THE WESTERN ONTARIO SERIES IN PHILOSOPHY OF SCIENCE

A SERIES OF BOOKS IN PHILOSOPHY OF SCIENCE, METHODOLOGY, EPISTEMOLOGY, LOGIC, HISTORY OF SCIENCE, AND RELATED FIELDS

VOLUME 82

More information about this series at http://www.springer.com/series/6686

John L. Bell

The Continuous, the Discrete and the Infinitesimal in Philosophy and Mathematics

Springer

John L. Bell
Department of Philosophy
University of Western Ontario
London, ON, Canada

ISSN 1566-659X ISSN 2215-1974 (electronic)
The Western Ontario Series in Philosophy of Science
ISBN 978-3-030-18706-4 ISBN 978-3-030-18707-1 (eBook)
https://doi.org/10.1007/978-3-030-18707-1

This Springer imprint is published by the registered company Springer Nature Switzerland AG.
The registered company address is: Gewerbestrasse 11, 6330 Cham, Switzerland

Preface

This book has a double purpose: first, to trace the historical development of the concepts of the continuous, the discrete, and the infinitesimal and, second, to describe the ways in which the first and last of these concepts are treated in contemporary mathematics. Accordingly, the first part of the book is largely philosophical, while the second is almost exclusively mathematical. In writing the book, I have found it necessary to thread my way through a wealth of sources, both philosophical and mathematical; and it is inevitable that a number of topics have not received the attention they deserve. Still, the thread itself, if tangled in places, has been luminous. "Only connect ... Live in fragments no longer," says E. M. Forster, and that is what I have tried to do here.

London, ON, Canada
March 2005/July 2018

John L. Bell

Introduction

Continuous as the stars that shine
And twinkle on the milky way,
They stretched in never-ending line
Along the margin of a bay:
Ten thousand saw I at a glance,
Tossing their heads in sprightly dance. (William Wordsworth)

To see a World in a Grain of Sand
And a Heaven in a Wild Flower,
Hold Infinity in the palm of your hand
And Eternity in an hour. (William Blake)

The homeland, friends, is a continuous act
As the world is continuous. (Jorge Luis Borges)

We are all familiar with the idea of *continuity*. To be continuous[1] is to constitute an unbroken or uninterrupted whole, like the ocean or the sky. A continuous entity—a *continuum*—has no "gaps." Opposed to continuity is *discreteness*: to be discrete[2] is to be separated, like the scattered pebbles on a beach or the leaves on a tree. Continuity connotes unity, while discreteness, plurality.

---

[1]The word "continuous" derives from a Latin root meaning "to hang together" or "to cohere"; this same root gives us the nouns "continent," an expanse of land unbroken by sea, and "continence," self-restraint in the sense of "holding oneself together." Synonyms for "continuous" include *connected, entire, unbroken,* and *uninterrupted.*

[2]The word "discrete" derives from a Latin root meaning "to separate." This same root yields the verb "discern," to recognize as distinct or separate, and the cognate "discreet," to show discernment, hence "well-behaved." It is a curious fact that, while "continuity" and "discreteness" are antonyms, "continence" and "discreetness" are synonyms. Synonyms for "discrete" include *separate, distinct, detached,* and *disjunct.*

The continuous and the discrete form a pair of *archetypal oppositions*.[3]

The domain of the discrete is a model of orderliness, a realm in which quality is entirely reduced to quantity and over which the concept of *number* reigns supreme. Since the units populating the realm of the discrete do not possess intrinsic qualities which might serve to distinguish them from one another, their difference is manifested through plurality alone. The simplicity of the principles governing discreteness has recommended it as a paragon of intelligibility, a realm within which reason can be realized to its fullest extent.[4]

By contrast, the realm of the continuous is a jungle, a labyrinth. It teems with such exotic and intractable entities as incommensurable lines, horn angles, space curves, and one-sided surfaces. The taming of this jungle by reduction to the discrete has been a principal task, if not the principal task, of mathematics. The origins of this effort lie in the practical procedure of mensuration, of comparing continuous magnitudes in a useful way.

If the constituency of the continuous is complex, it was founded in apparent simplicity. *Space* and *time* have traditionally been taken as founding members of that constituency. Certain philosophers have maintained that all natural processes occur continuously: witness, for example, Leibniz's famous apothegm *natura non facit saltus* – "nature makes no jump." In mathematics, the word is used in the same general sense but has come to be furnished with increasingly precise definitions. So, for instance, in the later eighteenth century, continuity of a function was taken to mean that infinitesimal changes in the value of the argument induce infinitesimal changes in the value of the function. With the abandonment of infinitesimals in the nineteenth century, this definition gave way to one employing the more precise concept of *limit*.

While it is the fundamental nature of a continuum to be *undivided*, it is nevertheless generally (although not invariably) held that any continuum admits of repeated or successive *division*, namely, that the process of dividing a continuum into ever smaller parts will never terminate in an *indivisible* or an *atom* – that is, a part which, lacking proper parts itself, cannot be further divided. In short, continua are *indefinitely* or *infinitely divisible*. The unity of a continuum thus conceals a potentially infinite plurality. In ancient Greece, this claim met with the objection that if one were to carry out *completely*, if only in imagination, the process of dividing an extended magnitude, such as a continuous line, then the magnitude would be reduced to a multitude of atoms – in this case, extensionless points – or even, possibly, to nothing at all. But then, it was held, no matter how many such points, or atoms, there may be – even if infinitely many – they cannot be "reassembled" or "juxtaposed" to restore the original continuum, for surely a sum

---

[3]See Bell (2016).

[4]But precisely because of the orderliness of the discrete, from a sociopolitical standpoint, it has been associated with regimented societies in which individuals are reduced to "numbers."

of extensionless elements still lacks extension.[5] The atoms or points within a continuum must not be separated but somehow "bonded" or "glued together" within it.

Such difficulties attended the birth in ancient Greece (fifth century BC) of the school of *atomism*.[6] The founders of this school, Leucippus and Democritus, claimed that matter, and, more generally, extension, is not infinitely divisible. Not only would the successive division of matter ultimately terminate in atoms, that is, in discrete particles incapable of being further divided, but matter had *in actuality* to be conceived as being compounded from such atoms. In attacking infinite divisibility, these atomists were at the same time mounting a claim that the continuous is ultimately reducible to the discrete, whether it be at the physical, theoretical, or perceptual level.

Atomism was later to flower into a general doctrine of the reducibility of the complex to the simple:[7] in addition to the physical atomism of the ancients, one can identify *epistemological* atomism, or the doctrine of units of perception; *linguistic* atomism, the alphabetic principle;[8] *logical* atomism, the positing of atomic or elementary propositions; and *biological* atomism, the postulation of discrete organic units such as cells or genes. A version of atomism can also be found in *mathematics*, namely, the doctrine – originating with the Pythagoreans of the sixth century BC – that all mathematical concepts are ultimately reducible to numbers, or, in the doctrine's contemporary form, *sets*.

The eventual triumph of the atomic theory in physics and chemistry in the nineteenth century paved the way for the idea of "atomism," as applying to matter, at least, to become widely familiar: it might well be said, to paraphrase Sir William

[5]Of course, this presupposes that there are no "gaps" between the elements or points, which is implicit in the assumption that the points have been obtained by complete division of a continuum.

[6]Atomism also emerged in India – independently and at about the same time in Jain and Buddhist philosophy. See Part I, 1.2 below.

[7]Whyte, *Essay on Atomism*, p. 12.

[8]It has been suggested that the emergence of atomism is connected with the alphabetic principle on which the great majority of natural (written) languages rest. In *Science and Civilisation in China*, vol. 4, 26(b), Joseph Needham notes the parallel between the limitless variety of words formable from the relatively few letters of the alphabet and the idea that a very small number of "elementary" particles could, in a multitude of combinations, engender the limitless variety of material bodies. But in China, atomism never really took root (see Part I, 1.2 below); in this connection, Needham observes, "the Chinese written character is an organic whole, a Gestalt, and minds accustomed to an ideographic language would perhaps hardly have been so open to the idea of an atomic constitution of matter." As Needham points out, however, the Chinese recognized the function of the atomic principle in numerous contexts, for example, the reduction of written characters to radicals, the composition of melodies from the notes of the pentatonic scale, and the representation of nature through the permutations and combinations of the broken and unbroken lines in the hexagrams of their ancient work of divination, the *I Ching*.

Harcourt's[9] famous observation in regard to the socialists of his day, "We are all atomists now." Nevertheless, only a minority of philosophers of the past espoused atomism at a metaphysical level, a fact which may explain why the analogous doctrine upholding continuity lacks a familiar name: that which is generally, if unconsciously, acknowledged requires no name. Peirce, that great wordsmith, coined the term *synechism* (from Greek *syneche,* "continuous") for his own philosophy – a philosophy permeated by the idea of continuity.[10] In what follows, I shall appropriate Peirce's term and use it in a sense shorn of its Peircean overtones, simply as a contrary to atomism. Thus, if "atomism" is the doctrine that the Real is *discrete*, then "synechism" is the doctrine that the Real is *continuous*. I shall also use the term "divisionism" for the doctrine that continua are infinitely divisible.

Closely associated with the concept of a continuum is that of *infinitesimal*.[11] An *infinitesimal magnitude* has been, traditionally, somewhat hazily conceived as a continuum "viewed in the small," an "ultimate part" of a continuum. In something like the same sense as a discrete entity is made up of its individual units, its "indivisibles", so, it was maintained, a continuum is "composed" of infinitesimal magnitudes, its ultimate parts. (It is in this sense, e.g., that mathematicians of the seventeenth century held that continuous curves are "composed" of infinitesimal straight lines.) Now the "coherence" of a continuum entails that each of its (connected) parts is also a continuum and, accordingly, divisible. Since points are indivisible, it follows that no point can be part of a continuum. Points are, in fact, no more than *locations* in a continuum. Infinitesimal magnitudes, as parts of continua, cannot, of necessity, be points: they are, in a word, *nonpunctiform*.

Magnitudes are normally taken as being *extensive* quantities, like mass or volume, which are defined over extended regions of space. By contrast, infinitesimal magnitudes have been conceived as *intensive* magnitudes resembling locally defined intensive quantities such as temperature or density. The effect of "distributing" or "integrating" an intensive quantity over an infinitesimal magnitude is to convert the former into an infinitesimal extensive quantity: thus, temperature is transformed into

[9]Sir William Harcourt (1827–1904) was a British lawyer, journalist, and liberal statesman, who served as home secretary and chancellor of the Exchequer under Gladstone. In a speech delivered in 1888, he remarked that "we are all socialists now." By this, he meant that capitalism had "progressed" sufficiently to render socialist aspirations, and their potential disturbance to the capitalist order, unnecessary. My paraphrase has a different purpose, namely, to indicate that, while atomism was the radical opposition to synechism (just as socialism was considered to be the radical opposition to capitalism), nevertheless philosophical synechists could, with good conscience, concede that – at the material level – atomism prevails.

[10]It should also be mentioned that the German philosopher, Johann Friedrich Herbart (1776–1841), introduced the term *synechology* for the part of his philosophical system concerned with the continuity of the real.

[11]According to the *Oxford English Dictionary*, the term *infinitesimal* was originally:

> *an* ordinal, *viz. the "infinitieth" in order; but, like other ordinals, also used to name fractions, thus* infinitesimal part *or* infinitesimal *came to mean unity divided by infinity* ($\frac{1}{\infty}$), *and thus an infinitely small part or quantity*

infinitesimal heat and density into infinitesimal mass. When the continuum is the trace of a motion, the associated infinitesimal/intensive magnitudes have been identified as *potential* magnitudes – entities which, while not possessing true magnitude themselves, embody a *tendency* to generate magnitude through motion, so manifesting "becoming" as opposed to "being."

An infinitesimal *number* has been conceived as a number so small that, while not coinciding with zero, is in some sense smaller than any finite positive number. An infinitesimal number is, so to speak, "greater than nothing but less than anything."[12] An infinitesimal number has been construed as a "number" which fails to satisfy *Archimedes' principle,* that is, as a nonzero "number" a such that, for any integer n, $n.a$ is less than any finite nonzero number. In "practical" approaches to the differential calculus, an infinitesimal is a number with the property that its square and all higher powers can be "neglected," that is, set to zero. In the theory of limits, the term "infinitesimal" is sometimes applied to any sequence whose limit is zero.

The concept of an *indivisible* is closely allied to, but to be distinguished from, that of an infinitesimal. An indivisible is, by definition, something that cannot be divided, which is usually understood to mean that it has no proper parts. Now a partless or indivisible entity does not necessarily have to be infinitesimal: souls, individual consciousnesses, and Leibnizian monads all supposedly lack parts but are surely not infinitesimal. But these have in common the feature of being unextended; extended entities such as lines, surfaces, and volumes prove a much richer source of "indivisibles." Indeed, if the process of dividing such an entity were to terminate, as the atomists maintained, this would necessarily issue in indivisibles having a qualitatively different nature from the original entity. In the case of a straight line, such indivisibles would, plausibly, be points; in the case of a circle, straight lines; and in the case of a solid cylinder divided by sections parallel to its base, circular areas. In each case, the indivisible in question is infinitesimal in the sense of *possessing one fewer dimension than its generating figure.* In the sixteenth and seventeenth centuries, indivisibles in this sense were used in the calculation of areas and volumes of curvilinear figures, a surface being thought of as the sum of linear indivisibles and a volume as the sum of planar indivisibles.

The concept of infinitesimal was beset by controversy from its beginnings. The idea makes an early appearance in the mathematics of the Greek atomist philosopher, Democritus c. 450 B.C., only to be banished c. 350 B.C. by Eudoxus in what was to become official "Euclidean" mathematics. The reappearance of infinitesimals as indivisibles in the sixteenth and seventeenth centuries has already been mentioned: in this form, they were systematically employed by Kepler; Galileo's student, Cavalieri; the Bernoulli clan; and a number of other mathematicians. In the guise of the delightfully named "linelets" and "timelets," infinitesimals played an essential role in Barrow's "method for finding tangents by calculation," which appears in his *Lectiones Geometricae* of 1670. As "evanescent quantities," infinitesimals were instrumental (although later abandoned) in Newton's development of the calculus

[12] Pyle (1997), p. 208.

and, as "inassignable quantities," in Leibniz's. The Marquis de l'Hôpital, who in 1696 published the first treatise on the differential calculus (entitled *Analyse des Infiniments Petits pour l'Intelligence des Lignes Courbes*), invokes the concept in postulating that "a curved line may be regarded as being made up of infinitely small straight line segments" and that "one can take as equal two quantities differing by an infinitely small quantity."

However useful infinitesimals may have been in practice, they could scarcely withstand logical scrutiny. Derided by Berkeley in the eighteenth century as "ghosts of departed quantities," in the nineteenth century execrated by Cantor as "cholera-bacilli" infecting mathematics, and in the twentieth roundly condemned by Bertrand Russell as "unnecessary, erroneous, and self-contradictory," the use of infinitesimals in the calculus and mathematical analysis was believed to have been finally supplanted by the limit concept which took rigorous and final form in the latter half of the nineteenth century. By the beginning of the twentieth century, the concept of infinitesimal had become, in analysis at least, a virtual "unconcept."

Nevertheless, the proscription of infinitesimals did not succeed in extirpating them; they were, rather, driven further underground. Physicists and engineers, for example, never abandoned their use as a heuristic device for the derivation of correct results in the application of the calculus to physical problems. Differential geometers of the stature of Lie and Cartan relied on their use in the formulation of concepts which would later be put on a "rigorous" footing. Even in mathematical analysis, they survived in Du Bois-Reymond's "orders of infinity. And, in a mathematically rigorous sense, they lived on in the algebraists' investigations of non-Archimedean fields.[13]

A new phase in the long contest between the continuous and the discrete has opened in the past few decades with the refounding on a solid basis of the concept of infinitesimal in analysis and differential geometry. This has been achieved in two essentially different ways: the one providing a rigorous formulation of the idea of infinitesimal *number* and the other of infinitesimal *magnitude.*

First, in the 1960s, Abraham Robinson, using methods of mathematical logic, created *nonstandard analysis,* an extension of mathematical analysis embracing both "infinitely large" and infinitesimal numbers in which the usual laws of the arithmetic of real numbers continue to hold, an idea which, in essence, goes back to Leibniz. Here by an infinitely large number we mean one which exceeds every positive integer; the reciprocal of any one of these is infinitesimal in the sense that, while being nonzero, it is smaller than every positive fraction $^1/_n$. Much of the usefulness of nonstandard analysis stems from the fact that within it, every statement of ordinary analysis involving limits has a succinct and highly intuitive translation into the language of infinitesimals. The elements of nonstandard analysis will be presented in Chap. 8.

[13]For a penetrating analysis of the emergence of non-Archimedean mathematics, see Ehrlich (2009).

The second development in the refounding of the concept of infinitesimal is the emergence in the 1970s of *smooth infinitesimal analysis*. This provides an image of the world in which the continuous is an autonomous notion, not explicable in terms of the discrete. Founded on the methods of category theory, it is a rigorous framework of mathematical analysis in which every function between spaces is smooth (i.e., differentiable arbitrarily many times, and so in particular continuous) and in which the use of limits in defining the basic notions of the calculus is replaced by *nilpotent infinitesimals*, that is, of quantities so small (but not actually zero) that some power – most usefully, the square – vanishes. Smooth infinitesimal analysis embodies a concept of intensive magnitude in the form of *infinitesimal tangent vectors* to curves. A tangent vector to a curve at point p on it is a short straight line segment **l** passing through the point and pointing along the curve. In fact, we may take **l** actually to be an infinitesimal *part* of the curve. Curves in smooth infinitesimal analysis are "locally straight" and accordingly may be conceived as being "composed of" infinitesimal straight lines in de l'Hôpital's sense, or as being "generated" by an infinitesimal tangent vector. Smooth infinitesimal analysis will be presented in considerable detail in Chap. 10.

* * *

The book is divided into two parts: The Continuous, the Discrete, and the Infinitesimal in the History of Thought and Continuity and Infinitesimals in Today's Mathematics. The bulk of the first part of the book is of a philosophical nature and can be read without requiring any specialized mathematical knowledge on the part of the reader. The second part is necessarily more mathematical and does demand some knowledge of contemporary mathematics for a proper understanding of its content.

Contents

Part I
The Continuous, the Discrete, and the Infinitesimal in the History of Thought

Chapter 1
The Continuous and the Discrete in Ancient Greece, the Orient, and the European Middle Ages

1.1 Ancient Greece

The Presocratics

The opposition between the Continuous and the Discrete played a significant role in ancient Greek philosophy. This probably derived from the still more fundamental opposition concerning the One and the Many, an antithesis lying at the heart of early Greek thought.[1]

The opposition between the One and the Many[2] seems to have been an animating principle in the thought of the Milesian philosophers **Thales** (*fl.* 585 B.C.), **Anaximander** (*fl.* 570 B.C.), and **Anaximenes** (*fl.* 550 B.C.). Monists all, they shared the belief that the world, manifold in appearance, could be reduced to a single underlying principle—although they disagreed as to what that principle was. The **Pythagoreans**[3] were, in essence, dualists, claiming that the world was built on the two ultimate principles of the limited and the unlimited, which in turn engender the whole series of opposites such as odd and even, one and many, still and moving. The Pythagoreans are also believed to have held that all things are made of number, from which it would seem to follow that they were atomists in some sense. But they can be considered genuine atomists only if the "numbers" they held to be constitutive of magnitude are indivisible atomic magnitudes in something like the sense of later atomism. Their discovery of incommensurable lines – in particular, of the incommensurability of the side and diagonal of a square[4] – provides an instant refutation of a narrow version of atomism in which it is claimed that any continuous line is

---

[1]See Stokes (1971).

[2]For a general account of Oppositions, see Bell (2016).

[3]Pythagoras himself is believed to have been active between 540 and 520 B.C.

[4]See, e,g., Bell (2016).

J. L. Bell, *The Continuous, the Discrete and the Infinitesimal in Philosophy and Mathematics*, The Western Ontario Series in Philosophy of Science 82,
https://doi.org/10.1007/978-3-030-18707-1_1

composed of a definite finite number of minimal indivisible unit lengths. Perhaps the alarm with which, according to tradition, the Pythagoreans reacted to this discovery is an indication that they did subscribe to this narrower atomism. But this was unclear even to Aristotle:

> *Nor is it in any way defined in what sense numbers are the causes of substances and of Being; whether as bounds, e.g. as points are the bounds of spatial magnitudes and as Eurytus determined which number belongs to which thing—e.g. this number to man, and this to horse—by using pebbles to copy the shapes of natural objects, like those who arrange numbers in the form of geometrical figures, the triangle and the square. Or is it because harmony is a ratio of numbers, and so is man and everything else?*[5]

Contemporary scholars are, appropriately perhaps, divided on the issue.

Heraclitus (*fl.* 500 B.C.), while essentially a monist, introduced into his monism two strikingly novel elements: these are the *Doctrine of Flux*, namely that all things are undergoing constant, if imperceptible change; and the *Unity of Opposites*: each object is constituted by opposing features. Heraclitus's "Flux" seems to mean continuous flux. The doctrine of the Unity of Opposites may derive from the observation that objects acquire contradictory attributes through a process of continuous change—as the ground, initially dry, becomes wet after rainfall. On this basis Heraclitus may be counted a forerunner of the synechists—a "protosynechist", perhaps.

The Greek debate over the continuous and the discrete seems to have been ignited by the efforts of the Eleatic philosophers **Parmenides** (*fl.* 515 B.C.), **Zeno** (*fl.* 460 B.C.) and **Melissus** (*fl.* 440 B.C.) to establish their doctrine of absolute monism. They were concerned to show that the divisibility of Being into parts leads to contradiction, so forcing the conclusion that the apparently diverse world is a static, changeless unity.[6] In his *Way of Truth* Parmenides asserts that Being is *homogeneous* and *continuous:*

> *It [Being] never was nor will be, since it is now, all together, one, continuous. ...Nor is it divided, since it all exists alike; nor is it more here and less there, which would prevent it from holding together, but it is all full of Being. So it is all continuous; for what is neighbours what is.*[7]

These passages suggest that Parmenides should be identified as a synechist. But in asserting the continuity of Being Parmenides is likely no more than underscoring its essential unity. For consider the later passage:

> *But look at things which, though far off, are securely present to the mind.; for you will not cut off for yourself what is from holding to what is, neither scattering everywhere in every way in order, nor drawing together.*[8]

[5] Aristotle (1996a, b), 1092b8.

[6] That this was the Eleatic position may be inferred from Plato's *Parmenides.*

[7] Kirk et al. (1983), pp. 249–50.

[8] *Ibid.*, p. 262.

Parmenides seems to be claiming that Being is more than merely continuous—that it is, in fact, a single whole, indeed an *indivisible* whole. The single Parmenidean existent is a continuum without parts, at once a continuum and an atom. If Parmenides was a synechist, his absolute monism prevented him from being a divisionist.

Parmenides' assertion that reality is a unique, partless continuum was reiterated by his disciple Melissus. However, the latter's observation:

> *If there were a plurality, things would have to be of the same kind as I say the One is,*[9]

which was intended as a *reductio ad absurdum* of belief in a plurality of things, seems to have opened the door to the emergence of atomism. In the atomists' hands, Melissus's assertion became, in effect,

> *there is a plurality of things, all of the same character as the One.*

This "plurality of things" are the indivisible atoms from which, according to the atomists, reality is constituted.

Zeno's arguments take the form of *paradoxes*[10] which are collectively designed to undermine the belief in motion, and hence in any notion of change.

The first two of Zeno's paradoxes, both of which rest on the assumption that space and time are continuous, are designed to prove that motion is impossible. This is done by showing that a finite motion generates an actual infinity.

The first paradox, the *Dichotomy*, goes as follows. Consider a body moving from one position to another. It is clear that, before the body can reach a given point in its path, it must first traverse half of the distance to that point. But before it can traverse half of that distance, it must traverse one quarter; and so on *ad infinitum*. So, for a body to pass from one point A to another, B, it must traverse an infinite number of divisions. But an infinite number of divisions cannot be traversed in a finite time, and so the goal cannot be reached.

The second paradox, *Achilles and the Tortoise*, is the best known. Achilles and a tortoise run a race, with the latter enjoying a head start. Zeno asserts that no matter how fleet of foot Achilles may be, he will never overtake the tortoise. For, while Achilles traverses the distance from his starting-point to that of the tortoise, the tortoise advances a certain distance, and while Achilles traverses this distance, the tortoise makes a further advance, and so on *ad infinitum*. Consequently Achilles will run forever without overtaking the tortoise.

This second paradox is formulated in terms of two bodies, but it has a variant involving, like the *Dichotomy*, just one. To reach a given point, a body in motion must first traverse half of the distance, then half of what remains, half of this latter, and so on *ad infinitum*, and again the goal can never be reached. This version of the *Achilles* exhibits a pleasing symmetry with the *Dichotomy*. For the former purports to show that a motion, once started, can never stop; the latter, that a motion, once stopped, can never have started.

---

[9] *Ibid.*, p. 399.

[10] For an account of Zeno's paradoxes, see, e.g., Bell (2016).

The third of Zeno's paradoxes, the *Arrow*, rests on the assumption of the *discreteness of time*. Here we consider an arrow flying through the air. Since time has been assumed discrete we may "freeze" the arrow's motion – as in a film frame – at an indivisible instant of time. For it to move during this instant, time would have to pass, but this would mean that the instant contains still smaller units of time, contradicting the indivisibility of the instant. So at this instant of time the arrow is at rest. Since the instant chosen was arbitrary, the arrow is at rest at any instant. In other words, it is always at rest, and so motion does not occur.

As pointed out, Zeno's *Dichotomy* and *Achilles* paradoxes both rest explicitly on the limitless divisibility of space and time. It has been supposed by some scholars, following Aristotle, that Zeno's *Arrow* paradox depends on the assumption that time, at least, is composed of atomic instants, but this view has been challenged. If indeed, as Barnes[11] and Furley[12] claim, none of Zeno's paradoxes of motion assume the atomic hypothesis, then it would be not unreasonable to number him among the divisionists. This is also consistent with the fact that he was a disciple of Parmenides.

The Eleatic arguments that plurality and change are illusions created an impasse – the *Eleatic problem* – for those philosophers (whom we may term protophysicists) concerned with the understanding of natural phenomena. They felt it essential to circumvent the Eleatic arguments, so preserving the multiplicity of the world evident to the senses, without deriving in the process a plurality from a pre-existing unity or allowing the generation or change of anything real. Two essentially different ways out of the impasse were found, one based on continuity, the other on discreteness. The first was the creation of **Anaxagoras** (500–428 B.C.), who conceived of matter, like magnitude, as being infinitely divisible, and who eliminated both generation and the derivation of plurality from unity by postulating *ab initio* an endless variety of primary substances in the form of infinitely divisible "seeds", all mixed together. Anaxagoras's theory of matter, the *homoimereia* or theory of homogeneous mixtures, was described by Lucretius with memorable scorn:

> *In speaking of the* homoimereia *of things Anaxagoras means that bones are formed of minute miniature bones, flesh miniature morsels of flesh, blood by the coalescence of many drops of blood; gold consists of grains of gold; earth is a conglomeration of little earths, fire of fires, moisture of moistures. And he pictures everything else as formed in the same way. At the same time he does not admit any vacuum in things, or any limit to the splitting of matter...*[13]

From this last phrase it may be inferred that Anaxagoras can be counted among the divisionists. Further evidence for this is provided by his own assertion, as reported by Simplicius:

> *Neither is there a smallest part of what is small, but there is always a smaller, for it is impossible that what is should ever cease to be.*[14]

---

[11]Barnes (1986).

[12]Furley (1967).

[13]Lucretius (1955), p. 51.

[14]Kirk et al. (1983), p. 360.

That Anaxagoras may even be considered a full-blown synechist follows from two other passages attributed to him:

> *But before these things were separated off, while all things were together, there was not even any colour plain; for the mixture of all things prevented it, of the moist and the dry, the hot and the cold, the bright and the dark.*[15]

> *The things in the one world-order are not separated one from the other nor cut off with an axe, neither the hot from the cold nor the cold from the hot.*[16]

The second attempt at escaping the Eleatic problem, *atomism*, was first and foremost a physical theory. It was mounted by **Leucippus** (*fl.* 440 B.C.) and **Democritus** (b. 460–457 B.C.) who, in contrast with Anaxagoras, maintained that matter was composed of indivisible, solid, homogeneous, spatially extended corpuscles, all below the level of visibility. Leucippus was regarded by Aristotle as the true founder of atomism. In *On Generation and Corruption*, Book I, he asserts:

> *For some of the ancients* [i.e., Parmenides and Melissus] *thought that what is must necessarily be one and motionless, since the void is nonexistent and there could be no motion without a separately existing void, and again there could not be a plurality without something to separate them. And if someone thinks that the universe is not continuous but consists of divided pieces in contact with each other, this is no different, they held, from saying that it is many, not one, and is void. For if it is divisible everywhere, there is no unit, and therefore no many, and the whole is void. If on the other hand it is divisible at one place and not another, this seems like a piece of fiction. For how far is it divisible, and why is one part of the whole like this—full—and another part divided? Again, it is necessary similarly that there be no motion... But Leucippus thought he had arguments which would assert what is consistent with sense-perception and not do away with coming into being or perishing or motion, or the plurality of existents. He agrees with the appearances to this extent, but he concedes, to those who maintain the One, that there would be no motion without void, and says that void is "non-existence", and nothing of "what is" is "not being"; for 'what is" in the strictest sense is a complete* plenum. *But this* plenum, *he says, is not one but many things of infinite number, and invisible owing to their minuteness. These are carried along in the void (for there* is *a void) and, when they come together, they cause coming-to-be and, when they dissolve, they cause passing-away. They act and are acted upon where they happen to come into contact (for there they are not one), and they generate when they are placed together and intertwined. From what is truly one no plurality could come into being, nor a unity from what is truly a plurality—this is impossible.*[17]

And in *Physics*, Book I:

> *Some gave in to both of these arguments—to the argument that all is one if what is signifies one thing, and to the argument from dichotomy—by positing atomic magnitudes.*[18]

As can be seen from the first passage, Leucippus's atomism—his theory of infinitely numerous invisible corpuscles moving in a void—is presented by Aristotle as an attempt in the first instance to reconcile the evidence of the senses with Eleatic

---

[15] *Ibid.* p. 358.
[16] *Ibid.* p. 371.
[17] *Ibid.* p. 407.
[18] *Ibid.* p. 408.

metaphysics. The senses tell us that the world is not a unity, but a plurality. In that case, where is unity to be sought? According to (Aristotle's) Leucippus, this unity is to be found in his postulated indivisible atomic magnitudes, each of which is, on a minute scale, an embodiment of the Eleatic One. Their combinations and dispersions underlie the phenomena of coming to be and passing away. The final sentence of the quotation indicates that Leucippus did not regard extended continua as true unities, since he presumably accepted the evident fact that such continua can be divided, thereby engendering (as observed above) a plurality. The second passage's mention of the argument from dichotomy has been taken by scholars[19] as indicating Leucippus's rejection of Zeno's putative divisionism.

Scholarly opinion is divided over the question of whether Leucippus and, especially, Democritus considered their atoms to be *physically*, but not *theoretically* indivisible. The traditional view[20] was that Democritus, at least, did not. Furley, in his *Two Studies of the Greek Atomists,*[21] on the other hand, argues that neither Democritus nor Aristotle made any distinction between physical and theoretical indivisibility, and so the former would think of his indivisible magnitudes as being theoretically as well as physically indivisible. As Furley points out, the hypothesis that Leucippus and Democritus postulated theoretically indivisible atoms is confirmed by Simplicius:

> *Leucippus and Democritus think that the cause of the indivisibility of the primary bodies is not merely their imperviousness but also their smallness and partlessness; Epicurus, later, does not think they are partless, but says that they are atomic because of their imperviousness.*[22]

Like the Eleatic One, Democritus's atoms were, Furley thinks, "absolutely solid, packed with being and nothing else". As plena atoms are impenetrable and so indivisible. But the universe as a whole is divisible since it consists of a multiplicity of existents separated by void.

Now if Democritus was a geometric atomist (and the case will in all likelihood never be proven), he was almost certainly aware of the difficulties attendant on the idea of theoretical indivisibles. For witness the following well-known passage from Plutarch, believed to be a quotation of Democritus's own words, and which has been termed by scholars the *cone dilemma:*

> *If a cone is cut by sections parallel to its base, are we to say that the sections are equal or unequal? If we suppose that they are unequal, they will make the surface of the cone rough and indented by a series of steps. If the surfaces are equal, then the sections will be equal and the cone will become a cylinder, being composed of equal, instead of unequal, circles. This is a paradox.*[23]

[19]See, e.g. Kirk et al. (1983), p. 408.

[20]As presented, e.g. in Heath (1981).

[21]Furley (1967).

[22]For Epicurus's views see below.

[23]Quoted in Sambursky (1963), p. 153.

A related fact is Archimedes' attribution (in *The Method*) to Democritus of the discovery[24] that the volume of a right circular cone is one third that of the circumscribed cylinder. Archimedes also claims that Democritus was unable to prove the result rigorously. While it is not known whether Democritus managed to resolve the cone dilemma, it is highly likely that he arrived at the volume formula by analyzing the cone into a collection of infinitesimally thin circular laminas. This use of infinitesimals anticipates Cavalieri's method of indivisibles.[25]

Even if Democritus did not uphold the actual existence of geometric indivisibles such as lines or surfaces, his *material* atomism may well have suggested the geometric analogy, which, while metaphysically problematic, proved to be mathematically most fruitful.

The Method of Exhaustion

Antiphon the Sophist, a contemporary of Socrates, is believed to have made one of the earliest attempts to rectify the circle. According to Simplicius, his procedure involved the inscribing in a circle of a regular polygon, for example a triangle or square, and then successively doubling the number of sides. In this way,

> *Antiphon thought that the area (of the circle) would be used up, and we should some time have a polygon inscribed in the circle the sides of which, owing to their smallness, coincide with the circumference of the circle. And, as we can make a square equal to any polygon... we shall be in a position to make a square equal to a circle.*[26]

As Simplicius notes, this infringes the principle that magnitudes are divisible without limit. If Antiphon truly thought that a circle could actually coincide with an inscribed polygon of a sufficiently large number of sides, then he has to be considered an atomist. Millennia later, the idea that a curve can be considered as an assemblage of infinitesimal straight lines came to play an important role in the development of the calculus.

If Antiphon and Democritus were in fact geometric atomists, they were exceptions among the mathematicians of ancient Greece. For Greek geometry rested on the assumption that magnitudes are divisible without limit, and so the very practice of Greek mathematicians would incline them towards divisionism. In particular the method of exhaustion worked out by **Eudoxus** (408–355 B.C.) the germ of which is stated in the proposition opening Book X of **Euclid**'s (325–265 B.C.) *Elements*, clearly presupposes that any magnitude can be divided without limit:

---

[24]But not the rigorous proof, which in his treatise *On the Sphere and Cylinder* Archimedes ascribes to Eudoxus.

[25]See Chap. 2.

[26]Quoted in Heath (1981), vol. I, p. 222.

If from any magnitude there be subtracted not less than its half, from the remainder not less than its half and so on continually, there will at length remain a magnitude less than any assigned magnitude of the same kind.

Eudoxus is also believed to have created the general theory of proportions presented in Book V of the Elements. In Definition 4 of that book, magnitudes are decreed to have a ratio to one another just when they "are capable, when multiplied, of exceeding one another". This prescription effectively excludes infinitesimal magnitudes from consideration.[27]

Archimedes (287–212 B.C.), the greatest mathematician of antiquity, made a number of important applications of the method of exhaustion. As a pivotal principle he employed what has come to be known as the *axiom of Archimedes,* an elaborated version of Definition 4 in Euclid's Book V:

Of unequal lines, unequal surfaces, or unequal solids, the greater exceeds the less by such a magnitude as is capable, if added (continually) to itself, of exceeding any magnitude of those which are comparable to one another.

As has been pointed out, a prescription of this sort excludes infinitesimal magnitudes.[28] Yet one of the central ideas in Archimedes' *Method* is that surfaces may be regarded as being composed of lines. How Archimedes intended this to be understood is not entirely clear. He does not speak of the number of lines in each figure as infinite, saying only that the figure is *made up of all* the lines in it. But this does suggest that he probably thought of these lines as indivisibles, infinitesimally narrow surface "elements". Further evidence for this is offered "by the highly suggestive fact that he was led to many new results by a process of balancing, in thought, elements of dissimilar figures, using the principle of the lever precisely as one would in weighing mechanically a collection of thin laminae or material strips."[29]

Important as the method of indivisibles was to Archimedes as a heuristic, however, he did not consider results discovered through its use as having been rigorously proved. Rigorous proof was invariably supplied by means of the method of exhaustion. For Archimedes, atomism pointed the way to (geometric) truth, but that truth could only be secured by rigorous derivation from synechist postulates.

Plato

Plato (429–328 B.C.) may have accepted the existence of indivisible magnitudes. Aristotle, in the *Metaphysics*, reports:

[27]Yet in Book III infinitesimals appear in the form of "horn" angles: angles between curved lines. Proposition 16 asserts that the angle between a circle and a tangent is less than any rectilineal angle.

[28]Arhcimedes' axiom, or primciple is now usually stated in the following form: if $0 < a < b$, then there is a natural number n such that $b < n.a$.

[29]Boyer (1959), p. 50.

Fig. 1.1

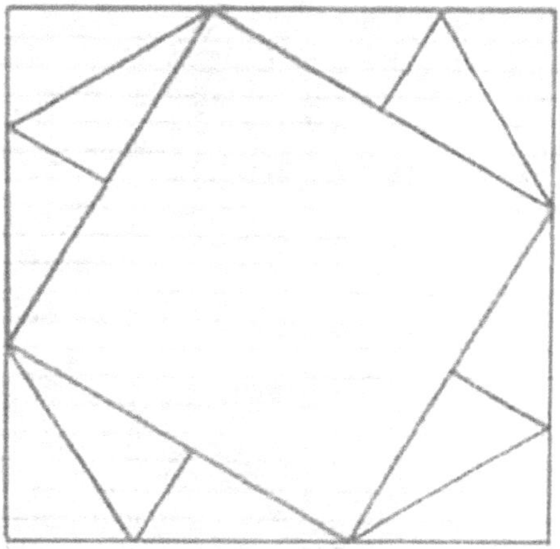

Plato steadily rejected this class of objects [i.e., points]*as a geometrical fiction, but he recognized "the beginning of a line," and he frequently assumed this latter class, which he called the "indivisible lines".*[30]

Similar views were held by Plato's pupil Xenocrates, who postulated the existence of atomic magnitudes in an attempt to avoid the pitfalls of Zeno's paradoxes.

In the *Timaeus* Plato postulates that the Empedoclean elements earth, water, air and fire are made up of two basic geometric units, both right-angled triangles. One is the right-angled isosceles triangle, the half-square, the other the Pythagorean "1, $\sqrt{3}$, 2" triangle, or half-equilateral triangle. The half-square is used to build up cubes, the characteristic shape of earth particles; the half-equilateral is used to build up regular pyramids, octahedra, and icosahedra, the characteristic shapes of the particles of fire, air and water, respectively.

While Plato says nothing concerning the indivisibility or otherwise of his triangular units, a number of scholars, including Cornford and Furley, have argued that the theory implicitly requires the use of indivisible magnitudes. Furley[31] points out that in Plato's theory, while fire, air and water can be transformed into one another—for instance, the half-equilaterals forming an icosahedron of water may re-form into pyramids and thus become fire—earth, composed of half-squares, cannot become anything else. But from this it would seem to follow that at least the elementary half-squares must be indivisible. For a (divisible) square can be divided into twelve half-equilaterals and a smaller square as in the diagram (Fig. 1.1). It follows that, unless the faces of a cube of earth are indivisible, the cube can be transformed into nine pyramids of fire and a smaller cube of earth, which is plainly at odds with Plato's assertion that earth is irreducible to anything else. From this it would seem to follow that Plato's theory involves the existence of indivisible triangles.

---

[30] Aristotle (1996a, b), 992a 20.

[31] Furley (1967), p. 108 f.

What sort of indivisibility would these triangles have? If they are material, then their indivisibility would be no more than physical. On the other hand, if they are not material, but some kind of geometric abstraction (as some scholars have claimed), then their indivisibility would perforce be of a theoretical order. This question, and indeed the whole issue of Plato's "atomism", remains unresolved.

Aristotle

It was **Aristotle** (384–322 B.C.) who first undertook the systematic analysis of continuity and discreteness. A thoroughgoing synechist, Aristotle maintained that physical reality is a continuous plenum, and that the structure of a continuum, common to space, time and motion, is not reducible to anything else. His answer to the Eleatic problem is a refinement of that of Anaxagoras, namely, that continuous magnitudes are potentially divisible to infinity, in the sense that they may be divided *anywhere*, though they cannot be divided *everywhere* at the same time.

Aristotle identifies continuity and discreteness as attributes applying to the category of Quantity.[32] As examples of continuous quantities, or *continua*, he offers lines, planes, solids (i.e., solid bodies), extensions, movement, time and space; among discrete quantities he includes number[33] and speech.[34] He also lays down definitions of a number of terms, including continuity:

> *Things are said to be* "together" *in place when the immediate and proper place of each is identical with that of the other and* "apart" (or *"severed"*) when this is not so. *They* are "in contact" *when their extremities are in this sense "together". One thing is* "in (immediate) succession" *to another if it comes after the point you start from in an order determined by position, or "form", or whatsoever it may be, and if there is nothing of its own kind between it and that to which it is said to be in immediate succession* ... "Contiguous" *means in immediate succession and in contact. Lastly, the* "continuous" *is a subdivision of the contiguous; for I mean by one thing being continuous with another that those extremities of the two things in virtue of which they are in contact with each other become one and the same thing and (as the very name indicates) are "held together", which can only be if the two limits do not remain two but become one and the same. From this definition it is evident that continuity is possible in the case of such things as can, in virtue of their natural constitution, become one by coming into contact; and the whole will have the same sort of union as that which holds it together, e.g. by rivet or glue or contact or organic union.*[35]

---

[32]In Book VI of the *Categories.* Quantity (ποσον) is introduced by Aristotle as the category associated with *how much.* In addition to exhibiting continuity and discreteness, quantities are, according to Aristotle, distinguished by the feature of being *equal* or *unequal.*

[33]Here it must be noted that for Aristotle, as for ancient Greek thinkers generally, the term "number" – *arithmos* – means just "plurality".

[34]Aristotle points out that (spoken) words are analyzable into syllables or phonemes, linguistic "atoms" themselves irreducible to simpler linguistic elements.

[35]Aristotle (1980), V, 3.

In effect, Aristotle here defines continuity as a *relation* between entities rather than as an *attribute* appertaining to a single entity; that is to say, he does not provide an explicit definition of the concept of *continuum*. At the end of this passage he indicates that a single continuous whole can be brought into existence by "gluing together" two things which have been brought into contact, which suggests that the continuity of a whole should derive from the way its *parts* "join up". That this is indeed the case is revealed by turning to the account of the difference between continuous and discrete quantities offered in the *Categories:*

> *Discrete are number and language; continuous are lines, surfaces, bodies, and also, besides these, time and space. For the parts of a number have no common boundary at which they join together. For example, ten consists of two fives, however these do not join together at any common boundary but are separate; nor do the constituent parts three and seven join together ay any common boundary. Nor could you ever in the case of number find a common boundary of its parts, but they are always separate. Hence number is one of the discrete quantities .... A line, on the other hand, is a continuous quantity. For it is possible to find a common boundary at which its parts join together—a point. And for a surface—a line; for the parts of a plane join together at some common boundary. Similarly in the case of a body one would find a common boundary—a line or a surface—at which the parts of the body join together. Time also and space are of this kind. For present time joins on to both past time and future time. Space again is one of the continuous quantities. For the parts of a body occupy some space, and they join together at a common boundary. So the parts of the space occupied by various parts of the body themselves join together at the same boundary as the parts of the body do. Thus space is also a continuous quantity, since its parts join together at one common boundary.*[36]

Accordingly, for Aristotle quantities such as lines and planes, space and time are continuous by virtue of the fact that their constituent parts "join together at some common boundary". By contrast *no* constituent parts of a discrete quantity can possess a common boundary.

One of the central theses Aristotle is at pains to defend in *Physics VI* is the irreducibility of a continuum to discreteness—that a continuum cannot be "composed" of indivisibles or atoms, parts which cannot themselves be further divided. He begins his reasoning as follows:

> *Now if the terms "continuous", "in contact", and "in immediate succession" are understood as defined above—things being "continuous" if their extremities are one, "in contact" if their extremities are together, and "in succession" if there is nothing of their own kind intermediate between them—nothing that is continuous can be composed of indivisibles: e.g. a line cannot be composed of points, the line being continuous and the point indivisible. For two points cannot have identical extremities, since in an indivisible there can be no extremity as distinct from some other part; and (for the same reason) neither can the extremities be together, for a thing which has no parts can have no extremity, the extremity and the thing of which it is the extremity being distinct. Yet the points would have to be either* continuous *or* contiguous *if they were to compose a continuum. And the same reasoning applies in the case of any indivisible. As to the impossibility of their being continuous, the proof just given will suffice; and one thing can be contiguous with another only if whole is in contact with whole or part with part or part with whole. But since indivisibles have no parts,*

[36] Aristotle (1996a), *Categories*, VI.

they must be in contact with one another as whole with whole. And if they are in contact with one another as whole with whole, they cannot compose a continuum, for a continuum is divisible into parts which are distinguishable from each other in the sense of being in different places.[37]

Aristotle is essentially saying here that, because points are partless, two given points must either coincide, i.e., be one, or must be totally separate, i.e., cannot overlap or even be in contact with one another. Thus a number of points, or indivisibles in contact with one another would teduce to a single indivisible, and so could not constitute a continuum, since a continuum is always divisible.

Having disposed of the possibility that a continuum could be made up of indivisibles either continuous or in contact with one another, Aristotle next turns to the question of whether a continuum such as length or time could be composed of indivisibles in succession. Once more he answers in the negative:

Again, one point, so far from being continuous or in contact with another point, cannot even be in immediate succession *to it, or one "now" to another "now", in such a way as to make up a length or a space of time; for things are "in succession" if there is nothing of their own kind intermediate between them, whereas two points have always a line (divisible at intermediate points) between them, and two "nows" a period of time (divisible at intermediate "nows"). Moreover, if a continuum such as length or time could thus be composed of indivisibles, it could itself be resolved into its indivisible constituents. But, as we have seen, no continuum can be resolved into elements which have no parts.*

While it has been shown that continua such as length and time cannot be composed of successive points or instants with *nothing of the same kind between them*, there remains the possibility that there might lie between them something of a *different* kind, e.g. stretches of emptiness or "void" such as certain Pythagoreans supposed to separate the distinct points composing a line. (Aristotle's arguments against the existence of void in general are presented in Book IV of the *Physics.*) Against this sort of picture Aristotle argues:

Nor can there be anything of any other kind between the points or between the moments: for if there could be any such thing it is clear that it must be indivisible or divisible, and if divisible, it must be divisible either into indivisibles or divisibles that are infinitely divisible, in which case it is a continuum. Moreover, it is plain that every continuum is divisible into parts that that are divisible without limit: for if it were divisible into indivisibles, we should have one indivisible in contact with another, since the extremities of things that are continuous with one another are one and are in contact.[38]

This somewhat cryptic argument deserves elucidation. Let us suppose with the Pythagoreans that a continuous line is composed of successive points separated by stretches of "void". Since any such void stretch, as a part of a continuum, cannot be indivisible, it is accordingly divisible either (*a*) into indivisible parts or (*b*) infinitely. Alternative (*a*) can be dismissed, for, if there were indivisible parts, in order to make up a continuum they would have to be in contact with one another, which we have already seen to be impossible. This leaves alternative (*b*), but in that case our stretch

[37] Aristotle (1980) VI, 1.

[38] *Ibid.*

of void is itself a (linear) continuum, and so the alleged "successive" points separated by it are not in fact successive, for there is now a line stretching between them which is itself divisible at intermediate points of the same kind.

Aristotle sometimes recognizes *infinite divisibility*—the property of being divisible into parts which can themselves be further divided, the process never terminating in an indivisible—as a consequence of continuity as he characterizes the notion in Book V. But on occasion he takes the property of infinite divisibility as *defining* continuity. It is this definition of continuity that figures in Aristotle's demonstration of what has come to be known as the *isomorphism thesis*, namely:

> *The same argument applies to magnitude, time and motion: either they are all composed of indivisible things and divided into indivisible things, or none of them is.*[39]

Briefly, the isomorphism thesis asserts that either magnitude, time and motion are all continuous, or they are all discrete. Aristotle's demonstration of this thesis rests on two key postulates concerning motion:

1. When motion is taking place, something is moving from here and *vice-versa*.
2. A moving object cannot simultaneously be in the act of moving towards a given point and in the state of being already at it.

Here in a nutshell is Aristotle's argument: given components *L, M, N*... of a motion, and assuming that each of these is itself a motion, then, by postulate 2, after *L* has started and before it has finished, the moving object *P* is, by postulate 1, past the start and short of the finish of the corresponding distance *A*. It follows that *A* is divisible in correspondence with *L*, and so likewise are the distances *B, C*, ... corresponding to *M, N*, ... Aristotle also shows how the assumption that the distances *A, B, C*, ... are indivisible leads to what he saw as absurdities concerning the motion. For if *L*, say, is a motion, then *P* would be moving over *A*, but since *A* lacks parts, *P's* movement over *A* leaves no "trace": in traversing *A* it "jumps" instantaneously from a state of rest to a state of rest. On the other hand, if *L* is not a motion then *P* would never be in motion but would accomplish the motion without moving. Accordingly, both distance and motion must be divisible. As for time, Aristotle argues that it is divisible if both distance and motion are, and *vice-versa*. For, he says, if the whole of the length *A* is traversed in time *T*, a part of it would be traversed (at equal speed) in less than *T*. On the other hand, if the whole time *T* were occupied in traversing *A*, then in part of the time a part of *A* would be covered.

While Aristotle held that magnitude, motion and time possessed a common (continuous) form, he had doubts as to whether *time* was an existent in the same sense as the first two. On this question we read in *Physics* Book IV, 10:

> *The following considerations might make one suspect that there is really no such thing as time, or at least that it has only an equivocal and obscure existence.*

[39]*Ibid.*

1. *Some of it is past and no longer exists, and the rest is future and does not yet exist; and time, whether limitless or any given length of time we take, is entirely made up of the no-longer and not-yet; and how can we conceive of that which is composed of non-existents sharing in existence in any way?*
2. *Moreover, if anything divisible exists, then, so long as it is in existence, either all its parts or some of them must exist. Now time is divisible into parts, and some of these were in the past and some will be in the future, but none of them exists. The present "now" is not part of time at all, for a part measures the whole, and the whole must be made up of the parts, but we cannot say that time is made up of "nows".*

Aristotle thus questions the existence of time on the grounds that none of its parts can be said to exist: the past no longer exists, the future does not yet exist, and the present, while it may exist, is a sizeless instant and so cannot be considered a *part* of time. These perplexities concerning the nature of time continued to puzzle Aristotle's successors. In *Physics* Book IV, 11, Aristotle had defined time as "the number of motion in respect of 'before' and 'after'"—a definition to which his pupil Strato later objected, not unreasonably, on the grounds that the use of the term "number", as a discrete quantity, is inappropriate in connection with time, which is continuous.

The question of whether magnitude is perpetually divisible into smaller units, or divisible only down to some atomic magnitude leads to the *dilemma of divisibility,*[40] a difficulty that Aristotle necessarily had to face in connection with his analysis of the continuum. In the dilemma's first, or *nihilistic* horn, it is argued as above that, were magnitude everywhere divisible, the process of carrying out this division completely would reduce a magnitude to extensionless points, or perhaps even to nothingness. The second, or *atomistic,* horn starts from the assumption that magnitude is not everywhere divisible and leads to the equally unpalatable conclusion (for Aristotle, at least) that indivisible magnitudes must exist.

Aristotle mounts his main attack on the atomistic horn of the dilemma in Book VI of the *Physics*, where, as we have seen, he repudiates the idea that a continuum can be composed of indivisibles. His refutation of the dilemma's nihilistic horn, presented in Book I of *On Generation and Corruption,* rests on two ideas: that the conception that it is the nature of a continuum to exist prior to its parts, and that a point is nothing more than a cut or division in a line, as the beginning or end—the limit—of a line segment. Precisely because points exist only as divisions or limits Aristotle denies them substantial reality; they are mere "accidents" arising from operations performed on substances or magnitudes. Points exist for Aristotle essentially in a *potential* mode, as marking out possible divisions in magnitudes. When a body moves continuously along a continuous path, he avers, the points over which it moves have no more than a potential existence and are only actualized by the body coning to a halt and starting to move again.[41] Similarly, a point in a straight line is

[40]Miller (1982).

[41]This is a crucial point for Aristotle in his refutation of Zeno's dichotomy paradox, since Aristotle concedes to Zeno only that in order to reach a goal a moving body must pass through a *potential* infinity of half-distances. If the body were to traverse an actual infinity of such distances, it would have to make an infinite number of stops and starts. In other words, only an impossibly *discontinuous* motion (in Aristotle's sense) would convert this potential infinity into an actual one.

brought into existence only by dissecting the line. Aristotle refutes the dilemma's nihilistic horn by showing that even though unlimited division of a magnitude is possible and a point exists everywhere *potentially*, it does not follow that magnitude reduces to points. A magnitude can be "divided throughout" only by a process in which a subsection is divided into further subsections. There is never a stage at which the division is completed and the line is reduced to indivisible constituents. For an actually existing point necessarily presupposes the existence of extended magnitudes which have been divided: until the division has actually been performed the point has no more than potential existence. Accordingly, the division must be successive rather than simultaneous, and it occurs "at every point" not in the sense of actually existing points but in the sense of points which could mark further subdivisions.

In Book VI of the *Physics* Aristotle brings some of these ideas to bear in his refutation of Zeno's paradoxes. As reported by Aristotle, in the first of these, the Dichotomy, the possibility of motion is denied

> *because, however near the moving object is to any given point, it will always have to cover the half, and then the half of that, and so on without limit until it gets there.*[42]

That is,

> *It is impossible for a thing to traverse or severally come into contact with illimitable things in a limited time.*

In repudiating this Aristotle argues that

> *there are two senses in which a distance or a period of time (or indeed any continuum) may be regarded as illimitable, viz., in respect of its divisibility or in respect of its extension. Now it is not possible to come into contact with quantitatively illimitable things in a limited time, but it is possible to traverse what is illimitable in its divisibility: for in this respect time itself is also illimitable. Accordingly, a distance which is (in this sense) illimitable is traversed in a time which is (in this sense) not limited but illimitable; and the contacts with the illimitable (points) are made at "nows" which are not limited but illimitable in number.*[43]

Here Aristotle traces the paradox as arising from Zeno's tacit assumption that in the course of the motion the number of contacts to be established accord in the case of the distance with its unlimited *divisibility* but in the case of the time with its limited *extension*. Aristotle dissolves the paradox by observing that actually in both cases these contacts accord with unlimited divisibility. A definite distance and a definite period of time are (by the isomorphism thesis) divisible in exactly the same way: the distance at points into shorter distances, the time at nows into shorter periods. A point marks off a stage in the journey, and the corresponding now marks off the time taken to accomplish that stage. Neither the points nor the nows are limitable; but they both exist only in a potential sense.

Aristotle takes the third paradox, that of the Arrow, as asserting the impossibility for a thing to be moving during a period of time, because it is impossible for it to be

---

[42]Aristotle (1980) VI, 9.

[43]*Ibid.*, 2.

moving at an indivisible instant. This is summarily dismissed on the grounds that time, as a continuum, cannot be composed of indivisible instants. Some modern scholars[44] claim that Aristotle has misunderstood the core of Zeno's argument (as plausibly reconstructed). In essence Aristotle takes the paradox as resting on the assumption that time is not infinitely divisible; but in fact Zeno's argument requires no assumption concerning the structure of time (or space). All he requires for the validity of his inference is that what is true of something (in this case, to be at rest) *at every moment* of a period of time (whether or not moments are indivisible instants) is true of it *throughout* the period.

Aristotle considered (uninterrupted) *movement*, or, more generally, *change*, to be a prime example of the continuous. In Book V of the *Physics* he answers the question, "what constitutes the unity of a movement?", by asserting:

> *Not its indivisibility (for every movement is potentially divisible without limit), but its uninterrupted continuity. Thus if a movement is strictly one, it must be continuous, and if continuous, one. ... For a movement to possess absolute unity and continuity (a) the movement must be specifically the same throughout the course, (b) the mobile must retain its numerical identity, and (c) the time occupied must [itself] be "one".*[45]

In the final section of Book VI Aristotle argues that no indivisible object can undergo movement or change in this unified sense, concluding:

> *It follows that a thing without parts cannot move, or indeed change at all. The only way in which it* could *move is if time were composed of "nows"; for in any "now" it would* have moved *or* have changed, *and so it would never move but always have moved. But the impossibility of this has already been demonstrated; time is not composed of "nows", nor a line of points, nor motion of jerks—for anyone who asserts that a partless thing can move is in fact saying that motion is composed of partless units, as if time were composed of "nows" or length were composed of points.*[46]

Another way of putting this is that an indivisible can move only in instantaneous "jumps" or "jerks".

As regards space, time, motion and extension, Aristotle was a thoroughgoing divisionist. But with regard to matter, or, at least, organic matter, his divisionism took a qualified form. This emerges in Book I, Ch. 4 of the *Physics* where, in criticizing Anaxagoras's theory of mixtures, with its arbitrarily small "seeds" of matter, he puts forward his view as to why the "natural parts" of a thing must have determinate size:

> *Flesh, bone and the like are the parts of animals It is clear then that neither flesh, bone, nor any such thing can be great or small without limit.*[47]

---

[44]E.g. Barnes (1986) and Kirk et al. (1983).

[45]Aristotle (1980) V, 4.

[46]*Ibid.* VI, 10.

[47]*Ibid.* I, 4; 187b 18–21.

He goes on to present his objections to Anaxagoras' assertion that everything contains all possible kinds of seed. Here as an example he takes water, in which according to Anaxagoras the seeds of flesh must be present:

> *For if flesh has been extracted from a given body of water and then more flesh is sifted from the remainder by repeating the process of separation: then, even if the successive extracts will continually decrease in quantity, still they cannot fall below a certain magnitude.*[48]

These quotations make it clear[49] that Aristotle does not admit the infinite divisibility of matter (or at least of organic matter) in an actual, physical sense. In a word, matter must, according to Aristotle, have *natural minima*. Aristotle does not develop this theory to any extent, but it assumed a more explicit form at the hands of later commentators such as Simplicius, Alexander of Aphrodosias (c. 200 A.D.), Themistius (4th cent. A.D.) and Philoponus (6th cent. A.D.).[50] The *doctrine of natural minima* that emerged was based on the following theses[51]:

1. *Qua* mathematical extension, quantity is (potentially) infinitely divisible; physically, it is not.
2. Each type of substance has its *natural minimum*, beyond which it cannot be further divided.

This doctrine, which came to exert a considerable influence on the scholars of the Middle Ages, bears a superficial resemblance to atomism, but is in fact quite distinct.[52] While natural minima certainly are, and atoms may be, mathematically divisible, natural minima of a given type may, unlike atoms, be physically divisible into other substances. Moreover, unlike atoms, minima do not exist in an objective, independent sense, they are only *potential* parts of substances.

Epicurus

As a thoroughgoing materialist, **Epicurus** (341–271 B.C.) could not accept the notion of potentiality on which Aristotle's theory of continuity rested, and so was propelled towards atomism in both its conceptual and physical senses. According to Simplicius,

> *Aristotle often refuted the doctrine of Democritus and Leucippus; because of these refutations, perhaps, as they were directed at the concept of the "partless", Epicurus, a later adherent of the doctrine of Democritus and Leucippus about the primary bodies, retained*

---

[48]This having been established above.

[49]Van Melsen (1952), p. 42.

[50]*Ibid.*, p. 47.

[51]Pyle (1997), pp. 216–217.

[52]*Ibid.*, p. 217.

their imperviousness but dropped their partlessness, since they had been refuted on this ground by Aristotle.[53]

Like Leucippus and Democritus before him, Epicurus felt it necessary to postulate the existence of physical atoms, but to avoid Aristotle's strictures he proposed that these should not be themselves conceptually indivisible but should *contain* conceptually indivisible parts. Aristotle had shown that a continuous magnitude could not be composed of *points,* that is, indivisible units lacking extension, but he had not shown that an indivisible unit must necessarily lack extension. Epicurus met Aristotle's argument that a continuum could not be composed of indivisibles by taking indivisibles to be partless units of magnitude possessing extension.

Epicurus's *Letter to Herodotus* contains a summary of his natural philosophy, and more particularly of his atomism. According to Epicurus the ultimate contents of the universe are bodies and space, or "void"; these are themselves irreducible and everything can be reduced to them. These ultimate bodies are

physically indivisible and unchangeable, if all things are not to be destroyed into non-being but are to remain durable in the dissolution of compounds—solid by nature, unable to be dissolved anywhere or anyhow. It follows that the first principles must be physically indivisible bodies.[54]

In other words, real things cannot be "destroyed into non-being"; but unless there were a limit to physical divisibility this is what would happen; accordingly, there must be a limit to physical divisibility.[55] Two millenia after Epicurus, the English philosopher Samuel Clarke, in his correspondence with Leibniz, put Epicurus' argument for the existence of atoms in the following way:

If there be no perfectly solid atoms, then there is no matter at all in the universe. For, the further the division and the subdivision of the parts of any body is carried, before you arrive at parts perfectly solid and without pores; the greater is the proportion of the pores to solid matter in that body. If, therefore, carrying on the division in infinitum, *you never arrive at parts perfectly solid and without pores; it will follow that all bodies consist of pores only, without any matter at all: which is a manifest absurdity.*[56]

The existence of atoms having been demonstrated, Epicurus proceeds to investigate their properties. In *Two Studies in the Greek Atomists,* David Furley provides the following paraphrase of Epicurus' analysis:

(A) In a finite body such as the atom, there cannot be an infinite number of parts with magnitude, however small they may be.
This implies:
(B1) the body cannot be divided into smaller and smaller parts to infinity (if we admitted this, we would put the whole world upon an insecure foundation; when we tried to get a firm mental grasp on the atoms we should find it impossible, because our mental picture of them would crumble away until nothing was left); and

---

[53]Quoted in Furley (1967) p. 111.

[54]Quoted *ibid.*, p. 7.

[55]*Ibid.*, p. 8.

[56]Alexander (1956), p. 54.

(B2) the process of traversing in the imagination from one side to the other of a finite body cannot consist of an infinite number of steps, not even with progressively diminishing steps.

We establish (A) on the following grounds:

(C1) If someone asserts that there is an infinite number of parts with magnitude in an object, then that object must itself be infinite in magnitude; this is true however small the parts may be. Moreover:

(C2) In the process of traversing an object in the imagination, one begins with the outermost distinguishable portion, and moves to the next; but this next must be similar to the first; hence it must be possible, in the view of one who asserts that there is an infinite number of such parts, to reach infinity in thought, when that object is totally grasped by the mind.

We establish (C2) by the following analogy:

(D1) The minimum perceptible *quantity is like larger perceptible quantities except that no parts can be perceived in it. (D2) The fact that it is like larger perceptible quantities, in which parts* can *be perceived, may suggest that we can distinguish one part from another in the minimum, too. But this is false. If we perceive a second quantity, it must be at least equal to the first, since the first was a minimum. (D3) We measure perceptible objects by studying these minima in succession, beginning from the first. They do not touch each other part to part (since they have no parts), nor do they coincide in one and the same place. They are arranged in succession, and they form the units of perceptible magnitude; more of them form a larger magnitude, fewer of them a smaller magnitude.*

(E) Similarly with the minimum in the atom—though it is much smaller than the perceptible minimum. The similarity is to be expected, since we have already argued that the atom has magnitude by an analogy with perceptible things, thus in effect projecting the atom on the larger scale of perceptible things.

(F) Furthermore, these minimal, partless extremities furnish the primary, irreducible unit of measurement, in terms of which we "see" the magnitude of atoms of different size when we study them in thought. So much can be inferred from the analogy with perceptible things (D3); but the latter of course are liable to change, and we must not be led by our analogy to think that atoms too are liable to change, in the sense of being put together out of separable parts.[57]

Epicurus was, as it were, faced with a choice between infinite divisibility and minimal parts. He must have seen that the former alternative would lead to positions inconsistent with experience: for instance, it would be necessary to be able to "reach infinity in thought". Aristotle, who also rejected actual infinity, had resolved the problem of infinite divisibility by introducing the subtle and somewhat elusive concept of potential infinity, and so contrived to avoid the postulation of minimal parts. But Epicurus, as a thorough-going materialist, rejected the idea of a potentiality which could never be actualized,[58] a Becoming never brought to full Being. He would have regarded it as contradictory to assert that a finite magnitude is potentially divisible to infinity, and yet to deny that it consists of an infinity of parts: an entity

[57]Furley (1967), pp. 8–10.

[58]Epicurus' position in this respect is similar to that of Cantor (see Chap. 4 below). Both can be said to have accepted the thesis that any potential infinity presupposes an actual infinity. But the consequences of this acceptance were quite different for the two. Epicurus, a finitist who repudiated actual infinity, was led necessarily to reject the potential infinite as well. But Cantor's whole world view reposed on the actual infinite, so for him the thesis served not to demonstrate the non-existence of the potential infinite, but rather to reveal it as a shadow cast by the substantial reality of the actual infinite.

just consists of those entities into which it is divided, whether potentially or actually. For him the potential had to be treated as if it was actual. This left him no option but to postulate of minimal units of magnitude.

Epicurus's physical atoms were *materially*, but not *conceptually* indivisible.[59] But while the atoms of Democritus *may* have been conceptually divisible *ad infinitum,* those of Epicurus have a finite number of minimal parts which *are* conceptually indivisible. Minimal parts of atoms may be considered as constituting the ultimate units of magnitude (the existence of which Aristotle, as a divisionist, had explicitly denied). This left Epicurus' theory vulnerable to another objection raised by Aristotle against atomism:

> *For if they [atoms] are all of one substance…why do not they become one when they come into contact, just as water does when it becomes water?*[60]

In the Epicurean theory atoms are conceived as being composed of a finite number of minimal parts in contact. However, when two atoms come into contact they must, to avoid Aristotle's objection, remain distinct, and accordingly physically separable. But only finitely many minimal parts, perhaps just two, of the respective atoms touch. If two minimal parts can be physically separated, so can any finite number. Since the atom is composed of finitely many minimal parts, it is separable. To avoid this difficulty it has been speculated[61] that the Epicureans adopted the view that the minimal parts of an atom are essentially constituents of *that* atom, and have no separate existence outside it. Thus minimal parts of two *different* atoms coming into contact are separable, but from this it no longer follows that minimal parts of the *same* atom are separable.[62]

On one point, however, the Epicurean theory is clear. The properties of atoms are reducible to the numbers and arrangements of ultimate units, and physics is thereby reduced to combinatorics. What, then, of geometry? As far as is known, Epicurus made no attempt to work out the consequences for geometry of his atomistic doctrine of magnitudes. Such an "Epicurean geometry", with its ultimate units of magnitude, would have to meet the challenge of the existence of incommensurable magnitudes (e.g. the side and diagonal of a square), and resolve the seeming absurdity that, if a square is built up from miniature tiles as "units", there are as many tiles along the diagonal as there are along the side, and so the diagonal should be equal in length to the side.[63] Furley suggests that it might be expected that Epicurus would regard geometry as irrelevant to the study of nature, since one of its basic principles—infinite divisibility—is contrary to the facts of nature. Some evidence for this

---

[59]For a penetrating discussion of the problem of material indivisibility, see Pyle (1997), Appendix 1.

[60]Aristotle (2000a), *On Coming-to-Be and Passing Away.*, I, 8., 326a 32–33.

[61]Pyle (1997), Appendix 1.

[62]In this respect Epicurean minimal parts may be said to resemble the quarks that are currently presumed to be the ultimate building-blocks of matter: just as Epicurean minimal parts have no separate existence, quarks appear only in groups of two, three, or five.

[63]These difficulties are similar to those encountered by the Islamic atomists in the ninth and tenth centuries: see below.

surmise is provided by Proclus, in whose *Commentary on Euclid* the Epicureans are identified as "those who criticize the principles of geometry alone".

Furley illustrates Epicurus' theory through the analogy of a drawing made on a piece of graph paper by shading some squares of the grid and leaving others blank: the shaded squares then represent units of matter, the unshaded squares units of "void". The squares are all considered as wholes, so there is no place for part of a square, or the diagonal of a square. If they are arranged in rows, the right edge of one is in contact with the left edge of the next. But the edges are not "parts" of the squares in the sense that one might fill in the edge of a void square while leaving the rest blank: the squares are indivisible wholes. The Epicurean atom was, according to Furley, supposed to exist within a three-dimensional grid of this kind. It is not necessary for the cells of the grid to be all of the same shape or size.

But why did Epicurus identify his minimum units of extension with *parts* of atoms, rather than with the atoms themselves? Furley's conjecture is that it was a response to Aristotle's analysis of motion, which had established that, if indivisible magnitudes actually exist, then the distance traversed by a moving body must be composed of indivisible minimal units. This made it necessary for Epicurus to consider, in addition to the moving atoms themselves, the places they successively occupied. It would then become clear that the units must all be equal, for otherwise absurd consequences would follow, such as that a (geometrically) indivisible space was too large or too small for a (geometrically) indivisible atom to fit into it. But from the equality of the minimal units, it would have to follow that either all atoms are identical in size, or else some atoms occupy more than one unit of spatial extension. Epicurus, Furley surmises, would have rejected the first alternative as not squaring with phenomena, and would accordingly have adopted the second.

The Stoics and Others

In opposition to the atomists, the Stoic philosophers **Zeno of Cition** (fl. 250 B.C.) and **Chrysippus** (280–206 B.C.) upheld the Aristotelian position that space, time, matter and motion are all continuous. And, like Aristotle, they explicitly rejected any possible existence of void within the cosmos. The cosmos is pervaded by a continuous invisible substance which they called *pneuma* (Greek: "breath"). This pneuma—which was regarded as a kind of synthesis of air and fire, two of the four basic elements, the others being earth and water—was conceived as being an elastic medium through which impulses are transmitted by wave motion. All physical occurrences were viewed as being linked through tensile forces in the pneuma and matter itself was held to derive its qualities form the "binding" properties of the pneuma it contains.

A major difficulty encountered by the Stoic philosophers was that of the nature of *mixture*, and, in particular, the problem of explaining how the pneuma mixes with material substances so as to "bind them together". The atomists, with their granular conception of matter, did not encounter any difficulty here, since they could regard the mixture of two substances as an amalgam of their constituent atoms into a kind of lattice or mosaic. But the Stoics, who regarded matter as continuous, had difficulty with the notion of mixture. For in order to mix fully two continuous substances, they would either have to interpenetrate in some mysterious way, or, failing that, these would each have to be subjected to an infinite division into infinitesimally small elements which would then have to be arranged, like finite atoms, into some kind of discrete pattern. The mixing of particles of finite size, no matter how small they may be, presents no difficulties. But this is no longer the case when we are dealing with a continuum, whose parts can be divided *ad infinitum*. Thus the Stoic philosophers were confronted with what was at bottom a *mathematical* problem.

Plutarch reports an attempted resolution of the cone dilemma by Chrysippus:

> *[Chrysippus] says that the surfaces will neither be equal nor unequal; the bodies, however, will be unequal, since their surfaces are neither equal nor unequal.*[64]

Sambursky[65] thinks that the first part of this quotation refers to the process of convergence to the limit; he also regards Chrysippus as having been the first to get a clear grasp of this idea. Sambursky suggests that, if we consider the infinite sequence of sections of the cone approaching the given one, "we have to discard the static concept of equal and unequal, taking into account that for each given difference in surfaces one can determine a distance which will yield a still smaller difference." It is Sambursky's contention that this is what Chrysippus intended to express by the phrase "neither equal nor unequal". Sambursky also contends that, provided one interprets "body" as the solid contained between parallel sections of the cone, the second part of the quotation is intended by Chrysippus to mean that, of the surfaces of three adjacent sections A_1, A_2, A_3, the volume defined by the surfaces A_1 and A_2 is unequal to that defined by A_2 and A_3, despite the fact that both $A_3 - A_2$, and $A_2 - A_1$ both converge to zero. Sambursky points out that, while it is most unlikely that Chrysippus formulated a rigorous proof of this proposition, it is necessary to ensure that in the limit process the volumes of adjacent sections do not become equal, which would lead to a cylinder instead of a cone and thus restore Democritus' dilemma.

Michael White[66] interprets Chrysippus as meaning that the two adjacent surfaces cannot be exactly equal, yet there is no discriminable quantity by which one exceeds the other. White suggests that the indiscriminably small difference between the surfaces may be represented as infinitesimals within Robinson's nonstandard analysis.[67]

---

[64]Quoted in Sambursky (1971), p. 93.

[65]*Ibid.*, p. 94 f.

[66]White (1992), Ch. 7.

[67]Another possibility is to formulate the problem within the framework of smooth infinitesimal analysis, where intuitionistic logic prevents infinitesimals from being in general equal or unequal to zero. See Chap. 10 below.

Chrysippus also considered the problem raised by Aristotle concerning the reality of time. Aristotle had pointed out that only the "now" actually existed, but as a mere boundary between past and future, a mathematical point, it cannot be counted a part of time. As Sambursky[68] points out, the reduction of the "now" to a mathematical point could have been avoided by postulating the existence of indivisible temporal atoms. While atomists such as Xenocrates and Epicurus would have regarded this move as unexceptionable, Chrysippus and his fellow-Stoics, as synechists, had no choice but to reject it. Instead they suggested what Sambursky calls a "dynamic solution" to the problem: as reported by Plutarch,

> *The Stoics do not admit the existence of a shortest element of time, nor do they concede that the "now" is indivisible, but that which someone might assume and think of as present is according to them partly future and partly past. Thus nothing remains of the Now, nor is there left any part of the present, but what is said to exist now is partly spread over the future and partly over the past.*[69]

Plutarch also quotes Chrysippus's view on time:

> *He states most clearly that no time is entirely present. For the division of continua goes on indefinitely, and by this distinction time, too, is infinitely divisible; thus no time is strictly present but is defined only loosely.*[70]

Sambursky interprets this "looseness" of definition of the present as "the result of a limiting process of convergence consisting in an infinite approach to the mathematical Now both from the direction of the past and from the future". So, according to his account, "the present is given by an infinite sequence of nested time intervals shrinking towards the mathematical "now", and it is therefore to be regarded as a duration of only indistinctly defined boundaries whose fringes cover the immediate past and future."[71]

In his celebrated work *De Rerum Natura,* Epicurus's Roman disciple **Lucretius** (*c.*100–55 B.C.) offers a systematic exposition of the former's materialist atomism, arguing against the views of various divisionists, including Anaxagoras and the Stoics. Lucretius formulates what appears to be a new argument for the existence of minimal parts:

> *If there are no such least parts, even the smallest bodies will consist of an infinite number of parts, since they can be halved and their halves halved again without limit. On this showing, what difference will there be between the whole universe and the very least of things? None at all. For, however endlessly infinite the universe may be, yet the smallest things will equally consist of an infinite number of parts. Since true reason cries out against this and denies that the mind can believe it, you must needs give in and admit that there are least parts which themselves are partless.*[72]

---

[68]Sambursky (1963), p. 151.

[69]Quoted in Sambursky (1963), p. 151.

[70]*Ibid.*

[71]*Ibid.*, pp. 151–2. Smooth infinitesimal analysis suggests another way of interpreting the Stoic conception of time.

[72]Lucretius (1955), p. 45

In appealing to "true reason" Lucretius would appear to be invoking the Euclidean axiom that the whole is always greater than the part. And indeed, divisionism does seem to lead to a violation of that hallowed principle. For, under the hypothesis of infinite divisibility, complete division into parts of, say, a line, and its half, would in both cases yield infinities manifesting no "difference".[73] Aristotle would have striven to avoid this unpalatable conclusion by taking refuge in potentiality and denying that complete division of a continuum could actually be carried out. But, as has already been observed, the materialist Epicurus rejected the notion of an unrealizable potentiality, and Lucretius followed suit. For them the complete division of a continuum must terminate after finitely many steps (as we would say), yielding minimal parts.

The neoplatonist philosopher **Damascius** (c. 462–540) was exercised by Aristotle's conundrum concerning the unreality of time. Arguing that the present is more than a mere instant, indeed has an extension, and so can be regarded as a *part* of time, he concluded that one of the parts of time does actually exist. Of the views of Damascius and the neoplatonist school on this question Simplicius reports:

> *I am impressed by how they solve Zeno's problem by saying that the movement is not completed with an indivisible bit, but rather progresses in a whole stride at once. The half does not always precede the whole, but sometimes the movement as it were leaps over both whole and part. But those who said that only an* indivisible *now existed did not recognize the same thing happening in the case of time. For time always accompanies movement and as it were runs along with it, so that it strides along together with it in a whole continuous jump and does not progress one now at a time ad infinitum. This must be the case because motion obviously occurs in things, and because Aristotle shows clearly that nothing moves or changes in a now but only* has *moved or changed, whereas things do change and move in time. At any rate, the leap in movement is a* part *of the movement which occurs in the course of moving and will not be taken in the now; nor, being present, will it occur in the non-present. So that in which the present movement occurs is the present time, and it is infinitely divisible, just as the movement is, for each is continuous, and every continuum is infinitely divisible.*[74]

Here "Zeno's problem" refers to the dichotomy paradox. Certain of Aristotle's synechist successors, including, apparently, the Stoics, were not content with Aristotle's resolution of the paradox by appeal to potentiality and introduced another device for circumventing it. They held that a motion could occur all at once, in a "leap", without the half-motion taking place before the whole. In that case the moving body vanishes from one position and reappears a little further on, without an intervening time lapse. The leap itself could still be thought of as infinitely divisible because the distance traversed would be infinitely divisible; another leap could be made across a shorter distance, in fact over any distance, however short. This idea of "motion by leaps" served to resolve Zeno's paradox: provided a body can leap in the manner indicated, it is not necessary for it to travel first the half-

[73]In the sense that the collections of parts of a line and its half can be put into one-one correspondence.

[74]Quoted in Sorabji (1982), pp. 74–5.

distance, and before that half of the half, ad infinitum, as Zeno's paradox threatens. Of course, this "resolution" raises difficulties of its own, not the least being that it involves a body being in two places at once.

Damascius's idea seems to have been that time itself embodies such "leaps". Being divisible, these temporal leaps are not atoms, however. Damascius defines time as "the *measure* of the flow of being"; Sorabji (1982) suggests that Damascius had in mind here a numerical, or discrete measure, like the hours in a day. If time is a discrete measure, then it will obviously contain leaps. On the other hand, the leaps can be called infinitely divisible, for the discrete stages recorded by the measure can be made arbitrarily close. Here we see the extension to the continuum of time of the idea, long familiar in the case of the linear continuum, of imposing a discrete measure in the form of an arbitrarily small unit.

1.2 Oriental and Islamic Views

China

Chinese natural philosophy seems to have inclined more to synechism than to atomism. In the *Chuang Tzu,* "The Book of Master Chuang", written around 290 B.C., are to be found a number of paradoxes, including some startlingly similar to those of Zeno. For example[75]:

> *That which has no thickness cannot be piled up, but it can cover a thousand square miles in area.*
>
> *The shadow of a flying bird has never yet moved.*
>
> *There are times when a flying arrow is neither in motion nor at rest.*
>
> *If a stick one foot long is cut in half every day, it will still have something left after ten thousand generations.*

The last of these would seem to be affirm a divisionist view and to deny the existence of atoms, at least of a material nature.

On the other hand,[76] the idea of a geometric point appears in the Mohist Canon of c. 330 B.C. There we find the following definition of a point:

> *The line is separated into parts, and that part which has no remaining part (i.e., cannot be divided into still smaller parts) and thus forms the extreme end of a line is a point.*

Further elaboration follows:

---

[75]Needham (1954–), vol. II, pp. 190–1.

[76]Needham, (1954-), 19(h).

If you cut a length continually in half, you go forward until you reach the position that the middle (of the fragment) is not big enough to be separated into halves; and then it is a point. Cutting away the front part (of a line) and cutting away the back part, there will eventually remain an indivisible point in the middle. Or if you keep cutting into half, you will come to a stage at which there is an almost nothing, and since nothing cannot be halved, this can no more be cut.

It is to be noted that this characterization as uncuttable applies equally well to an infinitesimal as to a point.

Like the Islamic philosophers (see below), the Mohists also[77] seem to have considered the idea of atoms or instants of time, as witness these passages from the Mohist canon:

The "beginning" means an (instant of) time.

Time sometimes has duration and sometimes not, for the "beginning" point of time has no duration.

The following passages on the nature of cohesion, contact, and coincidence would not seem out of place in Aristotle:

A discontinuous line includes empty spaces.

The meaning of "empty" is like the spaces between two opposed pieces of wood. In these spaces there is no wood; that is, surfaces cannot be absolutely smooth and cannot therefore fully cohere.

Contact means two bodies mutually touching.

Lines placed in contact with each other will not necessarily coincide, since one may be longer or shorter than the other. Points placed in contact with one another will coincide because they have no dimensions. If a line is placed in contact with a point, they may or may not coincide; they will do so if the point is placed at the end of the line, for both have no thickness; they will not do so if the point is placed at the middle of the line, for the line has length while the point has no length. If a hard white thing is placed in contact with another hard white thing, the hardness and whiteness will coincide mutually; since the hardness and whiteness are qualities diffused throughout the two objects, they may be considered to permeate the new larger object formed by the contact of the two smaller ones. But two material bodies cannot mutually coincide because of the mutual impenetrability of material solids.

Qualities such as hardness or whiteness being conceived by the Mohists as "diffused throughout" or "permeating" material objects, that is, continuously, it would seem naturally to follow that they saw matter itself as continuous. Indeed, according to Needham, Chinese natural philosophy as a whole was, like the world-view of the Stoics, "dominated throughout by the concept of waves rather than of atoms". The two fundamental forces in the universe, the Yin and the Yang, were conceived by the Chinese as exerting their influence in oscillatory succession, the one waxing as the other wanes. Needham sums up the Chinese view as follows:

---

[77]Needham (1954-), 26(b).

> *. . .the Chinese physical universe in ancient and medieval times was a perfectly continuous whole.* Chhi [matter-energy, similar to the pneuma of the Stoics] *condensed in palpable matter was not particulate in any important sense, but individual objects acted and reacted with all other objects in the world. Such mutual influences could be effective over very great distances, and operated in a wave-like or vibratory manner dependent in the last resort on the rhythmic alternation at all levels of the two fundamental forces, the Yin and the Yang. Individual objects thus had their intrinsic rhythms. And these were integrated like the sounds of individual instruments in an orchestra, but spontaneously, into the general pattern of harmony in the world.*[78]

This conception of the universe could not be further removed from atomism's flurry of particles in a void.

India

From very early on atomism played a role in Indian philosophical thought.[79] Generally speaking, it was subscribed to by thinkers of a realist tendency, such as the Jains, the Hinayana school of Buddhism, and the adherents of Nyaya-Vaisesika; and opposed by the idealists, most notably the Mahayana school of Buddhism and the Vedantists. The origins of atomism in India are shrouded in obscurity. Some scholars have claimed to find traces of atomism in the Upanishads. Others have sought to explain its emergence in India by drawing a parallel with the situation in ancient Greece: just as atomism emerged there as a response to Eleatic monism, so the analogous doctrine arose in India as a response to the doctrine of the eternal and immutable Brahman of the Upanishads.

The Indian idealists took issue with the atomist claim that atoms were both corporeal and yet also partless, holding this to be contradictory. For, they argued, the corporeal, being spatially extended, is composed of parts. Only the noncorporeal, for instance consciousness and sensation, is partless. For a dualist this would mean that reality is made up of two continua – a continuum of matter and a continuum of mind – with radically different properties: the one composed of parts and so divisible, and the other a partless unity, an Eleatic monad. But idealists, like materialists, are first and foremost monists, and the Indian idealists were no exception. The Vedantists in particular repudiated the idea that the world could be a plurality. In their unswerving pursuit of the ideal of unity, they took the unity of consciousness and the self as the ultimate reality, regarding as illusory, as what they tened *Maya,* not merely the external world with its apparent multiplicities, but also the received notion that there exists a multiplicity of minds or consciousnesses. For the Vedantists, there was only one Mind, and individual minds were held to be held to be illusions.

[78]Needham (1954-), 26(b).

[79]See Gangopadhyaya (1980).

Islamic Thought

Greek philosophy, and in particular Greek theories of the continuum, enjoyed a revival in Islamic thought from the seventh century A.D. Synechism and atomism once again did battle, with the latter eventually proving the more infuential within Islamic philosophy.

The dispute seems to have begun soon after 800 with the controversy between the philosophers **Nazzam** (died c. 846), a divisionist, and **Abu l-Hudahyl al-Allaf** (died c. 841), an atomist.[80] Like Damascius and the Stoics, Nazzam shielded his divisionist belief from Zenonian paradox by maintaining that motion takes place in divisible leaps. Nazzam had a number of interesting arguments against atomism. One of these had been earlier used by the Greeks for the same purpose. Since atomic movements take no time, they must all occur at the same speed. Now the Islamic atomists accounted for evident differences in (linear) speed by allowing an atom to linger for a varying number of time atoms in successive space atoms. But this raises diffculties in accounting for *rotatory* motion: the inner atoms of a rotating millstone, for example, must linger in their places, while the more rapidly moving outer atoms continue to progress, resulting in fragmentation or distortion of the millstone. Sorabji suggests how Nazzam's theory of divisible leaps may have overcome this difficulty. For, since all points in the millstone can leap simultaneously, it is never required that one point in the millstone remains still while others move. The divisibility of the leaps allows the points to rematerialize at precisely those distances required to preserve the millstone's shape. Of course, motion conceived as continuous also avoids this "fragmentation" problem, but, in Nazzam's eyes such motion was subject to Zeno's paradox. Leaping motion alone could avoid both Zeno's paradox and the fragmentation problem.

Another argument of Nazzam's against atomism is of particular interest, because of its influence on medieval European discussions. Consider a square and one of its diagonals. If atoms are sizeless, then, Nazzam contends, from every sizeless atom on the diagonal a straight line can be drawn at right angles until it joins a sizeless atom on one of the two sides. When all such lines have been drawn, they will be parallel and no gaps will lie between them. Thus to each atom on the diagonal there corresponds exactly one atom one one of the two sides, and vice-versa. So there must be the same number of atoms along the diagonal of a square as along the two adjoining sides. In that case the absurd conclusion is reached that the route along the diagonal should be no quicker than the route along the two sides.

These and other arguments against material atomism due to **Avicenna** (980–1037) appear in the *Metaphysics* of **Algazel** (1058–1111), through which they came to exert a considerable influence on the thinkers of medieval Europe.

The most forceful champions of atomism among the Islamic philosophers were the Mutakallemim, or professors of the Kalam, of the tenth and eleventh centuries. The views of this school are critically summarized in **Maimonides**' (1135–1204)

[80] Sorabji (1982), pp. 37–87.

Guide for the Perplexed. The summary takes the form of 12 propositions and commentaries, of which those concerning atomism are the first three—that all things are composed of atoms, that a vacuum exists, and that time is composed of time-atoms.

On the first of these propositions Maimonides comments:

> *"The Universe, that is, everything contained in it, is composed of very small parts [atoms} which are indivisible on account of their smallness; such an atom has no magnitude; but when several atoms combine, the sum has a magnitude, and thus forms a body." . . . All these atoms are perfectly alike; they do not differ from each other in any point. . .*[81]

On the second proposition he observes:

> *The original also believe that there is a vacuum, i.e. one space, or several spaces that contain nothing, which are not occupied by anything whatsoever, and are devoid of all substance. This proposition is to them an indispensable sequel to the first. For if the Universe were full of such atoms, how could any of them move? For it is impossible to conceive that one atom should move into another. And yet the composition, as well as the decomposition of things, can only be effected by the motion of atoms. Thus the Mutakallemim are compelled to assume a vacuum, in order that the atoms may combine, separate, and move in that vacuum that does not contain any thing or any atom.*[82]

On the existence of time-atoms and the consequences thereof Maimonides is at his most critical:

> *"Time is composed of time-atoms, i.e., of many parts, which on account of their short duration cannot be divided." This proposition also is a logical consequence of the first. The Mutakallemim undoubtedly saw how Aristotle proved that space, time and locomotion are of the same nature, that is to say, they can be divided into parts which stand in the same proportion to each other: if one of them is divided, the other is divided in the same proportion. They, therefore, knew that if time were continuous and divisible* ad infinitum, *their assumed atom of space would of necessity likewise be divisible. Similarly, if it were supposed that space is continuous, it would necessarily follow, that the time-element, which they considered to be indivisible, could also be divided. . . . Hence they concluded that space was not continuous, but was composed of elements that could not be divided; and that time could likewise be reduced to time-elements, which were indivisible. . .Time would thus be an object of position and order.*
>
> *The Mutakallemim did not at all understand the nature of time. . . Now mark what conclusions were accepted by the Mutakallemim as true. They held that locomotion consisted in the translation of each atom of a body from one point to the next one; accordingly the velocity of one body in motion cannot be greater than that of another body. When, nevertheless, two bodies are observed to move during the same time through different spaces, the cause of the difference is not attributed by them to the fact that the body which has moved through a larger distance had a greater velocity, but to the circumstance that motion which in ordinary language is called slow, has been interrupted by more moments of rest, while the motion which ordinarily is called quick has been interrupted by fewer moments of rest. . . .*

Maimonides derides the atomists' account of geometry:

[81] Maimonides (1956), p. 120.

[82] *Ibid.*, p. 121.

> *Nor must you suppose that the aforegoing theory concerning motion is less irrational than the proposition resulting from this theory, that the diagonal of a square is equal to one of its sides, and some of the Mutakallemim go so far as to declare that the square is not a thing of real existence. In short, the adoption of the first proposition would be tantamount to the rejection of all that has been proved in Geometry. The propositions in Geometry would, in this respect, be divided into two classes: some would be absolutely rejected; e.g., those which relate to properties of the incommensurability and the commensurability of lines and planes, to rational and irrational lines, and all other propositions in the tenth book of Euclid, and in similar works. Other propositions would appear to be only partly correct; e.g., the solution of the problem of dividing a line in two equal parts, if the line consists of an odd number of atoms; according to the theory of the Mutakallemim such a line cannot be bisected.*[83]

And the atomists' account of bodily rotation is scrutinized:

> *We ask them: "Have you observed a complete revolution of a millstone? Each point in the extreme circumference of the stone describes a large circle in the very same time in which a point nearer the centre describes a small circle; the velocity of the outer circle is therefore greater than that of the inner circle. You cannot say that the motion of the latter was interrupted by more moments of rest; for the whole moving body, i.e., the millstone, is one coherent body." They reply, "During the circular motion, the parts of the millstone separate from each other, and the moments of rest interrupting the motion of the portions nearer the centre are more than those which interrupt the motion of the outer portions." We ask again: "How is it that the millstone, which we perceive as one body, and which cannot be easily broken, even by a hammer, resolves into its atoms when it moves, and becomes again one coherent body, returning to its previous state as soon as it comes to rest, while no one is able to observe its breaking up?" Again their reply is based on the twelfth proposition, which is that the senses cannot be trusted, and that only the evidence of the intellect is admissible.*[84]

This last sentence bears witness to the philosophical gulf that had opened up between Epicureanism and Islamic philosophy: the Epicureans held that all knowledge derived from the senses, while their Islamic successors maintained the exact opposite. Yet for both atomism was a central tenet.

The commentaries on Aristotle by the Islamic philosopher **Averroes** (1126–1198) came to be widely disseminated in the West, where they exerted great influence. The doctrine of natural minima played a central role in Averroes' conception of continuous substance, in which the distinction between physical and mathematical divisibility is made quite clearly.[85] Witness, for example, this passage from his commentary on Aristotle's *Physics*:

> *A line as a line can be divided indefinitely. But such a division is impossible if the line is taken as made of earth*[86]

Averroes considered natural minima as actual parts of (continuous) substances, so possessing a kind of physical reality which makes them not dissimilar to atoms. The

[83] *Ibid.*

[84] *Ibid.*, p. 122.

[85] Van Melsen (1952), p. 59.

[86] Quoted in *ibid.*, p. 59.

doctrine of natural minima thus allowed the atomistic principle to gain a foothold in the continuum theory.

1.3 The Philosophy of the Continuum in Medieval Europe

The scholastic philosophers of Medieval Europe, in thrall to the massive authority of Aristotle, mostly subscribed in one form or another to the thesis, argued with great effectiveness by the Master in Book VI of the *Physics,* that continua cannot be composed of indivisibles. On the other hand, the avowed infinitude of the Deity of scholastic theology, which ran counter to Aristotle's thesis that the infinite existed only in a potential sense, emboldened certain of the Schoolmen to speculate that the actual infinite might be found even outside the Godhead, for instance in the assemblage of points on a line.

A few scholars of the time chose to follow Epicurus in upholding atomism and attempted to circumvent Aristotle's counterarguments. **Henry of Harclay** (c.1275–1317), for instance, claimed that a line is composed of an actual infinity of points in immediate juxtaposition, touching "whole to whole" without superposition. Even with the introduction of actual infinity, this view is open to the objection raised by Aristotle against the contiguity of points and was attacked on that and related grounds.

The anti-Aristotelian **Nicholas of Autrecourt** (c.1300–69) was also an atomist, holding that space is composed of points and time of instants. Like Henry, he claimed *contra* Aristotle that points, "having their own position and mode of being", can touch "whole to whole" without superposition, and thereby constitute an extended magnitude. In the section entitled "Indivisibles" of his *Universal Treatise* he attempts to answer a number of Aristotle's objections to the atomistic thesis. With regard to motion he reiterates the Mutakallemim theory that motion takes place through atomic "jerks", drawing the conclusion that a body moving without stopping has attained the upper limit to velocity. Rejecting Aristotle's contention that a continuum is divisible into parts only potentially, not actually, he sums up his own view of the composition of the continuum in the following terms:

> *First, the continuum is not composed of parts which can always be further divided (and in this there would be a departure from common opinion). Secondly, a continuum demonstrable to sense or imagination is not composed of a finite number of points (and in this a departure would be made from the opinion of all those who have posited that a continuum is composed of indivisibles). And on this basis other propositions are true. For example, "For every magnitude which is given or which is pointed out to sense or imagination, there is a smaller one (at least, nothing seems to prevent this), and yet, along with this, it will be true to say that there is in a thing some magnitude such that a smaller cannot be found."*[87]

[87]Nicholas of Autrecourt (1971), p. 82.

Nicholas's conclusions

> *For every magnitude pointed out to sense or imagination, there is a smaller one,*

and

> *there is in a thing some magnitude such that a smaller one cannot be found,*

seem contradictory at first sight. Andrew Pyle[88] has pointed out that the contradiction between these two propositions is only apparent, since the first asserts the reducibility of every sensible or imaginable magnitude, while the second denies that this holds for every magnitude *tout court.* The "irreducible" magnitude of the second proposition "does not come to sense or imagination as a finished being"[89] and is smaller than any reducible (that is, sensible or imaginable) magnitude. Pyle tentatively suggests that Nicholas's irreducible magnitude might be seen as an early instance of the idea of an *infinitesimal.* But Pyle, who holds the view that the infinitesimal is an incoherent concept, makes this suggestion only to underline what he sees as the ultimate untenability of Nicholas's program.[90] It seems to me, however, that, despite its obscurities, Nicholas's vision is closer in certain respects to the punctate account of the continuum given by Cantor and Dedekind in the nineteenth century.[91] For Nicholas actually asserts that "a continuum is composed of points; and a continuum, marked on a wall or elsewhere, whether sensed or imagined, is composed of *infinite* points."[92] (my emphasis). And he goes on to deny that the compounding of infinitely many points would necessarily lead to infinite extension. These are two key features of the Cantor-Dedekind theory.

The incipient atomism of the fourteenth century met with a determined synechist rebuttal. This was initiated by **John Duns Scotus** (c. 1266–1308). In his analysis[93] of the problem of "whether an angel can move from place to place with a continuous motion" he offers a pair of purely geometrical arguments against the composition of a continuum out of indivisibles. One of these arguments is a variant of Nazzam's: that, if the diagonal and the side of a square were both composed of points, then not only would the two be commensurable in violation of Book X of Euclid, they would even be equal. The other[94] starts from Euclid's second postulate that a circle of any diameter can be constructed with any point as centre. Two unequal circles are

---

[88]Pyle (1997), p. 208.

[89]Nicholas of Autrecourt (1971), p. 82.

[90]In Pyle's words:

> *We might almost claim that Nicholas was one of the founders of the doctrine of the* infinitesimal, *that curious creature greater than nothing yet less than anything, an infinity of which make up a magnitude. However great its* heuristic *value in the history of mathematics this doctrine is quite incoherent and the infinitesimal—as was already apparent in Zeno's day—an impossible entity.*

[91]See Chap. 4 below.

[92]Nicholas of Autrecourt (1971), p. 80.

[93]In his *Opus Oxoniense*. My source here is Murdoch's discussion of medieval atomism in §52 of Grant (1974).

[94]Grant (1974), p. 317.

Fig. 1.2

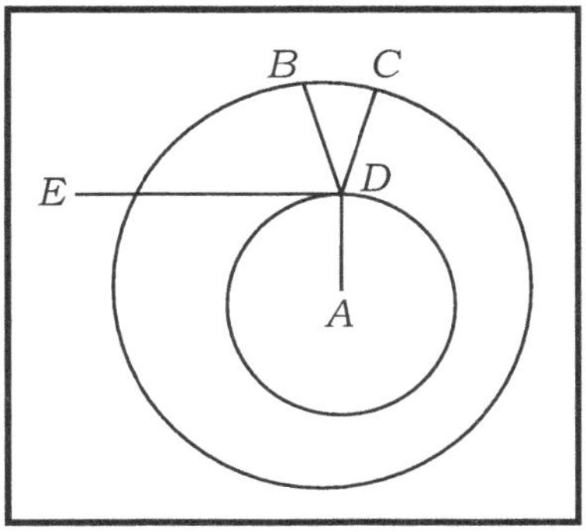

constructed about a common centre A. Supposing the larger circle to be composed of points, choose two contiguous such points B and C and draw straight lines from A to B and C (Fig. 1.2).

The lines then cut the circumference of the smaller circle at right angles, either at different points, or at a single point. If at different points, there will be just as many points on the larger circle as on the smaller, violating Euclid's fifth axiom that the whole is greater than the part. Suppose, on the other hand, that the straight lines AB and AC cut the smaller circle at a single point D. Then the tangent line ED to the smaller circle at D makes two right angles with AB, and also with AC. Hence $\angle ADE$ together with $\angle BDE$ is equivalent to two right angles, and likewise for $\angle ADE$ together with $\angle CDE$. By Euclid's third postulate, all right angles are equal, so if we subtract the angle these two pairs have in common, namely, ADE, the remainders will be equal; consequently $\angle BDE$ will be equal to $\angle CDE$ and the part again equal to the whole.

William of Ockham (c. 1280–1349) brought considerable dialectical subtlety[95] to his analysis of continuity; it has been the subject of much scholarly dispute.[96] For Ockham the principal difficulty presented by the continuous is the infinite divisibility of space, and in general, that of any continuum.[97] The treatment of continuity in the first book of his *Quodlibet* of 1322–7 rests on the idea that between any two points on a line there is a third—perhaps the first explicit formulation of the property of density—and on the distinction between a *continuum* "whose parts form a unity" from a *contiguum* of juxtaposed things.[98] Ockham recognizes that it follows from the property of density that on arbitrarily small stretches of a line infinitely many points must lie, but resists the conclusion that lines, or indeed any continuum, consists of points. Concerned, rather, to determine "the sense in which the line may be said to consist or to be made up of anything.",[99] Ockham continues:

[95]He seems to have refrained, however, from subjecting the continuum to his celebrated "razor".

[96]See, e.g. the papers of Murdoch and Stump in Kretzmann (1982).

[97]Burns (1916), p. 506.

[98]*Ibid.*, p. 510.

[99]*Ibid.*, p. 507.

> *Here we must express the true meaning of the problem, which is this: whether any parts of a line are indivisible: and if such be the meaning, I say that* no part of the line is indivisible, nor is any part of a continuum indivisible.[100]

In proving this he first notes that indivisible parts of a line would have to be at once points and minimum lengths; there can be only finitely many of these and so a line consists of finitely many points. The remainder of his proof proceeds along the lines of the similar demonstration in Duns Scotus.[101]

While Ockham does not assert that a line is actually "composed" of points, he had the insight, startling in its prescience, that a punctate and yet continuous line becomes a possibility when conceived as a dense array of points, rather than as an assemblage of points in contiguous succession.

The most ambitious and systematic attempt at refuting atomism in the fourteenth century was mounted by **Thomas Bradwardine** (c. 1290–1349). The purpose of his *Tractatus de Continuo* (c. 1330) was to "prove that the opinion which maintains continua to be composed of indivisibles is false."[102] This was to be achieved by setting forth a number of "first principles" concerning the continuum—akin to the axioms and postulates of Euclid's *Elements*—and then demonstrating that the further assumption that a continuum is composed of indivisibles leads to absurdities. In his stimulating analysis[103] of this work John Murdoch enumerates what he sees as its successes in regard to what its author hoped to establish. Among these Murdoch includes Bradwardine's improved Aristotelian definition of continuum; his strict definition of indivisible rendering it independent of the idea of extension; his rigorous demonstration on the basis of the two previous definitions that a continuum cannot be created through the juxtaposition or superposition of indivisibles; and his demonstration that the primary assumptions of Euclidean geometry presuppose the infinite divisibility of the geometric line. On the other hand Bradwardine does not seem to have grasped the possibility, suggested by Ockham, that a continuous line could be composed of a dense array of indivisibles.

Nicole Oresme (c. 1325–1382), the foremost French thinker of the fourteenth century, made a number of significant contributions to mathematics, introducing in particular the idea of representing uniformly accelerated motion by means of linear graphs. He also made translations from Latin into French, with commentaries, of several of the works of Aristotle, including *De Caelo*.[104] In his commentary on Book I of Aristotle's work Oresme observes that the terms "magnitude", "continuous body" and "continuum" are synonymous. He then comments on Aristotle's assertion "the continuous is that which is divisible into parts, which themselves are continuously divisible." Like Averroes, Oresme draws a distinction between the physical

---

[100] *Ibid.*, p. 507.

[101] *Ibid.* p. 507–9.

[102] Murdoch (1957), p. 54.

[103] *Op. cit.*

[104] Oresme (1968).

and mathematical divisibility of continua, but places a stronger emphasis on the potentially infinite nature of the latter form of divisibility:

> Divisible *is used in two ways: one way it means the real separation of the parts of anything, and the other way it means division conceptually in the mind. It is not to be thought that every magnitude or continuum is divisible in the first sense, for it is naturally impossible to divide the heavens as one divides a wooden log, separating one part from another. In dividing a log or a stone or another material or destructible object, one can reach a part so small that further division would destroy its substance. But any continuum or magnitude is continually divisible conceptually in the human mind, just as astrologers divide the heavens into degrees, the degrees into minutes, the minutes into seconds, the seconds into thirds, fourths, and then fifths. The imagination can proceed thus endlessly. In the same way, any object such as earth, water, a stone, a log, etc., has many parts, and each of its parts has many parts, and so on and on; just as each body has two halves and each half has two halves, proceeding thus endlessly even though by these divisions we arrive at parts so small that they are imperceptible to the senses. This is true of all continuous things like a line, a surface, a solid body, motion, time and similar things; for each of these has parts and we cannot say nor think a number of parts so great that it could not be greater, even a hundred or a thousand times greater, beyond any ratio, without any end or limit, however small the thing may be, even the thousandth part of a grain of millet.*[105]

The later emergence of the mathematical concept of *function* owes much to Oresme. The function concept is closely tied up with the idea of continuity, more exactly, with the idea of one attribute varying continuously, but not necessarily uniformly, with another. The ancient Greek thinkers had grasped that motion, for instance, could be described as a continuous variation of distance or position with time. Thus the idea of a variable *quantity*, of one quantity depending on another quantity, was accepted in Greek philosophy, as is indicated by the fact that Greek mathematicians such as Hippias and Archimedes employed the idea in the generation of curves. On the other hand, motion itself was considered a quality and as such unquantifiable; indeed Aristotle had explicitly repudiated the idea of instantaneous velocity.[106] And the notion of a variable quality, that is, of a quality (continuously) correlated with a quantity, seems not to have been regarded as an admissible concept in Greek science. But the thirteenth century in Europe saw the emergence of a theory of variable qualities in which the germ of the concept of function can be discerned—the so-called doctrine of the *latitude of forms*, Here the term *form* "refers in general to any quality which admits of variation and which involves the intuitive idea of intensity—that is, to such notions as velocity, acceleration, density".[107] A form is accordingly what later became known as an *intensive quantity*. The latitude of a form was "the degree to which the latter possessed a certain quality", and the central concern of the theory was the study of the manner in which these qualities could be intensified or diminished—the *intensio* or *remissio* of the form. As forms or intensive quantities the Scholastics considered not only what we would today call instantaneous velocity (although they lacked an exact definition of the notion), but

[105]*Ibid.* pp. 45–47.

[106]Aristotle (1980), 234a.

[107]Boyer (1959), p. 73.

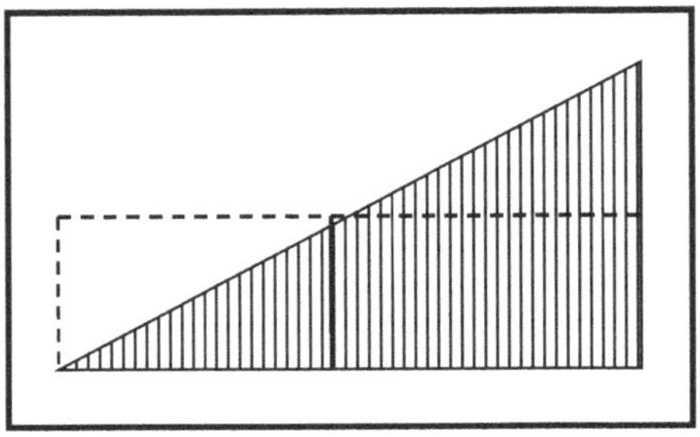

Fig. 1.3

also brightness, temperature, and density. They also distinguished between uniform and nonuniform rates of change, rates of rates of change, and the like.

Oresme held that everything measurable is imaginable as continuous quantity, and he hit upon the brilliant idea of drawing a picture or graph of the way a measurable form could vary. In doing so he ushered in the idea of a function being represented by (continuous) curve, although he was unable to make effective use of it except in the case of linear functions. In particular he drew what amounts to a velocity-time graph for a body moving with uniform acceleration (Fig. 1.3).[108] In this diagram points on the base of the triangle represent instants of time, and the length of each vertical line the velocity of the body at the corresponding instant. Oresme realized that the distance covered by the body is represented by the area of the triangle and, since this latter coincides with the area of the indicated rectangle, was able to infer the rule that, under uniformly accelerated motion, the average velocity is the arithmetic mean of the terminal and initial velocities.

The views on the continuum of **Nicolaus Cusanus** (1401–64), a champion of the actual infinite, leave a somewhat contradictory impression. In his treatise *Of Learned Ignorance* of 1440, he contrasts the indivisibility of the infinite line (a rather mystical conception) with the divisibility of any finite line:

> *A finite line is divisible, whereas an infinite line, in which the maximum is at one with the minimum, has no parts and is in consequence indivisible. The finite line, however, cannot be divided into anything but a line, for, as we have already seen, in dividing an extended object, we never reach a minimum point which is the smallest that can exist.*[109]

On the other hand, in *De Mente Idiotae* of 1450, in answer to the question "What dost thou understand by an atom?" Cusanus responds:

> *Under mental consideration that which is continuous becomes divided into the ever divisible, and the multitude of parts progresses to infinity. But by actual division we arrive at an actually indivisible part which I call an atom. For an atom is a quantity, which on account of its smallness is actually indivisible.*[110]

[108]Boyer and Merzbach (1989), p. 264 f.

[109]Cusanus (1954), p. 36.

[110]Quoted in Stones (1928).

Here Cusanus seems not to be contrasting the "mental" or "geometric" continua, which are infinitely divisible, and the "physical" continua of extended matter, which are not. Rather, he is saying that *any* continuum, be it geometric, perceptual, or physical, is subject to *two* types of successive division, the one *ideal*, the other *actual*. Ideal division "progresses to infinity"; actual division terminates in atoms after finitely many steps. This distinction is similar to that between physical and mathematical divisibility found in Oresme.

Cusanus's realist conception of the actual infinite is reflected in his quadrature of the circle.[111] He took the circle to be an *infinilateral* regular polygon, that is, a regular polygon with an infinite number of (infinitesimally short) sides. By dividing it up into a correspondingly infinite number of triangles, its area, as for any regular polygon, can be computed as half the product of the apothegm (in this case identical with the radius of the circle), and the perimeter. The idea of considering a curve as an infinilateral polygon was employed by a number of later thinkers, for instance, Kepler, Galileo and Leibniz.

[111]Boyer (1959), p. 91. The argument may well be of Greek origin.

Chapter 2
The Sixteenth and Seventeenth Centuries. The Founding of the Infinitesimal Calculus

2.1 The Sixteenth Century

From Stevin to Kepler

The early modern period saw the spread of knowledge in Europe of ancient geometry, particularly that of Archimedes, and a loosening of the Aristotelian grip on thinking. In regard to the problem of the continuum, the focus shifted away from metaphysics to technique, from the problem of "*what* indivisibles were, or *whether* they composed magnitudes" to "the new marvels one could *accomplish* with them"[1] through the emerging calculus and mathematical analysis. Indeed, tracing the development of the continuum concept during this period is tantamount to charting the rise of the calculus.

Traditionally, geometry is the branch of mathematics concerned with the continuous and arithmetic (or algebra) with the discrete. The infinitesimal calculus that took form in the sixteenth and seventeenth centuries, which had as its primary subject matter *continuous variation*, may be seen as a kind of synthesis of the continuous and the discrete, with infinitesimals bridging the gap between the two. The widespread use of indivisibles and infinitesimals in the analysis of continuous variation by the mathematicians of the time testifies to the affirmation of a kind of mathematical atomism which, while logically questionable, made possible the spectacular mathematical advances with which the calculus is associated. It was thus to be the infinitesimal, rather than the infinite, that served as the mathematical stepping stone between the discrete and the continuous.

The "dynamic" as opposed to "static" form this mathematical atomism assumed in the work of **Simon Stevin** (1548–1620) has been identified by historians of

[1]Murdoch (1957), p. 325.

J. L. Bell, *The Continuous, the Discrete and the Infinitesimal in Philosophy and Mathematics*, The Western Ontario Series in Philosophy of Science 82,
https://doi.org/10.1007/978-3-030-18707-1_2

Fig. 2.1

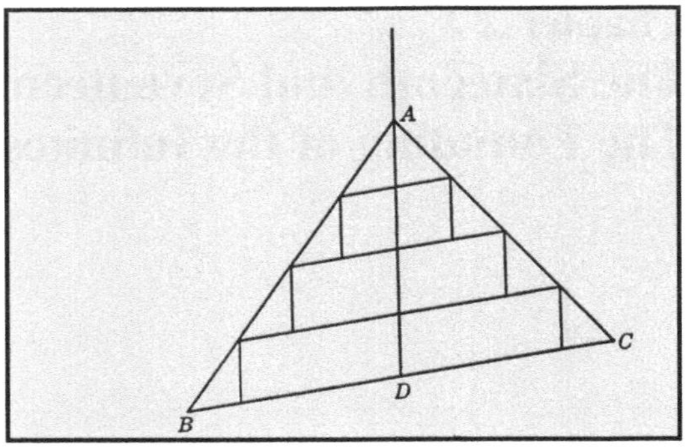

mathematics as an anticipation of the theory of limits.[2] It is instructive to see how Stevin proved (in his work on statics of 1586) that the centre of gravity of a triangle lies on its median. Here is Boyer's account (see Fig. 2.1):

> *Inscribe in the triangle ABC a number of parallelograms of equal height...The center of gravity of the inscribed figure will lie on the median by the principle that bilaterally symmetrical figures are in equilibrium.... However, we may inscribe in the triangle an infinite number of such parallelograms, for all of which the center of gravity will lie on AD. Moreover, the greater the number of parallelograms thus inscribed, the smaller will be the difference between the inscribed figure and the triangle ABC. If, now, the "weights" of the triangles ABD and ACD are not equal, they will have a certain fixed difference. But there can be no such difference, inasmuch as each of these triangles can be made to differ by less than this from the sums of the parallelograms inscribed within them, which are equal. Therefore the "weights" of ABD and ACD are equal, and hence the centre of gravity of the triangle ABC lies on the median AD.*[3]

Here Stevin implicitly employs the principle that any pair of magnitudes, the difference between which can be shown to be less than any assigned magnitude, are themselves equal.[4] The "infinity of parallelograms" inscribable in the triangle seems to mean a potential, not an actual infinity, while the parallelograms themselves are varying elements of area—and so of the same dimension as the figure to which their sum approximates—rather than fixed "indivisible" lines of a lower dimension. Stevin applied the same idea to determine the centres of gravity of a number of plane curvilinear figures and solids.

In contrast with Stevin's cautious approach, **Johann Kepler** (1571–1630) made abundant use of infinitesimals in his calculations. In his *Nova Stereometria* of 1615, a work actually written as an aid in calculating the volumes of wine casks, he regards curves as being infinilateral polygons, and solid bodies as being made up of infinitesimal cones or infinitesimally thin discs.[5] A particularly fetching application

---

[2]A similar, but independent approach was taken by **Luca Valerio** (1552–1618).

[3]Boyer (1959), pp. 99–100.

[4]Baron (1987), p. 97

[5]*Ibid.*, pp. 108–116; Boyer (1969), pp. 106–110.

Fig. 2.2

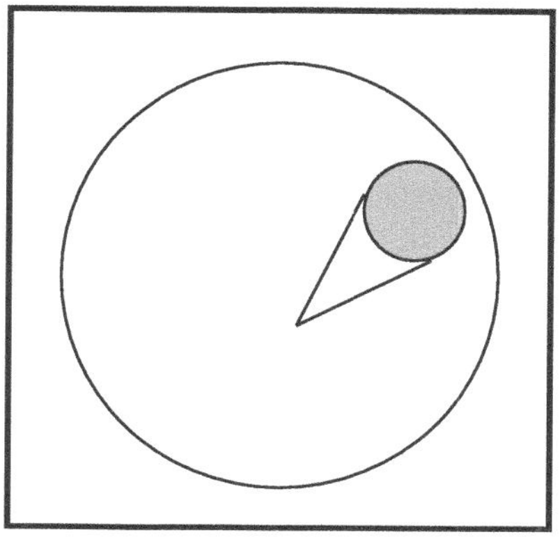

of the former idea is Kepler's determination of the volume of a sphere.[6] The sphere may be regarded as being made up of infinitesimal cones, each with its base on the sphere's surface and its apex at the centre (Fig. 2.2). From the fact that the volume of a cone is 1/3 × height × area of base, it follows that the sphere's volume is 1/3 × radius × surface area.

Such uses are in keeping with Kepler's customary use of infinitesimals of the same dimension as the figures they constitute; but he also used indivisibles on occasion. He spoke, for example, of a cone as being composed of circles,[7] and in his *Astronomia Nova* of 1609, the work in which he states his famous laws of planetary motion, he takes the area of an ellipse to be the "sum of the radii" drawn from the focus.

It seems to have been Kepler who first introduced the idea, which was later to become a reigning principle in geometry, of *continuous change of a mathematical object*, in this case, of a geometric figure.[8] In his *Astronomiae pars Optica* of 1604 Kepler notes that all the conic sections are continuously derivable from one another both through focal motion and by variation of the angle with the cone of the cutting plane.

Galileo and Cavalieri

Galileo Galilei (1564–1642) advocated a form of mathematical atomism in which the influence of both the Democritean atomists and the Aristotelian scholastics can be discerned. This emerges when one turns to the First Day of his *Dialogues Concerning Two New Sciences* (1638), more particularly to the extended discussion

[6]Baron (1987), p. 110.
[7]Boyer (1959), p. 107.
[8]Kline (1972) p. 299.

therein of the composition of material continua. Salviati, Galileo's spokesman, proposes an atomic account of matter similar in spirit to that of Democritus: bodies are composed of "infinitely small indivisible particles",[9] themselves infinite in number, interspersed with an infinity of infinitesimally small vacua. Salviati, that is, Galileo, also maintains, contrary to Bradwardine and the Aristotelians, that continuous magnitude is made up of indivisibles, indeed an infinite number of them:

> *Since lines and all continuous quantities are divisible into parts which are themselves divisible without end, I do not see how it is possible to avoid the conclusion that these lines are built up of an infinite number of indivisible quantities because a division and a subdivision which can be carried on indefinitely presupposes that the parts are infinite in number, since otherwise the subdivision would reach an end*[10]*; and if the parts are infinite in number, we must conclude that they are not finite in size, because an infinite number of finite quantities would give an infinite magnitude. And thus we have a continuous quantity built up of an infinite number of indivisibles.*[11]

Salviati/Galileo recognizes that this infinity of indivisibles will never be produced by successive subdivision, but claims to have a method for generating it all at once, thereby removing it from the realm of the potential into actual realization:

> *I am willing, Simplicio, at the outset, to grant to the Peripatetics the truth of their opinion that a continuous quantity is divisible only into parts which are still further divisible so that however far the division and subdivision be continued no end will be reached; but I am not so certain that they will concede to me that none of these divisions of theirs can be a final one, as is surely the fact, because there always remains "another"; the final and ultimate division is rather one which resolves a continuous quantity into an infinite number of indivisible quantities, a result I grant can never be reached by successive division into an ever-increasing number of parts. But if they employ the method which I propose for separating and resolving the whole of infinity, at a single stroke... I think that they would be contented to admit that a continuous quantity is built up out of absolutely indivisible atoms, especially since this method, perhaps better than any other, enables us to avoid many intricate labyrinths...*[12]

And what is this "method for separating and resolving, at a single stroke, the whole of infinity"? Simply the act of bending a straight line into a circle:

> *If now the change which takes place when you bend a line at right angles so as to form now a square, now an octagon, now a polygon of forty, a hundred or a thousand angles, is sufficient to bring into actuality the four, eight, forty, hundred, and thousand parts which, according to you, existed at first only potentially in the straight line, may I not say, with equal right, that, when I have bent the straight line into a polygon having an infinite number of sides, i.e., into a circle, I have reduced to actuality that infinite number of parts which you claimed, while it was straight, were contained in it only potentially.*[13]

---

[9]Galilei (1954), p. 55.

[10]Here Galileo is giving expression to the conviction, which he shares with Cantor, that no potential infinity is unaccompanied by an actual one.

[11]Galilei (1954), pp. 33–34

[12]*Ibid.*, p. 48

[13]*Ibid.*, p. 47.

Here Galileo finds an ingenious "metaphysical" application of the idea of regarding the circle as an infinilateral polygon. When the straight line has been bent into a circle Galileo seems to take it that that the line has thereby been rendered into indivisible parts, that is, points. But if one considers that these parts are the sides of the infinilateral polygon, they are better characterized not as indivisible points, but rather as *unbendable* straight lines, each at once part of and tangent to the circle.[14] Galileo does not mention this possibility, but nevertheless it does not seem fanciful to detect the germ here of the idea of considering a curve as an assemblage of infinitesimal "unbendable" straight lines.[15]

While Galileo was firm in his conviction that continua are *composed* of an infinity of indivisibles in an actual sense, he considers that the *number* of such indivisibles must be regarded as a potential infinity:

> *. . .if we consider discrete quantities, I think there is, between finite and infinite quantities, a third intermediate term which corresponds to every assigned number; so that if asked, as in the present case, whether the finite parts of a continuum are finite or infinite in number the best reply is that they are neither finite nor infinite but correspond to every assigned number.*[16]

In the last analysis both the infinite and the infinitesimal are essentially beyond the grasp of intuition:

> *. . .infinity and indivisibility are in their very nature incomprehensible to us; imagine then what they are when combined. Yet if we wish to build up a line out of indivisible points, we must take an infinite number of them, and are, therefore, bound to understand both the infinite and the indivisible at the same time. . .*[17]

The mysterious "merging of parts into unity"[18] to form a continuum is likened to the liquefaction of a solid:

> *Having broken up a solid into many parts, having reduced it to the finest of powder and having resolved it into its infinitely small indivisible atoms why may we not say that this solid has been reduced to a single continuum, perhaps a fluid like water or mercury or even a liquefied metal? And do we not see stones melt into glass and the glass itself become under strong heat more fluid than water?*
>
> *. . .I am not able to find any better means [than that substances become fluid in virtue of being resolved into their infinitely small indivisible components] of accounting for certain phenomena of which the following is one. When I take a hard substance such as stone or*

---

[14]Hermann Weyl makes a similar suggestion in connection with Galileo's "bending" procedure:

> *If a curve consists of infinitely many straight "line elements", then a tangent can simply be conceived as indicating the direction of the individual line segment; it joins two "consecutive" points on the curve.* (Weyl 1949, p. 44.)

[15]This conception was to prove fruitful in the later development of the calculus and to achieve fully rigorous formulation in the synthetic differential geometry of the later twentieth century. See Chap. 10 below.

[16]Galilei (1954), p. 37

[17]*Ibid.*, p. 30.

[18]Boyer (1959), p. 116

metal and when I reduce it by means of a hammer or hard file to the most minute and impalpable powder, it is clear that its finest particles, although when taken one by one are, on account of their smallness, imperceptible to our sight and touch, are nevertheless finite in size, possess shape, and the capability of being counted. It is also true that when once heaped up they remain in a heap; and if an excavation be made within limits the cavity will remain and the surrounding particles will not rush in to fill it; if shaken the particles come to rest immediately after the external disturbing agent is removed; the same effects are observed in all piles of larger and larger particles, of any shape, even if spherical, as is the case with piles of millet, wheat, lead shot, and every other material. But if we attempt to discover such properties in water we do not find them; for when once heaped up it immediately flattens out unless held up by some vessel or other external retaining body; when hollowed out nit quickly rushes in to fill the cavity; and when disturbed it fluctuates for a long time and sends out its waves through great distances.

Seeing that water has less firmness than the finest of powder, in fact has no consistence whatsoever, we may, it seems to me, very reasonably, conclude that the smallest particles into which it can be resolved are quite different from finite and divisible particles; indeed the only difference I am able to discover is that the former are indivisible.[19]

Despite Galileo's uncertainty concerning the exact nature of indivisibles, he employed them in analyzing rectilinear motion and the motion of projectiles, in so doing authorizing their use by his successors.

It was Galileo's pupil and colleague **Bonaventura Cavalieri** (1598–1647) who refined the use of indivisibles into a reliable mathematical tool; indeed the "method of indivisibles" remains associated with his name down to the present day. Cavalieri nowhere explains precisely what he understands by the word "indivisible", but it is apparent that he conceived of a surface as composed of a multitude of equispaced parallel lines and of a volume as composed of equispaced parallel planes, these being termed the indivisibles of the surface and the volume respectively.[20] While Cavalieri recognized that these "multitudes" of indivisibles must be unboundedly large, indeed was prepared to regard them as being actually infinite, he avoided following Galileo into ensnarement in the coils of infinity by grasping that, for the "method of indivisibles" to work, the precise "number" of indivisibles involved *did not matter*. Indeed, the essence of Cavalieri's method was the establishing of a correspondence between the indivisibles of two "similar" configurations,[21] and in the cases Cavalieri considers it is evident that the correspondence is suggested on solely geometric grounds, rendering it quite independent of number. The very statement of *Cavalieri's principle* embodies this idea: if plane figures are included between a pair of parallel lines, and if their intercepts on any line parallel to the including lines are in a fixed ratio, then the areas of the figures are in the same ratio. (An analogous principle holds for solids.) As an application, consider a circle of radius a inscribed in an ellipse with major semi-axis b (see Fig. 2.3). Since the intercepts of the lines parallel to the tangents are in the fixed ratio b: a, Cavalieri's principle entails that the area of the

[19]Galilei (1954), pp. 39–40.

[20]Boyer (1969), p. 117.

[21]*Ibid.*, p. 118.

Fig. 2.3

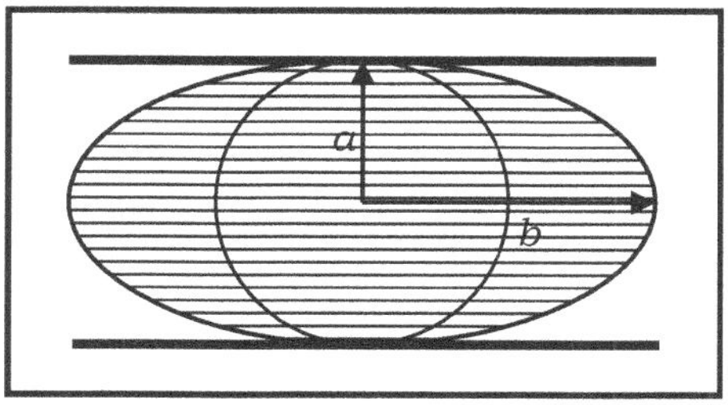

ellipse is $^b/_a \times$ the area of the circle. Clearly the number of parallel lines is irrelevant to the argument.

Cavalieri's method is in essence that of *reduction of dimension*: solids are reduced to planes with comparable areas and planes to lines with comparable lengths. While this method suffices for the computation of areas or volumes, it cannot be applied to rectify curves, since the reduction in this case would be to points, and no meaning can be attached to the "ratio" of two points. For rectification a curve has, it was later realized, to be regarded as the sum, not of indivisibles, that is, points, but rather of infinitesimal straight lines, its "microsegments".

Cavalieri recognized that his method differed from Kepler's in taking continua to be sums of *heterogenea,* that is, parts of a lower dimension, rather than *homogenea,* parts of an unreduced dimension. He was challenged by critics that this feature made his whole approach unsound: for how can a sum of planes without thickness produce a volume, or breadthless lines a surface? Cavalieri failed to provide a convincing answer to this question, sidestepping it with the response that, while the indivisibles are correctly regarded as lacking thickness or breadth, nevertheless one "could substitute for them small elements of area and volume in the manner of Archimedes".[22] In a number of places Cavalieri suggests that surfaces and volumes could be regarded as being generated by "the flowing of indivisibles".[23] Although he failed to develop this idea into a geometric method, the suggestion is natural enough, given that continua had from the first been regarded as being generated by motion: a line can be considered as the trace of a moving point, a surface as a moving line, a solid as a moving surface.

[22] *Ibid.*, p. 121.
[23] *Ibid.*, p. 122.

2.2 The 17th Century

The Cartesian Philosophy

The mathematical work of **René Descartes** (1596–1650), though unquestionably of major importance, has been described as "only an episode in his philosophy".[24] He, too, employed infinitesimals in his early mathematical work, and was acquainted with the views concerning them of his predecessors and contemporaries.[25] Later, however, he came to avoid infinitesimals, and even to deplore their use. He developed purely algebraic methods for determining tangents to curves, directing some of his sharpest criticism at those such as Fermat who employed infinitesimals for this purpose.

As a philosopher Descartes may be broadly characterized as a synechist. His philosophical system rests on two fundamental principles: the celebrated Cartesian dualism—the division between mind and matter—and the less familiar identification of matter and spatial extension. In the *Meditations* Descartes distinguishes mind and matter on grounds similar to that of the Indian idealists—that the corporeal, being spatially extended, is divisible, while the mental is partless:

> *...there is a vast difference between mind and body, in respect that body, from its nature, is always divisible, and that mind is entirely indivisible. For in truth, when I consider the mind, that is, when I consider myself in so far only as I am a thinking thing, I can distinguish no parts, but I very clearly discern that I am somewhat absolutely one and entire; and although the whole mind seems to be united to the whole body, yet, when a foot, an arm, or any other part is cut off, I am conscious that nothing has been taken from my mind; nor can the faculties of willing, perceiving, conceiving, etc., properly be called its parts, for it is the same mind that is exercised [all entire] in willing , in perceiving, and in conceiving, etc. But quite the opposite holds in corporeal or extended things; for I cannot imagine any one of them [however small soever it may be], which I cannot easily sunder in thought, and which, therefore, I do not know to be divisible. This would be sufficient to teach me that the mind or soul of man is entirely different from the body, if I had not already been apprised of it on other grounds.*[26]

In *The Principles of Philosophy* Descartes identifies space and matter:

> *Space or internal place, and the corporeal substance which is comprised in it, are not different in reality, but merely in the mode by which they are wont to be conceived by us. For, in truth, the same extension in length, breadth and depth, which constitutes space, constitutes body; and the difference between them lies only in this, that in body we consider extension as particular; whereas in space we attribute to extension a generic unity...Nothing remains in the idea of body, except that it is something extended in length, breadth and depth; and this something is comprised in our idea of space, not only of that which is full of body, but even of what is called void space.*[27]

[24] *Ibid.*, p. 166.

[25] *Ibid.*, p. 165

[26] Descartes (1927), p. 139.

[27] *Ibid.*, pp. 203–4.

And from this it follows that matter is continuous and divisible without limit:

> *We likewise discover that there cannot exist any atoms or parts of matter that are of their own nature indivisible. For however small we suppose those parts to be, yet because they are necessarily extended, we are always able in thought to divide any one of them into two or more smaller parts, and may accordingly admit their divisibility. For there is nothing we can divide in thought which we do not recognize to be divisible; and, therefore, were we to judge it indivisible our judgment would not be in harmony with the knowledge we have of the thing; and although we should even suppose that God had reduced any particle of matter to a smallness so extreme that it did not admit of being further divided, it would nevertheless be improperly styled indivisible, for though God had rendered the particle so small that it was not in the power of any creature to divide it, he could not however deprive himself of the ability to do so, since it is absolutely impossible for him to lessen his own omnipotence…Wherefore, absolutely speaking, the smallest extended particle is always divisible, since it is such of its very nature.*[28]

Since extension is the sole essential property of matter and, conversely, matter always accompanies extension, matter must be ubiquitous. Descartes' space is accordingly, as it was for the Stoics, a plenum pervaded by a continuous medium. But Descartes parted from the Stoics in postulating that this medium was primordially divided, in the manner of a Rubik cube, into material corpuscles of equal size. These corpuscles, being themselves divisible, are not atoms. Nor could they be spherical in form, at least initially, because spheres cannot be packed together without leaving some "empty" space. Now Descartes also posited that his corpuscles were originally in continual motion, a motion which, in such a densely packed universe, could only be circular. As a result of this motion certain corpuscles would be ground down, like the pebbles on a beach, into a quasispherical form, the resulting intermediary space becoming filled with the residue left by the process of abrasion. The circular motions of the particles are manifested as whirlpools of material particles varying in size and velocity. This is the basis of Descartes' celebrated theory of *vortices*.

The Cartesian view of the world was firmly endorsed by the authors of the *Port-Royal Logic* (1652), the philosopher-theologians **Antoine Arnauld** (1612–94) and **Pierre Nicole** (1625–95). In that work the Cartesian infinite divisibility of matter, with its worlds within worlds, is described in almost rhapsodic terms:

> *How to understand that the smallest bit of matter is infinitely divisible and that one can never arrive at a part that is so small that not only does it not contain several others, but it does not contain an infinity of parts; that the smallest grain of wheat contains in itself as many parts, although proportionately smaller, as the entire world; that all the shapes imaginable are actually to be found there; and that it contains in itself a tiny world with all its parts—a sun, heavens, stars, planets, and an earth—with admirably precise proportions; that there are no parts of this of this grain that do not contain yet another proportional world?*[29]

And yet these marvels must necessarily be realized, since the infinite divisibility of matter has been provided with geometric demonstration, "proofs of it as clear as

[28] *Ibid.*, p. 209.

[29] Arnauld and Nicole (1996), p. 230.

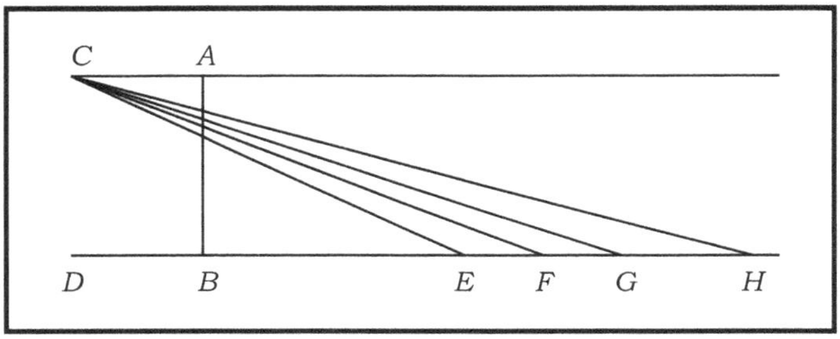

Fig. 2.4

proofs of any of the truths it reveals to us.". Three proofs are initially provided. The first draws on the demonstrated existence of incommensurable lines, the second on the theorem that no whole number exists whose square is double that of another whole number. The third pivots on the observation that the unextended cannot generate extension:

> *Finally, nothing is clearer than this reasoning, that two things having zero extension cannot form an extension, and that every extension has parts. Now taking two of these parts that are supposed to be indivisible, I ask whether they do or do not have any extension. If they have some extension, then they are divisible, and they have several parts. If they do not, they therefore have zero extension, and hence it is impossible for them to form an extension.*[30]

For good measure Arnauld and Nicole throw in another proof of infinite divisibility, this time based on the indefinite extensibility of lines:

> *Certainly, while we might doubt whether extension can be infinitely divided, at least we cannot doubt that it can be increased to infinity, and that we can join to a surface of a hundred thousand leagues another surface of a hundred thousand leagues, and so on to infinity. Now this infinite increase in extension proves its infinite divisibility. To understand this we have only to imagine a flat sea that is increased infinitely in extent, and a vessel at the shore which leaves the port in a straight line. It is certain that when the bottom of this vessel is viewed from the port through a lens or some other transparent body, the ray that ends at the bottom of this vessel will pass through a certain point of the lens, and the horizontal ray will pass through another point of the lens higher than the first. Now as the vessel goes further away, the point of intersection with the lens of the ray ending at the bottom of the vessel will continue to rise and will divide space infinitely between these two points. The further away the vessel goes, the more slowly it will rise, without ever ceasing to rise. Nor can it reach the point of the horizontal ray, because these two lines that intersect at the eye will never be parallel nor the same line. Thus this example provides at the same time proofs of the infinite divisibility of extension and of an infinite decrease in motion.*[31]

This argument was to turn up in a number of other places, notably in Kant's *Physical Monadology* of 1756. Its precise geometric form is the following (see Fig. 2.4). Given two parallel lines *CA* and *CB* and suppose that each is extended to infinity. Draw *AB* perpendicular to *CA* and *DB*. A line drawn from *C* to a point *E* lying to the

[30] *Ibid.*, p. 232.

[31] *Ibid.*

right of B will cut the lie AB at some point, and lines drawn from C to points F, G, H, ... proceeding still further to the right will cut AB at points closer and closer to A. Since DB extends to infinity, this process may be continued indefinitely, generating points on AB that become ever closer to A, but never actually reach it. Thus AB is infinitely divisible.[32]

Of Descartes' contemporaries, it was the philosopher and mathematician **Pierre Gassendi** (1592–1665) who mounted the strongest attack on the Cartesian vision of the world. Gassendi was chiefly responsible for the revival of the physical atomism of Democritus' and Epicurus' view that the physical world consists of indivisible corpuscles moving in empty space. Rejecting Descartes' identification of matter and extension, Gassendi was enabled to accept that pure extension was infinitely divisible, and at the same time maintain that matter was not. In defending Epicurean atomism he denies that Epicurus claimed that every entity is reducible to points:

> *Let us now investigate what is usually objected to in Epicurus: it is really a wonder not only that there should have been some in antiquity who attacked Epicurus as if he had believed that the division of magnitude is terminated in certain mathematical points; but also that there have been learned men from more recent times who inveighed against him in whole tomes as if he had said that bodies were constituted from surfaces, surfaces from lines, lines from points, and accordingly bodies and indeed all things, from points, into which, accordingly bodies and all things were resolved. This is a wonder, I say, since if they had been willing to pay the least attention, they would have been able to observe that the indissectibles into which Epicurus held divisions to be terminated are not mathematical points, but the meagerest bodies; since, moreover, he not only gave them magnitude, when it is acknowledged that there is nothing of this kind in a point; but he also gave them incomprehensibly variable shapes, such as cannot be conceived in a point, which lacks magnitude and parts.*[33]

Here is Gassendi on continuity of magnitude, in which he maintains that atoms are the only absolutely continuous entities:

> *... it will be evident that each body must be said to be continuous insofar as its parts are conjoined, cohering with, and unseparated from one another, and are such that, even though they are only contiguous with each other, their joints cannot be distinguished by any of the senses. That is to say, magnitude or as [the Schoolmen] call it, continuous quantity, differs from multiplicity, i.e., discrete quantity, in the fact that the parts of a continuous quantity can indeed be separated, but are not in fact separated, whereas the parts of a discrete quantity are actually or really separated. This is not to say that the parts of a multiplicity are not also contiguous with each other, a, for example, many stones in one pile; but that they do not mutually grip each other, bind together, and hold onto one another by their own angles or little hooks. ... Thus in a word, all bodies that may be broken apart by the force of heat, or by something else, have parts which are only in mutual contact, and which separate when this bond is severed and their continuity is broken.*
>
> *Thus it is that, if we are asked what is so continuous that it does not on any account consist of the contiguous, the only thing we can specify in reply is the Atom. And we should understand Democritus to have been talking about this when, according to Aristotle, he wrote: "And neither can one become two, nor two, one," since, of course, one atom is not dissectible in*

[32]It is interesting to note that this argument fails for geometries in which lines are not indefinitely extensible, for instance in elliptic geometries.

[33]Quoted in Leibniz (2001), p. 361.

such a way that two should emerge, nor are two penetrable by each other in such a way that they might coalesce into one. But this does not prevent every body that is not actually divided into parts from being called continuous, in accordance with common usage, and inasmuch as the senses cannot reach as far as atoms or their joints.

Aristotle, however, says that a continuum is "a thing whose parts are conjoined at a common boundary." This is physically true, to the extent that it has no two assignable parts which are not conjoined at some intermediate part, either a sensible part—as in a magnitude of three feet, the outermost feet are conjoined by the middle one—or an insensible one—as in a magnitude of two feet, these two feet have an item lying between them that evades the senses ... I ... declare that it is impossible for any continuous thing to be dissected with such great subtlety that middle-sized conglomerates of innumerable atoms are not expelled from it.[34] *Of course, accepting with Aristotle that the common boundary is a mathematical individual (for he asserts that the parts of a line are conjoined by a point, the parts of a surface by a line, and the parts of a body by a surface), this insensible boundary cannot be a physical reality, insofar as there do not exist in the nature of things indissectible things of this kind, which are only imagined or supposed ...*[35]

Gassendi also upheld the Mutakallemim theory that, while motion appears continuous to the senses, it actually occurs in jerks, with alternating periods of movement and rest.

Infinitesimals and Indivisibles

The use of infinitesimals and indivisibles was widespread in seventeenth century mathematics. **Grégoire de St.-Vincent** (1584–1667), like Kepler, regarded figures as being made up of infinitely many infinitesimal elements of the same dimension as the figure, but, with Stevin, conceived of these elements as being obtained by continuing subdivision.[36] **Gilles Personne de Roberval** (1602–75) upheld the infinite divisibility of matter, and of every mathematical continuum.[37] His method of determining areas and volumes, which he termed the "method of infinities", rested on the conception of figures being built up from small surface or volume elements, but treated *as if* they were indivisibles. It was Roberval who first, in 1634, determined the area under an arch of the cycloid. **Evangelista Torricelli** (1608–1647), while not wholly convinced of the logical soundness of the method of indivisibles, was nevertheless virtuosic in its application. The most striking use he made of the method was his discovery, in 1641, that the volume of an infinitely long solid, obtained by revolving a segment of the equilateral hyperbola about its asymptote, is finite.[38] **André Tacquet** (1612–60), who also employed indivisibles in his determinations of areas and volumes, did not regard the method as legitimate, maintaining

---

[34]In this connection see the discussion at the end of Appendix A,

[35]*Ibid.*, pp. 361–2.

[36]Boyer (1959), p. 137.

[37]Walker (1932), p. 34.

[38]Boyer (1959), pp. 125–6.

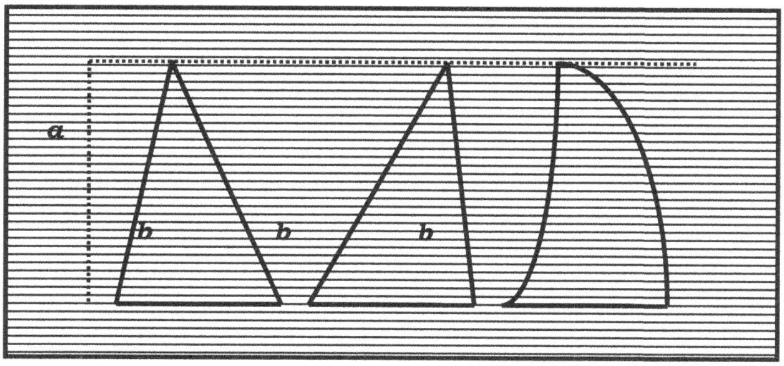

Fig. 2.5

that a geometrical magnitude is made up only of *homogenea*, as opposed to *heterogenea*.[39] **Blaise Pascal** (1623–1662) employed the intuition that addition to a figure of an indivisible of lower dimension had no effect on its size to justify the neglect of terms of lower degree in calculating curvilinear areas. Pascal's "arithmetical" interpretation of indivisibles was to be a major influence on Leibniz, who "repeatedly states that he was led to the invention of the calculus by a study of the works of Pascal".[40]

John Wallis (1616–1703) arithmetized Cavalieri's indivisible methods, using what amounts to limits of sums of integral powers of numbers. He went so far as to introduce the symbol "∞" for the infinitely many lines or parallelograms making up a surface.[41] Unlike Tacquet, Wallis was unconcerned with the question of whether the indivisibles constituting a figure should be taken as *homogenea* or *heterogenea*. On the other hand, he insisted that a line, for example, is to be regarded as possessing sufficient thickness so as to enable it, by infinite multiplication, to acquire an altitude equal to that of the figure in which it is inscribed.

Thus, for example, to determine the area of a triangle with base b and altitude a, one writes ∞ for the total number of lines in the triangle (see Fig. 2.5). Taking the lengths of the lines to be in arithmetic progression, the total length of the lines is then $\frac{1}{2}\infty b$, that is, the sum of the extreme terms 0 and b multiplied by one half the number of terms. Since the altitude of each inscribed parallelogram is $a/\infty = a.\ 1/\infty$, the triangle's area is $\frac{1}{2}\infty b \times a/\infty = \frac{1}{2}ab$.

The concept of infinitesimal had arisen with problems of a geometric character and infinitesimals were originally conceived as belonging solely to the realm of continuous magnitude as opposed to that of discrete number. But from the algebra

[39] *Ibid.*, p. 139.

[40] Baron (1987), p. 205n.

[41] *Ibid.*, pp. 206–7. Baron reports (p. 213) that in Wallis's *A Treatise of Algebra* of 1685 he lists the stages in the development of infinitesimal methods as follows: 1. Method of Exhaustion (Archimedes); 2. Method of Indivisibles (Cavalieri); 3. Arithmetick of Infinites (Wallis); 4. Method of Infinite Series (Newton).

and analytic geometry of the 16th and 17th centuries there issued the concept of *infinitesimal number.* Wallis, for example, treats $1/\infty$ as such a number. However, the idea first appears in the work of **Pierre de Fermat** (1601–65) on the determination of maximum and minimum (extreme) values, published in 1638. Fermat had been struck by the observation, traceable to Pappus, that in a problem which in general has two solutions, an extreme value gives just a single solution.[42] So, for example, in determining the maximum area of a rectangle of semiperimeter a, it is required to maximize the quantity $A = x(a - x)$. In general, there are two values of x corresponding to each value of A, say x and $x + E$. In that case,

$$(*) \qquad x(a - x) = (x + E)(a - x - E),$$

whence

$$0 = aE - 2xE - E^2$$

so that, dividing by E and rearranging,

$$2x = a - E.$$

Now by Pappus' observation, for a maximum value of A, x and $x + E$ must coincide, so that $E = 0$. It then follows that $x = a/2$.

There is a glaring logical flaw in this argument (of which Fermat seems to have been aware), namely, that E is initially assumed to differ from 0, and is then finally set equal to 0. Simultaneously equal and unequal to zero: a strange sort of "number" indeed! In answer to criticisms, Fermat later modified his method by observing that at a maximum point the two values of A set equal in $(*)$ above are not actually equal but "should be" equal.[43] He accordingly introduces the relation of "adequalitas" or "near-equality" between the two values, a relation which, on setting $E = 0$, coincides with genuine equality.[44] This amounts to treating E as an infinitesimal number, a quantity "nearly equal", but not necessarily coinciding with, zero. This move does not, of course, do away with the logical difficulties; what it does seem to indicate is that Fermat thought of infinitesimals as formal, algebraic conceptions rather than as variable quantities.

Fermat's treatment of maxima and minima contains the germ of the fertile technique of "infinitesimal variation", that is, the investigation of the behaviour of a function by subjecting its variables to small changes. Fermat applied this method in determining tangents to curves and centres of gravity.

Another of Fermat's major contributions to infinitesimal analysis is to be found in his work on the *rectification of curves.* While it had been long known that a

---

[42] Boyer (1959), p. 155.

[43] Here Fermat seems to recognize that the function $f(x) = x(a - x)$ is locally one-one.

[44] Boyer (1959), p. 156.

curvilinear figure could be equal in area to a rectilinear one (Archimedes, for instance, had shown that the area ofa parabolic segment is $\frac{4}{3}$ times thevarea of the triangle with the same base and vertex), it was by no means clear whether a curved line could be exactly equal in length to a straight one. Aristotle had held a negative view[45] on the matter, a view which Fermat and most of his contemporaries shared. Nevertheless, shortly before 1660 a number of mathematicians, including Christopher Wren, Heinrich van Heuraet, and John Wallis, produced such rectifications. Fermat's response was to carry out a rectification of the semicubical parabola $ky^2 = x^3$ by reducing the problem to the quadrature of a parabola.[46]

Barrow and the Differential Triangle

It has been claimed that **Isaac Barrow** (1630–77) was "the first inventor of the Infinitesimal Calculus".[47] While this may be an exaggeration, there is no disputing the importance of his role in its emergence. Barrow was one of the first mathematicians to grasp the reciprocal relation between the problem of quadrature and that of finding tangents to curves—in modern parlance, between integration and differentiation. Indeed, in his *Lectiones Geometricae o*f 1670, Barrow observes, in essence, that if the quadrature of a curve $y = f(x)$ is known, with the area up to x given by $F(x)$, then the subtangent to the curve $y = F(x)$ is measured by the ratio of its ordinate to the ordinate of the original curve.[48]

Barrow's approach, in 1669, to the problem of finding tangents to curves[49] is particularly instructive in its use of the *characteristic, differential* or *infinitesimal* triangle, a device which played an important role in the early development of the calculus. Starting with the triangle *PRQ* generated by the increment *PR* (Fig. 2.6), the fact that this triangle is similar to the triangle *PMN* enables one to claim that the slope *QR/PR* of the tangent is equal to *PM/MN*. Now, Barrow asserts, when the arc *PP'* is sufficiently small (which follows upon taking the increment *PR* sufficiently small, see Fig. 2.7) we may safely identify it with the segment *PQ* of the tangent at *P*. (Here Barrow is evidently thinking of the curve as an infinilateral polygon.) The triangle *PRP'*, in which *PP'* is regarded both as an arc of the curve and as part of the

[45]Aristotle (1980), VII, 4. See also Heath (1949), pp. 140–142.

[46]Boyer (1959), pp. 162–163.

[47]Child (1916), p. viii. The quotation continues:

> *Newton got the main idea of it from Barrow by personal communication; and Leibniz also was in some measure indebted to Barrow's work, obtaining confirmation of his own original ideas, and suggestions for their further development, from the copy of Barrow's book he purchased in 1673.*

[48]Article "Infinitesimal Calculus", *Encyclopedia Britannica,* 11th edition. But, while Barrow recognized the fact, he failed to put it to systematic use.

[49]Kline (1972), p. 346.

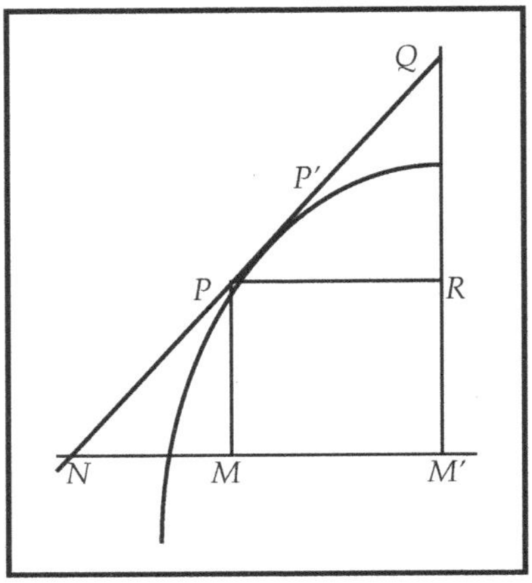

Fig. 2.6

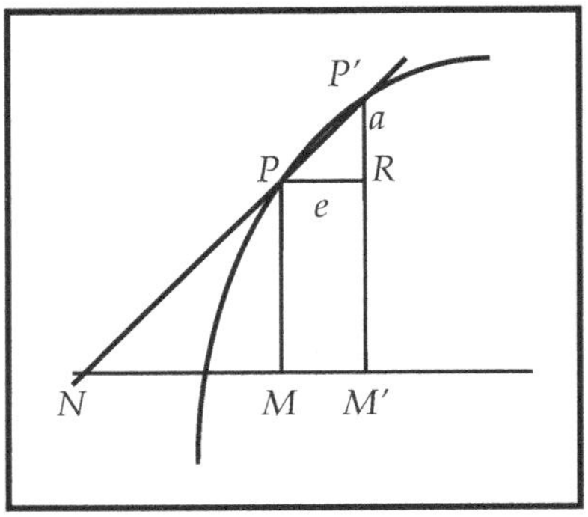

Fig. 2.7

tangent, is the characteristic triangle. The characteristic triangle associated with an infinitesimal abscissal increment at a point on a curve thus represents the additional increment in area under the curve over what would have been generated had the curve been straight from that point on.[50]

Barrow's explicit determination of tangents proceeds along lines similar to that followed by Fermat.[51] Starting with the equation of a curve, say $y^2 = px$, he replaces x by $x + e$ and y by $y + a$, obtaining

$$y^2 + 2ay + a^2 = px + pe.$$

Form this $y^2 = px$ is subtracted, yielding

[50]In smooth infinitesimal analysis, the area of a characteristic triangle always reduces to 0. See Chap. 10 below.

[51]Kline (1972), pp. 346–7.

$$2ay + a^2 = pe.$$

Then, in a typical move, he discards powers of a and e above the first, which amounts to replacing the first figure above by the second. It then follows that

$$\frac{a}{e} = \frac{p}{2y}.$$

But $a/e = PM/NM$, so that

$$\frac{PM}{NM} = \frac{p}{2y}.$$

Since PM is y, the position of N is determined.

Barrow regarded the conflict between divisionism and atomism as a live issue, as may be inferred from a remark in his *Lectiones Mathematicae* of 1683:

> *I am not unaware, for indeed nobody is unaware, that this doctrine of the perpetual divisibility of quantity is admitted by some only reluctantly, and simply rejected by others, and that the controversy over composition of magnitudes (whether it be from indivisibles, or from homogeneous parts) is everywhere conducted with great obstinacy.*[52]

As presented in the 1683 *Lectiones,* Barrow's case against geometric atomism pivots on the fact that it is in contradiction with most of the propositions of Euclidean geometry:

> *. . .there is the necessary agreement of all mathematicians [to the thesis of infinite divisibility]; for although they hardly ever suppose it openly, they often assume it covertly, and if it were not true, many of their demonstrations would fall to the ground.*[53]

As can be seen from the 1670 *Lectiones*, Barrow conceived of continuous magnitudes as being generated by motions, and so necessarily dependent on time:

> *Every magnitude can be either supposed to be produced, or in reality can be produced, in innumerable ways. The most important is that of "local movements". In motion, the matters chiefly to be considered are the mode of motion and the quantity of the motive force. Since quantity of motion cannot be discussed without Time, it is necessary first to discuss Time. Time denotes not an actual existence, but a certain capacity or possibility for a continuity of existence; just as space denotes a capacity for intervening length. Time does not imply motion, as far as its absolute and intrinsic nature is concerned; not any more than it implies rest; whether things move or are still, whether we sleep or wake, Time pursues the even tenor of its way. Time implies motion to be measurable; without motion we could not perceive the passage of Time. . . .*
>
> *Time has many analogies with a line, either straight or circular, and therefore may be conveniently represented by it; for time has length alone, is similar in all its parts, and can be looked upon as constituted from a simple addition of successive instants or as from a*

---

[52]Quoted in Jesseph (1993), p. 63n.

[53]Quoted *ibid,* p. 63.

continuous flow of one instant; either a straight or a circular line has length alone, is similar in all its parts, and can be looked upon as being made up of an infinite number of points or as the trace of a moving point.[54]

Barrow followed Cavalieri in regarding figures as being composed of indivisibles, and, like Wallis, was indifferent to the question of whether the indivisibles constituting a figure should be taken as *homogenea* or *heterogenea*:

To every instant of time, or indefinitely small particle of time, (I say instant or indefinite particle, for it makes no difference whether we suppose a line to be composed of points or indefinitely small linelets; and so in the same manner, whether we suppose time to be made up of instants or indefinitely minute timelets); to every instant of time, I say, there corresponds some degree of velocity, which the moving body is supposed to possess at the instant; to this degree of velocity there corresponds some length of space described.. Hence, if through all the points of a line representing time are drawn.. parallel lines, the plane surface that results as the aggregate of the parallel straight lines, when each represents the degree of velocity corresponding to the point through which it is drawn, exactly corresponds to the aggregate of the degrees of velocity, and thus most conveniently can be adapted to represent the space traversed also. Indeed, this surface, for the sake of brevity, will be called the aggregate of the velocity or the representative of the space. It may be contended that rightly to represent each separate degree of velocity retained during any timelet, a very narrow rectangle ought to be substituted for the right line and applied to the given interval of time. Quite so, but it comes to the same thing whichever way you take it; but as our method seems to be simpler and clearer, we will in future adhere to it.[55]

Barrow's conceptions formed the starting point for the groundbreaking work on the infinitesimal calculus of his illustrious pupil **Isaac Newton** (1642–1727).

Newton

Newton's meditations on the subject during the plague year 1665–66 issued in the invention of what he called the "Calculus of Fluxions", the principles and methods of which were presented in three tracts published many years after they were written[56]: *De analysi per aequationes numero terminorum infinitas; Methodus fluxionum et serierum infinitarum;* and *De quadratura curvarum.* Newton's approach to the calculus rests, even more firmly than did Barrow's, on the conception of continua as being generated by motion. In the introductory paragraph to *De quadratura curvarum* he adopts the kinematic view explicitly, contrasting it with the conception of magnitudes as being composed of infinitesimal parts:

I don't here consider Mathematical Quantities as composed of Parts extremely small, but as generated by a continual motion. Lines are described, and by describing are generated, not by any apposition of Parts, but by a continual motion of Points. Surfaces are generated by

[54] Child (1916), pp. 35–7.

[55] *Ibid.,* pp. 38–9.

[56] *De analysi,* written 1666, published 1711; *Methodus fluxionum,* written 1671, published 1736; *Quadratura,* written c. 1676, published 1704.

> *the motion of Lines, Solids by the motion of Surfaces, Angles by the rotation of their Legs, Time by a continual flux, and so in the rest.*[57]

Newton's exploitation of the kinematic conception went much deeper than had Barrow's. In *De Analysi,* for example, Newton introduces a notation for the "momentary increment" (*moment*)—evidently meant to represent a moment or instant of time—of the abscissa or the area of a curve, with the abscissa itself representing time. This "moment"—effectively the same as the infinitesimal quantity Fermat had denoted by *E* and Barrow by *e*—Newton denotes by *o* in the case of the abscissa, and by *ov* in the case of the area. From the fact that Newton uses the letter *v* for the ordinate, it may be inferred that Newton is thinking of the curve as being a graph of velocity against time. By considering the moving line, or ordinate, as the moment of the area Newton established the generality of and reciprocal relationship between the operations of differentiation and integration, a fact that Barrow had grasped but had not put to systematic use. Before Newton, quadrature or integration had rested ultimately "on some process through which elemental triangles or rectangles were added together",[58] that is, on the method of indivisibles. Newton's explicit treatment of integration as inverse differentiation was the key to the integral calculus.

In the *Methodus fluxionum* Newton makes explicit his conception of variable quantities as generated by motion and introduces his characteristic notation. He calls the quantity generated by a motion a *fluent*, and its rate of generation a *fluxion.* The fluxion of a fluent x is denoted by $x^{\bullet}$, and its moment, or "infinitely small increment accruing in an infinitely short time o ", by $x^{\bullet}\, o$. The problem of determining a tangent to a curve is transformed into the problem of finding the relationship between the fluxions $x^{\bullet}$ and $y^{\bullet}$ when presented with an equation representing the relationship between the fluents x and y. (A quadrature is the inverse problem, that of determining the fluents when the fluxions are given.) Thus, for example, in the case of the fluent $y = x^n$, Newton first forms $y + y^{\bullet}\, o = (x + x^{\bullet}o)^n$, expands the right-hand side using the binomial theorem, subtracts $y = x^n$, divides through by o, neglects all terms still containing o, and so obtains $y^{\cdot} = nx^{n-1}\, x^{\bullet}$.

Newton later became discontented with the undeniable presence of infinitesimals in his calculus and dissatisfied with the doubtful procedure of "neglecting" them. In the preface to the *De quadratura curvarum* he remarks that there is no necessity to introduce into the method of fluxions any argument about infinitely small quantities. In their place he proposes to employ what he calls the method of *prime and ultimate ratio*. This method, in many respects an anticipation of the limit concept, receives a number of allusions in Newton's celebrated *Principia mathematica philosophiae naturalis* of 1687. For example, the first section of Book One is entitled "The method of first and last quantities, by the help of which we demonstrate the propositions that follow"; its first Lemma reads:

[57]Quoted in Jesseph (1993), p. 144.

[58]Baron (1987), p. 268.

> *Quantities, and the ratios of quantities, which in any finite time converge continually to equality, and before the end of that time approach nearer to each other than by any given difference, become ultimately equal*[59]*;*

its third:

> *...the ultimate ratio of the arc, chord, and tangent, any one to any other, is the ratio of equality.*[60]

Later Newton compares the method of indivisibles with his procedure and attempts to meet any objections to its use.

> *For demonstrations are shorter by the method of indivisibles; but because the hypothesis of indivisibles seems somewhat harsh, and therefore that method is reckoned less geometrical, I chose rather to reduce the demonstrations of the following Propositions to the first and last sums and ratios of nascent and evanescent quantities, that is, to the limits of those sums and ratios, and so to premise, as short as I could, the demonstrations of those limits. For hereby the same thing is performed as by the method of indivisibles; and now those principles being demonstrated, we may use them with greater safety. Therefore, if hereafter I should happen to consider quantities as made up of particles, or should use little curved lines for right ones, I would not be understood to mean indivisibles, but evanescent divisible quantities; not the sums and ratios of determinate parts, but always the limits of sums and ratios...*
>
> *Perhaps it may be objected, that there is no ultimate proportion of evanescent quantities; because the proportion, before the quantities have vanished, is not the ultimate, and when they are vanished, is none. But by the same argument it may be alleged that a body arriving at a certain place, and there stopping, has no ultimate velocity; because the velocity, before the body comes to the place, is not its ultimate velocity; when it has arrived, there is none. But the answer is easy; for by the ultimate velocity is meant that with which the body is moved, neither before it arrives at its last place and the motion ceases, nor after, but at the very instant it arrives; that is, that velocity with which the body arrives at its last place, and with which the motion ceases. And in like manner, the by the ultimate ratio of evanescent quantities is to be understood the ratio of the quantities not before they vanish, nor afterwards, but with which they vanish. In like manner the first ratio of nascent quantities is that with which they begin to be....*
>
> *It may also be objected, that if the ultimate ratios of evanescent quantities are given, their ultimate magnitudes will also be given; and so all quantities will consist of indivisibles, which is contrary to what Euclid has demonstrated concerning incommensurables, in the tenth Book of his Elements. But this objection is founded on a false supposition. For those ultimate ratios with which quantities vanish are not truly the ratios of ultimate quantities, but limits towards which the ratios of quantities decreasing without limit do always converge; and to which they approach nearer than by any given difference, but never go beyond, nor in effect attain to, till the quantities are diminished in infinitum. ... Therefore if in what follows, for the sake of being more easily understood, I should happen to mention quantities as least, or evanescent, or ultimate, you are not to suppose that quantities of any determinate magnitude are meant, but such as are conceived to be always diminished without end.*[61]

[59]Newton (1962), p. 29.

[60]*Ibid.*, p. 32.

[61]*Ibid.*, pp. 38–9.

Here Newton presents, in kinematic dress, the idea of a continuously variable quantity approaching a limit.

Newton thus developed three approaches for his calculus, all of which he regarded as leading to equivalent results, but which varied in their degree of rigour.[62] The first employed infinitesimal quantities which, while not finite, are at the same time not exactly zero. Finding that these eluded precise formulation, Newton focussed instead on their ratio, which is in general a finite number. If this ratio is known, the infinitesimal quantities forming it may be replaced by any suitable finite magnitudes—such as velocities or fluxions—having the same ratio. This is the method of fluxions. Recognizing that this method itself required a foundation, Newton supplied it with one in the form of the doctrine of prime and ultimate ratios, which is, as we have observed, a kinematic form of the theory of limits.

While Newton the mathematician was unquestionably a synechist, his view of the ultimate constitution of matter inclined towards atomism. At the beginning of Book III of the *Principia* he observes:

> *. . .that the divided but contiguous particles of bodies may be separated from one another, is a matter of observation; and, in the particles that remain undivided, our minds are able to distinguish yet lesser parts, as is mathematically demonstrated. But whether the parts so distinguished, and yet not divided, may, by the powers of Nature, be actually divided and separated from one another, we cannot certainly determine. Yet, had we the proof of but one experiment that any undivided particle, in breaking a hard and solid body, suffered a division, we might conclude by virtue of this rule that the undivided as well as the divided particles may be divided and actually separated to infinity.*[63]

Newton's point here is that, while every particle of matter is divisible in theory, only *observable* particles are known to be physically divisible. Short of an *experimentum crucis* demonstrating the actual divisibility of all material particles, whether observable or not, the existence of physically indivisible material atoms remains a possibility. Newton defended material atomism with greater vigour in his *Opticks*, as can be seen from the following celebrated passage:

> *It seems probable to me, that God in the Beginning form'd Matter in solid, massy, hard, impenetrable, moveable Particles, of such Sizes and Figures, and with such other Properties, and in such proportion to Space, as most conduced to the End for which He formed them.*[64]

These particles, or atoms, are, according to Newton, perfectly solid plena of homogeneous matter. Each is inseparable in itself, but a number may come into contact to form compound bodies. In doing so, however, they

> *do not conjoin into solidity, but are in mutual contact only at some mathematical point, the remaining spaces being left in each case between them vacant.*[65]

---

[62]Boyer (1959), p. 200.

[63]Newton (1962), p. 399,

[64]Newton (1952), p. 400.

[65]Quoted from Pyle (1997), p.415.

As Andrew Pyle observes,[66] in replacing Epicurean minimal parts by mathematical points Newton avoids the difficulty arising in the Epicurean theory from the conception of atoms as being composed of a finite number of minimal parts in contact. For if two minimal parts can be physically separated after coming into contact, so can any finite number, leading to the conclusion that the atom is physically fissionable. Newton's account of the contact of atoms does not lead to this conclusion, because an atom contains, not a finite number of mathematical points (as opposed to minimal parts), but an infinite number; "splitting the atom" would then require the separation of an infinite number of such points, and hence the application of an infinite force.

Leibniz

The philosopher-mathematician **G. W. F. Leibniz** (1646–1716) was greatly preoccupied with the problem of the composition of the continuum—the "labyrinth of the continuum", as he called it.[67] Indeed, we have it on his own testimony that his philosophical system—*monadism*—grew from his struggle with the problem of just how, or whether, a continuum can be built from indivisible elements. Leibniz asked himself: if we grant that each real entity is either a simple unity or a multiplicity, and that a multiplicity is necessarily an aggregation of unities, then under what head should a geometric continuum such as a line be classified? Now a line is extended and Leibniz held that extension is a form of repetition:

> *All repetition…is either discrete, as in numbered things are discriminated; or continuous, where the parts are indeterminate and can be assumed in infinite ways.*[68]

Since a line is divisible into parts, it cannot be a (true) unity. It is then a multiplicity, and accordingly an aggregation of unities. But of what sort of unities? Seemingly, the only candidates for geometric unities are points, but points are no more than extremities of the extended, and in any case, as Leibniz knew, solid arguments going back to Aristotle establish that no continuum can be constituted from points. It follows that a continuum is neither a unity nor an aggregation of unities. Leibniz concluded that continua are *not real entities* at all; as "wholes preceding their parts" they have instead a purely ideal character. In this way he freed the continuum from the requirement that, as something intelligible, it must itself be simple or a compound of simples. Accordingly,

---

[66] *Ibid.*

[67] Actually, as pointed out in Wilson (2015), the metaphor of the labyrinth was a familiar one in seventeenth century writing. In connection with the continuum it had already been employed, for example, by Galileo in *Two New Sciences.*

[68] Russell (1958), p. 245.

> *...continuous quantity is something ideal, which belongs to possibles, and to actuals considered as possibles. For the continuum involves indeterminate parts, while in actuals there is nothing indefinite—indeed in them all divisions which are possible are actual...But the science of continua, i.e. of possibles, contains eternal truths, which are never violated by actual phenomena, since the difference is always less than any assignable given difference.*[69]

> *From the fact that a mathematical solid cannot be resolved into primal elements it follows immediately that it is nothing real but merely an ideal construct designating only a possibility of parts.*[70]

> *Properly speaking, the number ½ in the abstract is a mere ratio, by no means formed by the composition of other fractions...And we may say as much of the abstract line, composition being only in concretes, or masses of which these abstract lines mark the relations. And it is thus also that mathematical points occur, which are also only modalities, i.e. extremities. And as everything is indefinite in the abstract line, we take notice of it of everything possible, as in the fractions of a number, without concerning ourselves concerning the divisions actually made, which designate these points in a different way. But in substantial actual things, the whole is a result or assemblage of simple substances, or of a multiplicity of real units. And it is the confusion of the ideal and the actual which has embroiled everything and produced the labyrinth concerning the composition of the continuum. Those who compose a line of points have sought first elements in ideal things or relations, otherwise than was proper; and those who have found that relations such as number, and space...cannot be formed of an assemblage of points, have been mistaken in denying, for the most part, the first elements of substantial realities, as if they had no primitive units, or as if there were no simple substances.*[71]

Leibniz held that space and time, as continua, are ideal, and anything real, in particular matter, is discrete, compounded of simple unit substances he termed *monads:*

> *Matter is not continuous but discrete, and actually infinitely divided, though no assignable part of space is without matter. But space, like time, is something not substantial, but ideal, and consists in possibilities, or in an order of coexistents that is in some way possible. And thus there are no divisions in it but such as are made by the mind, and the part is posterior to the whole. In real things, on the contrary, units are prior to the multitude, and multitudes only exist through units. (The same holds of changes, which are not really continuous.)*[72]

> *Space, just like time...is something indefinite, like every continuum whose parts are not actual, but can be taken arbitrarily, like the parts of unity, or fractions... Space is something continuous but ideal, mass is discrete, namely an actual multitude, or being by aggregation, but composed of an infinite number of units. In actuals, single terms are prior to aggregates, in ideals the whole is prior to the part. The neglect of this consideration has brought forth the labyrinth of the continuum.*[73]

> *Within the ideal or continuum the whole precedes the parts... The parts are here only potential; among the real [i.e. substantial] things, however, the simple precedes the*

[69] *Ibid.*, p. 246.

[70] Quoted in Weyl (1949) p.41.

[71] Russell (1958), p. 246.

[72] *Ibid.*, p. 245.

[73] *Ibid.*

aggregates, and the parts are given actually and prior to the whole. These considerations dispel the difficulties regarding the continuum—difficulties which arise only when the continuum is looked upon as something real, which possesses real parts before any division as we may devise, and when matter is regarded as a substance.

In actuals there is nothing but discrete quantity, namely the multitude of monads or simple substances.[74]

Leibniz explains how he came to develop his doctrine:

At first, when I had freed myself from the yoke of Aristotle, I took to the void and the atoms, for that is the view which best satisfies the imagination. But having got over this, I perceived, after much meditation, that it is impossible to find the principles of a real unity in matter alone, or in that which is only passive, since it is nothing but a collection or aggregation of parts ad infinitum. Now a multiplicity can derive its reality only from true unities , which come from elsewhere and are quite other than mathematical points, which are only extremities of the extended... of which it is certain that the continuum cannot be composed. Therefore to find these real unities I was compelled to have recourse to a formal atom, since a material being cannot be both material and perfectly indivisible or endowed with a true unity. It was necessary, hence, to recall and, so to speak, rehabilitate the substantial forms so decried today, but in a way which would make them intelligible and which would separate the use we should make of them from the abuse that has been made of them. I thence found that their nature consists in force, and that from that there ensues something analogous to feeling and appetite; and that accordingly they must be conceived in imitation of the idea we have of Souls.[75]

That Leibniz had come to abandon atomism sometime before 1675 is attested to by the following passage written at that time, in which he repudiates the existence of minimal parts of continua:

A minimum time (minimum space) is part of a greater time (space) between whose boundaries it lies—from the notion that we have of whole and part. Therefore a minimum time is a minimum part of time, and a minimum space is a minimum part of space. There is no such thing as a minimum part of space. Because otherwise there would be as many minima in the diagonal as the in the side, and thus the diagonal would equal the side, since two things all of whose parts are equal, are themselves equal. In the same way it is demonstrated that there is no such thing as a minimum of time. If a minimum is a minimum of anything, then it will be a minimum of those things that are in space, or rather of the parts of space, seeing as you distinguish them from bodies. Nor can we speak otherwise about the matter. If, then, we suppose a minimum, both a moment and time will entail a contradiction. Everything greater is composed out of something smaller. Therefore every minimum is part of the greater thing within whose boundaries it lies.

If a continuum is something other than the sum of supposed minima (if there are minima in it), it follows that there is a part that is left over when the sum of the minima has been taken away; therefore this part is greater than a minimum, since it is neither smaller nor equal, therefore there are also minima in it. But this is absurd, since we have already removed all the minima. Therefore if there are minimum parts in the continuum, it follows that the continuum is composed of them. But it is absurd for the continuum to be composed out of minima, as I have demonstrated; therefore it is also absurd for there to be minima in the continuum, or for minima to be parts of the continuum. To be in something (i.e. to be within

[74]*Ibid.*, pp. 245–6.

[75]Leibniz (1951), pp. 107–108.

its boundaries) and to something which cannot be understood without something else, is to be a part. Therefore there is no such thing [as a minimum in a continuum]. Hence if there are instants in time, there will be nothing but instants, and time will be but the sum of instants.[76]

Yet in Leibniz's philosophy the concept of *point*, or indivisible, plays a key role. Indeed, despite having thrown off the Aristotelian yoke, Leibniz continued to adhere to the Aristotelian doctrine that mathematical points are extremities or positions and that they can never, by themselves, constitute a continuum. Thus:

A point is not a certain part of matter, nor would an infinite number of points make an extension.[77]

...The continuum is infinitely divisible. And this appears in the straight line, from the mere fact that its part is similar to the whole. Thus when the whole can be divided, so can the part, and similarly any part of the part. Points are not part of the continuum, but extremities, and there is no more a smallest part of a line than a smallest fraction of unity.[78]

As to indivisibles, while they are understood as the simple extremities of time or of line, they cannot be conceived as containing new extremities of either actual or potential parts. Whence, points are neither big nor small, and no jump is necessary to pass through them. However, the continuous, though it everywhere has such indivisibles, is definitely not composed of them.[79]

Extremities of a line and units of matter do not coincide. Three continuous points in the same straight line cannot be conceived. But two are conceivable: the extremity of one straight line and the extremity of another, out of which one whole is formed. As, in time, are the two instants, the last of life and the first of death.[80]

As Russell observes,[81] Leibniz actually distinguished three kinds of point or indivisible: *metaphysical* points, or monads, from which actual entities such as bodies are compounded[82]; *mathematical points,* or positions in space; and *physical* points, which Russell quite plausibly identifies with "an infinitesimal extension of the kind used in the Infinitesimal Calculus."[83] Thus:

Atoms of matter are contrary to reason...only atoms of substance, i.e. unities which are real and absolutely destitute of parts, are sources of actions and the absolute first principles of the composition of things and, as it were, the last elements of the analysis of substances. They might be called metaphysical points; they possess a certain vitality and a kind of

---

[76]Quoted in Leibniz (2001), pp. 37, 39.

[77]Russell (1958), p. 242.

[78]*Ibid.*, p. 248.

[79]Leibniz (1951), p. 99.

[80]Russell (1958), p. 247.

[81]*Ibid.*, p. 104.

[82]Strictly speaking, for Leibniz, as Russell points out (p. 106), matter or extended mass is nothing more than "a well-founded phenomenon", not a substantial unity but a plurality engendered by indifferent monadic aggregation. Monads are not parts of phenomena but rather constitute their foundation. The monads themselves are the sole substantial realities: with the Eleatics Leibniz avers (Russell, p. 242) "*What is not truly* one *being is also not truly a* being."

[83]Russell (1958), p. 105.

perception, and mathematical points are their points of view to express the universe. But when corporeal substances are compressed, all their organs form only a physical point to our sight. Thus physical points are only indivisible in appearance; mathematical points are exact, but they are merely modalities; only metaphysical points [i.e., monads] ...are exact and real, and without them there would be nothing real, for without true unities there would be no multiplicity.[84]

Like William of Ockham, Leibniz held that density of point-distribution was sufficient to ensure continuity:

There is continuous extension whenever points are assumed to be so situated that there are no two between which there is not an intermediate point.[85]

And like Cusanus, he accepted the existence of the actual infinite:

I am so much for the actual infinite that instead of admitting that nature abhors it, as is commonly said, I hold that it affects nature everywhere in order to indicate the perfections of its Author. So I believe that every part of matter is, I do not say divisible, but actually divided, and consequently the smallest particle should be considered as a world full of an infinity of creatures...[86]

Among the best known of Leibniz's doctrines is the *Principle* or *Law of Continuity*. In a somewhat nebulous form this principle had been employed on occasion by a number of Leibniz's predecessors, including Cusanus and Kepler, but it was Leibniz who gave the principle its first clear formulation. In a letter to Bayle of 1687, Leibniz stated the principle as follows:

In any supposed transition, ending in any terminus, it is permissible to institute a general reasoning in which the final terminus may be included.[87]

This would seem to indicate that Leibniz considered "transitions" of any kind as continuous. Certainly he held this to be the case in geometry and for natural processes, where it appears as the principle *Natura non facit saltum* ("Nature makes no jump"). It is the Law of Continuity that allows geometry and the evolving methods of the infinitesimal calculus to be applicable in physics:

You ask me for some elucidation of my Principle of Continuity. I certainly think that this principle is a general one and holds good not only in Geometry but also in Physics. Since geometry is but the science of the continuous, it is not surprising that this law is observed everywhere in it, for Geometry by its very nature cannot admit any break in its subject matter. In truth we know that everything in that science is interconnected and that no single instance can be adduced of any property suddenly vanishing or arising without the possibility of our determining the intermediate transition, the points of inflection and singular points, with which to render the change explicable, so that an algebraic equation which represents one state exactly virtually represents all the other states which may properly occur in the same subject[88]

---

[84]*Ibid.*, p. 105.

[85]Ibid.,, p. 247.

[86]Reply to Foucher, 1693, Leibniz (1951), p. 99.

[87]Quoted in Boyer (1959), p. 217.

[88]Letter to Varignon, 1702, Leibniz (1951), pp. 184–5.

The Principle of Continuity also furnished the chief grounds for Leibniz's rejection of material atomism:

> *My axiom that nature never acts by a leap has a great use in Physics. It destroys atoms, small lapses of motion, globules of the second element, and other similar chimeras...You are right in saying that all magnitudes may be subdivided. There is none so small in which we cannot conceive an inexhaustible infinity of subdivisions. But I see no harm in that or any necessity to exhaust them. A space infinitely divisible is traversed in a time also infinitely divisible. I conceive no physical indivisibles short of a miracle, and I believe nature can reduce bodies to the smallness Geometry can consider.*[89]

> *Matter, according to my hypothesis, would be divisible everywhere and more or less easily with a variation which would be insensible in passing from pone place to another neighbouring place; whereas, according to the atoms, we make a leap from one extreme to another, and from a perfect incohesion, which is in the place of contact, we pass to an infinite hardness in all other places. And these leaps are without example in nature.*[90]

The Principle of Continuity also played an important underlying role in Leibniz's mathematical work, especially in his development of the infinitesimal calculus.[91] Leibniz's essays *Nova Methodus* of 1684 and *De Geometri Recondita* of 1686 may be said to represent the official births of the differential and integral calculi, respectively. Characteristically, Leibniz's approach to the calculus has combinatorial roots, traceable to his early work on derived sequences of numbers. Let us attempt to describe, in modern terms, how he arrived at his conceptions.

Let $P = \{(x_0, y_0), (x_1, y_1), \ldots (x_n, y_n)\}$be a finite sequence of pairs of (real) numbers with $x_0 < x_1 < \ldots < x_n$. (See Fig. 2.8.) For each $i = 0, \ldots, n-1$ write $\mathrm{d}x_i = x_{i+1} - x_i$, $\mathrm{d}y_i = y_{i+1} - y_i$. Then the quantity

$$\mathrm{D}_i P = \frac{\mathrm{d}y_i}{\mathrm{d}x_i}$$

represents the slope of the line passing through the points in the plane with coordinates (x_j, y_j) and (x_{j+1}, y_{j+1}).Also for each $j = 0, \ldots, n-1$, the quantity

$$\int^j P = \sum_{i-0}^{j} y_i \mathrm{d}x_i$$

---

[89]Letter to Foucher, 1692, Leibniz (1951), p. 71.

[90]Quoted in Russell (1958), p. 235.

[91]Leibniz's abilities as a mathematician have been memorably characterized by E. T. Bell:

> *The union in one mind of the highest ability in the two broad, antithetical domains of mathematical thought, the analytical and the combinatorial, or the continuous and the discrete, was without precedent before Leibniz and without sequent after him. He is the one man in the history of mathematics to have both qualities of thought in a superlative degree. (Bell, E.T. 1965, p. 128).*

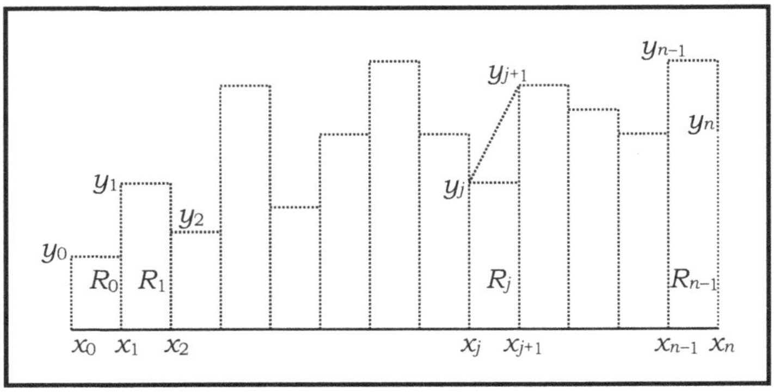

Fig. 2.8

represents the sum of the areas of the rectangles R_0, R_1, …, R_j. Now define $\int P$, the *integral* of P by

$$\int P = \left\{\left(x_0, \int^0 P\right), \ldots \left(x_{n-1}, \int^{n-1} P\right)\right\},$$

and DP, the *derivative* or *differential* of P by

$$\mathrm{D}P = \{(x_0, \mathrm{D}_0 P), \ldots, \{(x_{n-1}, \mathrm{D}_{n-1} P)\}.$$

An easy calculation then shows that, for each $j = 0, \ldots, n - 1$,

$$\mathrm{D}_j \int P = y_j \quad \int^j \mathrm{D}P = y_j - y_0.$$

The first of these means that

$$\mathrm{D} \int P = P. \tag{2.1}$$

Provided we set the "constant of integration" y_0 to 0, the second means that

$$\int \mathrm{D}P = P. \tag{2.2}$$

That is, for finite sequences, *differentiation and integration are mutually inverse operations.*

Now consider P to be a finite set of points in the plane, and let C be a continuous curve passing through the points of P (Fig. 2.9). Then $\mathrm{D}_j P$ is a rough approximation to the slope of the tangent to C at the point (x_j, y_j) and $\int^{n-1} P$ is a rough

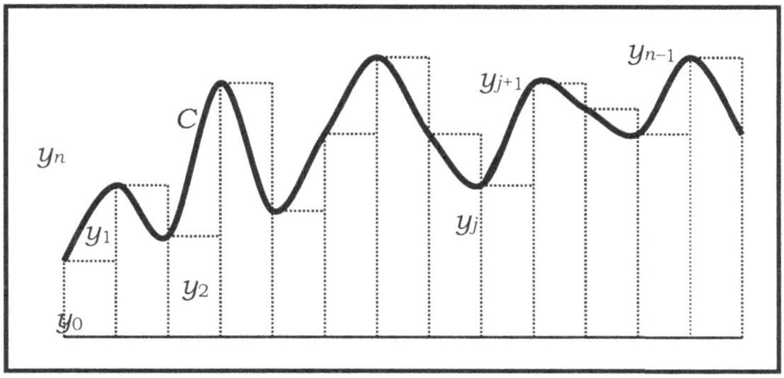

Fig. 2.9

approximation to the area under *C*. Clearly, these approximations improve as the number of points increases and the distance between successive points decreases. Leibniz saw that if the number *n* of points was permitted to increase to infinity and the differences $\mathrm{d}x_i$ and $\mathrm{d}y_i$ to become infinitesimal the corresponding "approximations" to tangent slope and area of *C* could be taken as *exact.* He wrote $\mathrm{d}x$ and $\mathrm{d}y$ for such infinitesimal differences, or *differentials*, and $\frac{\mathrm{d}y}{\mathrm{d}x}$ for the ratio of the two, which he then took to represent the slope of the curve at the corresponding point. This suggestive, if highly formal procedure led Leibniz to evolve rules for calculating with differentials, which was achieved by appropriate modification of the rules of calculation for ordinary numbers. Determining the *integral* for the curve, however, raised the thorny problem of attempting to sum infinitely many infinitesimal quantities. Leibniz's way out of this difficulty was to appeal to his Principle of Continuity, which suggested that the Eqs. (2.1) and (2.2) holding in the discrete case would also be preserved in the transition to continuity, with the result that "integration" and "differentiation" remain mutually inverse operations. As a result, it became *unnecessary* to provide a satisfactory account of the integral: as long as one had some way of calculating derivatives, of "differentiating", integrals could be determined by mere inversion of the procedure of differentiation.[92]

Although the use of infinitesimals was instrumental in Leibniz's approach to the calculus, in 1684 he introduced the concept of differential without mentioning infinitely small quantities, almost certainly in order to avoid foundational difficulties. He states without proof the following rules of differentiation:

If a is constant, then $da = 0$ *and* $d(ax) = adx$

[92]Indeed, Jakob and Johann Bernoulli, who made many contributions to the Leibnizian calculus and who introduced the term "integral", actually defined the operation of integration as the inverse of differentiation.

$$d(x + y - z) = dx + dy - dz$$

$$d(xy) = xdy + ydx$$

$$\mathrm{d}\left(\frac{x}{y}\right) = \frac{x\mathrm{d}y - y\mathrm{d}x}{y^2}$$

$$d(x^p) = px^{p-1}dx, \textit{also for fractional } p.$$

But behind the formal beauty of these rules—an early manifestation of what was later to flower into differential algebra—the presence of infinitesimals makes itself felt, since Leibniz's definition of tangent employs both infinitely small distances and the conception of a curve as an infinilateral polygon:

> *We have to keep in mind that to find a tangent means to draw a line that connects two points of a curve at an infinitely small distance, or the continued side of a polygon with an infinite number of angles, which for us takes the place of a curve.*[93]

In thinking of a curve as an infinilateral polygon (as in Fig. 2.9), the abscissae x_0, $x_1, \ldots$ and the ordinates y_0, y_1, are to be regarded as lying infinitesimally close to one another; the symbols *x, y* are then conceived as variables ranging over sequences of such abscissae or ordinates. Leibniz also conceived the differentials d*x*, d*y* as variables ranging over differences. This enabled him to take the important step of regarding the symbol d as an *operator* acting on variables, so paving the way for the *iterated application* of d, leading to the higher differentials $\mathrm{dd}x = \mathrm{d}^2x$, $\mathrm{d}^3x = \mathrm{dd}^2x$, and in general $\mathrm{d}^{n+1}x = \mathrm{dd}^n x$.[94]

Leibniz supposed that the first-order differentials d*x*, d*y*, ... were incomparably smaller than, or infinitesimal with respect to, the finite quantities *x, y*, ..., and, in general that an analogous relation obtained between the $(n + 1)^{th}$ –order differentials $\mathrm{d}^{n+1}x$ and the nth –order differentials $\mathrm{d}^n x$. He also assumed that the nth power $(\mathrm{d}x)^n$ of a first-order differential was of the same order of magnitude as an nth –order differential $\mathrm{d}^n x$, in the sense that the quotient $\mathrm{d}^n x/(\mathrm{d}x)^n$ is a finite quantity.

For Leibniz the incomparable smallness of infinitesimals derived from their failure to satisfy Archimedes' principle; and quantities differing only by an infinitesimal were to be considered equal:

> *...only those homogeneous quantities are comparable, of which one can become larger than the other if multiplied by a number, that is, a finite number. I assert that entities, whose difference is not such a quantity, are equal...This is precisely what is meant by saying that the difference is smaller than any given quantity.*[95]

But while infinitesimals were conceived by Leibniz to be incomparably smaller than ordinary numbers, the Law of Continuity ensured that they were governed by the same laws as the latter:

---

[93] Quoted in Mancosu (1996), p. 156.

[94] *Ibid.*, p. 156.

[95] Quoted from Bos (1974), p. 14.

Meanwhile, we conceive of the infinitely small not as a simple and absolute zero, but as a relative zero ... that is, as an evanescent quantity which yet retains the character of that which is disappearing.[96]

Leibniz's attitude toward infinitesimals and differentials seems to have been that they furnished the elements from which to fashion a formal grammar, an algebra, of the continuous. Since he regarded continua as purely ideal entities, it was then perfectly consistent for him to maintain, as he did, that infinitesimal quantities themselves are no less ideal—simply useful fictions, introduced to shorten arguments and aid insight.

Supporters and Critics of Leibniz

Although Leibniz himself did not credit the infinitesimal or the (mathematical) infinite with objective existence, a number of his followers did not hesitate to do so. Among the most prominent of these was **Johann Bernoulli** (1667–1748). A letter of his to Leibniz written in 1698 contains the forthright assertion that "inasmuch as the number of terms in nature is infinite, the infinitesimal exists *ipso facto*."[97] One of his arguments for the existence of actual infinitesimals begins with the positing of the infinite sequence $\frac{1}{2}, \frac{1}{3}, \frac{1}{4}, \ldots$. If there are ten terms, one tenth exists; if a hundred, then a hundredth exists, etc.; and so if, as postulated, the number of terms is infinite, then the infinitesimal exists.[98]

Leibniz's calculus gained a wide audience through the publication in 1696, by **Guillaume de L'Hôpital** (1661–1704), of the first expository book on the subject, the *Analyse des Infiniments Petits Pour L'Intelligence des Lignes Courbes*. L'Hôpital begins his exposition by laying down two definitions:

I. *Variable quantities are those that continually increase or decrease; and constant or standing quantities are those that continue the same while others vary.*
II. *The infinitely small part whereby a variable quantity is continually increased or decreased is called the differential of that quantity.*[99]

Following Leibniz, L'Hôpital writes dx for the differential of a variable quantity x.

Two postulates are next introduced:

I. *Grant that two quantities, whose difference is an infinitely small quantity, may be taken (or used) indifferently for each other: or (what is the same thing) that a quantity, which is increased or decreased only by an infinitely small quantity, may be considered as remaining the same.*
II. *Grant that a curve line may be considered as the assemblage of an infinite number of infinitely small right lines: or (what is the same thing) as a polygon with an infinite*

[96]Quoted in Boyer (1959), p. 219.
[97]*Ibid.*, p. 239.
[98]*Ibid.*
[99]Quoted in Mancosu (1996), p. 151.

number of sides, each of an infinitely small length, which determine the curvature of the line by the angles they make with each other.[100]

L'Hôpital applies Postulate I (and Definition II) in determining the differential of a product *xy:*

$$d(xy) = (x + dx)(y + dy) - xy = ydx + xdy + dxdy = ydx + xdy.$$

Here the last step is justified by Postulate I, since $dxdy$ is infinitely small in comparison to $ydx + xdy$.

A typical application of Postulate II is in determining the length of the subtangent to the parabola $x = y^2$. In Fig. 2.7 above, the infinitesimal abscissal increment *e* is dx and the corresponding increment in the ordinate, *a, is* dy. The triangles *PNM* and the infinitesimal triangle (which, by postulate II, *is* a triangle) *PRP'* are similar, so that $NM = ydx/dy$. Now the differential equation of the parabola is $dx = d(y^2) = 2ydy$, which yields $NM = 2y^2 = 2x$.

Leibniz's calculus of differentials, resting as it did on somewhat insecure foundations, soon attracted criticism. The attack mounted by the Dutch physician **Bernard Nieuwentijdt** (1654–1718) in works of 1694–6 is of particular interest, since Nieuwentijdt offered his own account of infinitesimals which conflicts with that of Leibniz and has striking features of its own.[101] Nieuwentijdt postulates a domain of quantities, or numbers, subject to an ordering relation of greater or less. This domain includes the ordinary finite quantities, but it is also presumed to contain infinitesimal and infinite quantities—a quantity being infinitesimal, or infinite, when it is smaller, or, respectively, greater, than any arbitrarily given finite quantity. The whole domain is governed by a version of the Archimedean principle to the effect that zero is the only quantity incapable of being multiplied by some quantity to equal any given quantity. Infinitesimal quantities may be characterized as quotients b/m of a finite quantity b by an infinite quantity m. In contrast with Leibniz's differentials, Nieuwentijdt's infinitesimals have the property that the product of any pair of them vanishes; in particular squares and all higher powers of infinitesimals are zero. This fact enables Nieuwentijdt to show that, for any curve given by an algebraic equation, the hypotenuse of the differential triangle generated by an infinitesimal abscissal increment *e* coincides with the segment of the curve between x and $x + e$. That is, a curve truly *is* an infinilateral polygon.[102]

It is instructive to see how Nieuwentijdt's calculus worked in practice.[103] For example, consider again the determination of the subtangent to the parabola $x = y^2$ (Fig. 2.10). Let e be an infinitesimal abscissal increment and a the corresponding increment in the ordinate. Thus $PN = y$, $PR = e$, $RP' = a$. Then the triangles *PNM* and the infinitesimal triangle *PRP'* are similar, so that $a = ey/NM$. Since the point

[100]Quoted *ibid.*, p. 152.

[101]*Ibid.*, p. 159 *et seq.*

[102]*Ibid.*, p. 159.

[103]*Ibid.*, pp. 153–4.

Fig. 2.10

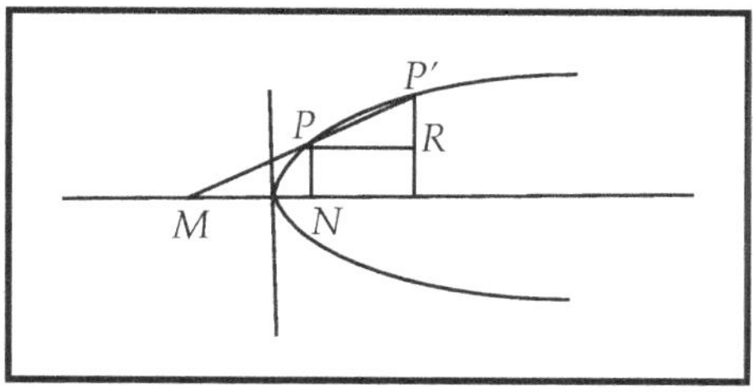

$(x + e, y + a)$ lies on the curve, it follows that $x + e = (y + a)^2 = y^2 + 2ay + a^2$. Since a is infinitesimal, $a^2 = 0$, so subtracting the original equation gives $e = 2ya$, whence $a = e/2y$. Accordingly $e/2y = ey/NM$, so that $NM = ey/(e/2y) = 2y^2 = 2x$.

The major differences between Nieuwentijdt's and Leibniz's calculi of infinitesimals are summed up in the following table:

Leibniz	Nieuwentijdt
Infinitesimals are variables	*Infinitesimals are constants*
Higher-order infinitesimals exist	*Higher-order infinitesimals do not exist*
Products of infinitesimals are not absolute zeros	*Products of infinitesimals are absolute zeros*
Infinitesimals can be neglected when infinitely small with respect to other quantities	*(first-order) infinitesimals can never be neglected*

In responding to Nieuwentijdt's assertion that squares and higher powers of infinitesimals vanish, Leibniz remarked that "it is rather strange to posit that a segment dx is different from zero and at the same time that the area of a square with side dx is equal to zero."[104]. Yet this oddity is in fact a consequence—apparently unremarked by Leibniz—of one of his own key principles, namely that curves may be considered as infinilateral polygons (L'Hôpital's Postulate II).[105] For consider the curve $y = x^2$ (Fig. 2.11). Given that the curve is an infinilateral polygon, an infinitesimal stretch of the curve between the abscissae 0 and dx must coincide with the tangent to the curve at the origin—in this case, the axis of abscissae—between those two points. But then the point (dx, dx^2) must lie on the axis of abscissae, which means that d$x^2 = 0$.

A related argument surfaces in the debate concerning the infinitesimal calculus that flared up at the start of the eighteenth century among the mathematical luminaries of the Paris Academy of Sciences.[106] The algebraist **Michel Rolle**

[104]*Ibid.*, p. 161.

[105]Bell (1998), p. 9. In fact the "nilsquare" property of infinitesimals and L'Hôpital's Postulate II (for algebraic curves) are equivalent. As remarked above, Nieuwentijdt saw that the first assertion implies the second.

[106]Mancosu (1996), pp. 165 *et seq.*

Fig. 2.11

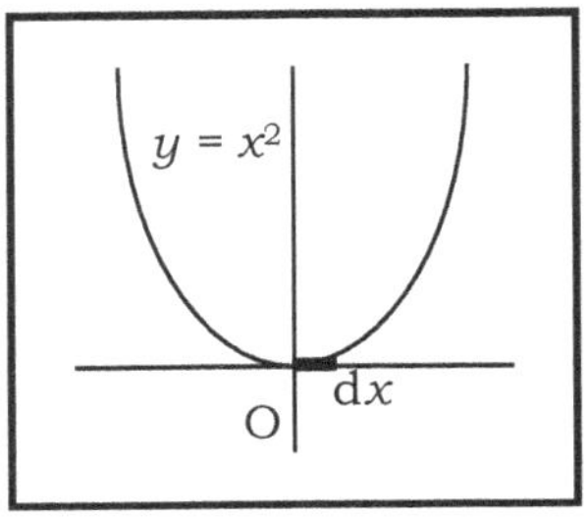

(1652–1719) raised a number of objections to the calculus, among which was the claim that talk of differentials was nonsense, since "it could be proved that differentials were absolute zeros". Rolle later used the following argument to back up his claim.[107] Starting with the parabola $y^2 = ax$, L'Hôpital's rules yield the differential equation $a\mathrm{d}x = 2y\mathrm{d}y$. Assuming that the point $(x + \mathrm{d}x, y + \mathrm{d}y)$ lies on the parabola, one obtains $ax + a\mathrm{d}x = y^2 + 2y\mathrm{d}y + \mathrm{d}y^2$, whence $\mathrm{d}y^2 = (ax - y^2) + (a\mathrm{d}x - 2y\mathrm{d}y) = 0 + 0 = 0$. Since $\mathrm{d}y^2 = 0$, Rolle infers that $\mathrm{d}y = 0$, from which it also follows that $\mathrm{d}x = 0$. Here the crucial step is from $\mathrm{d}y^2 = 0$ to $\mathrm{d}y = 0$, a step that Nieuwentijdt, who accepted the first assertion but not the second, would not have taken.

Nevertheless, a number of the Paris academicians embraced the infinitesimal (and the infinite) with enthusiasm. One such was the polymath **Bernard de Fontenelle** (1657–1757), who put the fact on record with the publication of his *Éléments de la Géométrie de l'Infini* in 1727. Dismissing the notion that the infinite is in any way mysterious,[108] Like Wallis he writes ∞ as the last term of the infinite sequence 0, 1, 2, 3, . . . and proceeds to treat this symbol with great latitude, allowing it to be raised not merely to integral but even to fractional and infinite powers as in $\infty^{\frac{3}{4}}$and ∞^{∞^3}. Infinitesimals of various orders are obtained as reciprocals of powers of ∞.

The insistence that infinitesimals obey precisely the same algebraic rules as finite quantities forced Leibniz and the defenders of his differential calculus into treating infinitesimals, in the presence of finite quantities, *as if* they were zeros, so that, for example, $x + \mathrm{d}x$ is treated as if it were the same as x. This was justified on the grounds that differentials are to be taken as variable, not fixed quantities, decreasing continually until reaching zero. Considered only in the "moment of their evanescence",[109] they were accordingly neither something nor absolute zeros.

Thus differentials (or infinitesimals) $\mathrm{d}x$ were ascribed variously the four following properties:

---

[107] *Ibid.*, p. 167.

[108] Boyer (1959), p. 242.

[109] Mancosu (1996), p. 167.

$$1.\mathrm{d}x \approx 0 \quad 2.\textit{neither } \mathrm{d}x = 0 \textit{ nor } \mathrm{d}x \neq 0 \quad 3.\mathrm{d}x^2 = 0 \quad 4.\mathrm{d}x \to 0$$

where "≈ "" stands for "indistinguishable from", and "→ 0″ stands for "becomes vanishingly small". Of these properties only the last, in which a differential is considered to be a variable quantity tending to 0, survived the nineteenth century refounding of the calculus in terms of the limit concept.[110]

Bayle

Finally, the sceptical views of the philosopher **Pierre Bayle** (1647–1706) concerning the continuous should be noted. In his *Dictionnaire* article on Zeno of Elea,[111] Bayle argues that the idea of extension is incoherent whatever its mode of composition. It cannot be composed of unextended mathematical points, since extension cannot be produced by compounding the extensionless. Nor can it consist of Epicurean atoms, extended but indivisible corpuscles, since anything extended is divisible. The sole remaining possibility is that extension is composed of parts that are themselves divisible to infinity. To refute this Bayle provides a number of arguments, among the most interesting of which are that extension conceived as infinitely divisible would make it impossible for parts of extended substances to touch one another, and at the same time necessary that they penetrate one another. Bayle's conclusions in regard to the existence of extended substance are similar to those of Leibniz:

> *We must acknowledge with respect to bodies what mathematicians acknowledge with respect to lines and surfaces...They frankly admit that a length and breadth without depth is something that cannot exist outside of our minds. Let us say the same of the three dimensions. They can exist only in our minds. They can exist only ideally.*[112]

While Bayle, like Leibniz, held that extension is a purely ideal notion, he viewed time quite differently. He affirmed its reality and, further, claimed that it was atomic in composition, with successive "nows" as atoms:

> *Time is not divisible ad infinitum. ... I will make it more obvious by an example. Thus I say that what suits Monday and Tuesday with regard to succession suits every part of time whatsoever. Since it thus impossible that Monday and Tuesday exist together, and since it must be the case necessarily that Monday cease before Tuesday begins to exist, there is no other part of time, whatever it may be, that can coexist with another. Every one has to exist alone. Every one must begin to exist, when the other ceases to do so. Every one must cease to exist, before the following one begins to be. From which it follows that time is not divisible to*

[110]But the other properties have resurfaced in the theories of infinitesimals which have emerged over the past several decades. Appropriately defining the relation, ≈, property 1 holds of the differentials in *nonstandard analysis*, while properties 1, 2 and 3 hold of the differentials in *smooth infinitesimal analysis*. See Chaps. 8 and 10 below.

[111]Bayle (1965), pp. 350–94.

[112]Bayle (1965), p. 363.

> *infinity, and that successive durations of things are composed of moments properly so called, of which each is simple and indivisible, perfectly distinct from the past and future, and contains only the present time.*[113]

Bayle's views later had a significant influence on Hume's philosophy of space, time, and mathematics.

[113] *Op. cit.*

Chapter 3
The Eighteenth and Early Nineteenth Centuries: The Age of Continuity

3.1 The Mathematicians

Euler

The leading practitioner of the calculus, indeed the leading mathematician of the eighteenth century, was **Leonhard Euler** (1707–83). While Euler's genius has been described as being of "equal strength in both of the main currents of mathematics, the continuous and the discrete",[1] philosophically he was a thoroughgoing synechist. Rejecting Leibnizian monadism, he favoured the Cartesian doctrine that the universe is filled with a continuous ethereal fluid and upheld the wave theory of light over the corpuscular theory propounded by Newton.

Euler's philosophical views may be gleaned from his *Letters to a German Princess*, written during 1760–62. He writes:

> *Two things, then, must be admitted: first, the space through which the heavenly bodies move is filled with a subtile matter; secondly, rays are not an actual emanation from the sun and other luminous bodies, in virtue of which part of their substance is violently emitted from them, according to the doctrine of* Newton.
>
> *That subtile matter which fills the whole space in which the heavenly bodies revolve is called* ether. *Of its extreme subtilty no doubt can be entertained... It is also, without doubt, possessed of elasticity, by means of which it has a tendency to expand itself in all directions, and to penetrate into spaces where there would otherwise be a vacuum: so that if by some accident the ether were forced out of any space, the surrounding fluid would instantly rush in and fill it again.*[2]

[1]Bell (1965), Vol I, p. 152

[2]Euler (1843), Vol 1, pp. 83–84.

J. L. Bell, *The Continuous, the Discrete and the Infinitesimal in Philosophy and Mathematics*, The Western Ontario Series in Philosophy of Science 82,
https://doi.org/10.1007/978-3-030-18707-1_3

Euler rejected the Newtonian doctrine that forces, in particular gravitation, could act at a distance and upheld the idea that the effect of such forces is transmitted continuously in some way through the ether. In an early letter he writes:

> *Gravity, then, is not an intrinsic property of body: it is rather the effect of a foreign force, the source of which must be sought for out of the body. This is geometrically true, though we know not the foreign forces which occasion gravity...*
>
> *I have already remarked that these forces may very probably be caused by the subtile matter which surrounds all the heavenly bodies, and fills the whole space of the heavens... This opinion, however, that attraction is essential to all matter, is subject to so many other inconveniences, that it is hardly possible to allow it a place in a rational philosophy. It is certainly much safer to proceed on the idea, that what is called attraction is a power contained in the subtile matter which fills the whole space of the heavens; though we cannot tell how.*[3]

In a sequence of later letters Euler mounts a determined attack against both material atomism and monadism. He argues that it is an inherent property of extension, established on geometric grounds, to be infinitely divisible; this being granted, it follows that bodies too, as instances of the extended, must be likewise. But, says Euler, certain philosophers deny this conclusion, insisting that the divisibility of bodies

> *extends only to a certain point, and that you may come at length to particles so minute that, having no magnitude, they are no longer divisible. These ultimate particles, which enter into the composition of bodies, they denominate* simple beings *and* monads.[4]

Indeed, he observes wryly,

> *There was a time when the dispute respecting monads employed such general attention, and was conducted with so much warmth, that it forced its way into company of every description, that of the guard-room not excepted. There was scarcely a lady at court who did not take a decided part in favour of monads or against them. In a word, all conversation was engrossed by monads—no other subject could find admission.*[5]

The partisans of monads are

> *obliged to affirm that bodies are not extended but have only an appearance of extension. They imagine that by this they have subverted the argument adduced in support of the divisibility in infinitum. But if body is not extended, I should be glad to know from whence we derived the idea of extension; for if body is not extended, nothing in the world is, since spirits are still less so. Our idea of extension, therefore, would be altogether imaginary and chimerical.*[6]

The monadists' case ultimately rests, says Euler, on the *principle of sufficient reason*, but applied in a question-begging way:

---

[3] *Ibid.*, pp. 254–5.

[4] Euler (1843), Vol. II, p. 39.

[5] *Ibid.*, p. 39–40.

[6] *Ibid.*, p. 41.

> *Bodies, say they, must have their sufficient reason somewhere; but if they were divisible to infinity, such reason could not take place; and hence they conclude, with an air altogether philosophical,* that as every thing must have its sufficient reason, it is absolutely necessary that all bodies should be composed of monads—*which was to be demonstrated.... It were greatly to be wished that a reasoning so slight could elucidate to us questions of this importance; but I frankly confess I understand nothing of the matter.*[7]

Euler argues that the monadists' seemingly reasonable basic premise to the effect that every compound being is made up of simple beings, is in fact fallacious since it leads to contradictions. He writes:

> *In effect, they* [the monadists] *admit that bodies are extended; from this point [they] set out to establish the proposition that they are compound beings; and having hence deduced that bodies are compounded of simple beings, they are obliged to allow that simple beings are incapable of producing real extension, and consequently that the extension of bodies is mere illusion.*
>
> *An argument whose conclusion is a direct contradiction of the premises is singularly strange: the reasoning sets out with advancing that bodies are extended; for if they were not, how could it be known that they are compound beings—and then comes to the conclusion that they are not so. Never was a fallacious argument, in my opinion, more completely refuted than this has been.*[8] *The question was,* Why are bodies extended? *And, after a little turning and winding, it is answered,* Because they are not so. *Were I to be asked, Why has a triangle three sides? and I should reply that it is a mere illusion—would such a reply be deemed satisfactory.*

Euler rejected the concept of infinitesimal in its sense as a quantity less than any assignable magnitude and yet unequal to 0, arguing:

> *There is no doubt that every quantity can be diminished to such an extent that it vanishes completely and disappears. But an infinitely small quantity is nothing other than a vanishing quantity and therefore the thing itself equals 0. It is in harmony also with that definition of infinitely small things, by which the things are said to be less than any assignable quantity; it certainly would have to be nothing; for unless it is equal to 0, an equal quantity can be assigned to it, which is contrary to the hypothesis.*[9]

In that case differentials must be zeros, and $\mathrm{d}y/\mathrm{d}x$ the quotient 0/0. Since for any number α, $\alpha \cdot 0 = 0$, Euler maintained that the quotient 0/0 could represent any real number whatsoever Or, to put it another way, real numbers are just the *ratios of infinitesimals.*[10] For Euler *qua* formalist the calculus was essentially a procedure for determining the value of the expression 0/0 as a real number in the manifold situations it arises as the ratio of evanescent increments.

But in the mathematical analysis of natural phenomena, Euler, along with a number of his contemporaries, did employ what amount to infinitesimals in the

---

[7] *Ibid.*, pp. 50–51.

[8] But here Euler seems to have missed the point that, if from the assumption that bodies are extended it follows logically that they are unextended, then it can be concluded that bodies are unextended. This is an instance of the logical tautology $(p \rightarrow \neg p) \rightarrow \neg p$.

[9] Quoted from Euler's *Institutiones* of 1755 in Kline (1972), p 429.

[10] This is a reigning principle of smooth infinitesimal analysis, see Chap. 10 below. See also Lawvere (2011).

Fig. 3.1

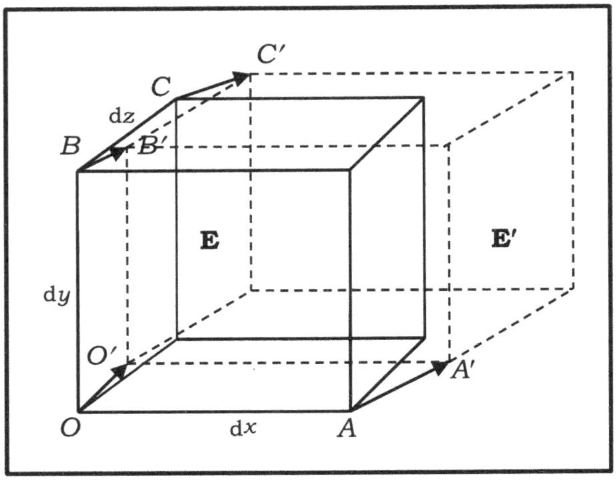

form of minute, but more or less concrete "elements" of continua. This is exemplified in his work on fluid flow, where his virtuosity in employing the principle of gaining knowledge of the external world from the behaviour of its infinitesimal parts[11] can be seen in full flower. In his fundamental papers on the subject[12] Euler derives the *equation of continuity* for a fluid free of viscosity but of varying density flowing smoothly in space. At any point $O = (x, y, z)$ in the fluid and at any time t, the fluid's density ρ and the components u, v, w of the fluid's velocity are given as functions of x, y, z, t. Euler considers the elementary volume element **E**—an infinitesimal parallelepiped—with origin O and edges OA, OB, BC of infinitesimal lengths dx, dy, dz (see Fig. 3.1). Fluid flow during the infinitesimal time dt transforms the volume element **E** into the infinitesimal parallelepiped **E**′ with vertices O', A', B', C'. Euler calculates the lengths of the sides $O'A'$, $O'B'$, $B'C'$ to be respectively

$$dx\left(1 + dt\frac{\partial u}{\partial x}\right),\quad dy\left(1 + dt\frac{\partial v}{\partial y}\right),\quad dz\left(1 + dt\frac{\partial w}{\partial z}\right),$$

ignoring infinitesimal terms of higher order than the second. The volume of **E'** is then the product of these quantities, which, ignoring terms in dt of second and higher order is seen to be

[11]Weyl (1950), p. 92.

[12]Euler, *Principia motus fluidorum* and *Principes généraux du mouvement des fluides,* 1755. Summarized in Dugas (1988), pp. 301–304.

$$\mathrm{d}x\mathrm{d}y\mathrm{d}z\left(1 + \mathrm{d}t\frac{\partial u}{\partial x} + \mathrm{d}t\frac{\partial v}{\partial y} + \mathrm{d}t\frac{\partial w}{\partial z}\right).$$

Since the coordinates of O' are $(x + u\mathrm{d}t,\ y + v\mathrm{d}t,\ z + w\mathrm{d}t)$, the fluid density ρ' there at time $t + \mathrm{d}t$ is

$$\rho + \mathrm{d}t\frac{\partial \rho}{\partial t} + u\mathrm{d}t\frac{\partial \rho}{\partial x} + v\mathrm{d}t\frac{\partial \rho}{\partial y} + w\mathrm{d}t\frac{\partial \rho}{\partial z}.$$

Euler now invokes the principle of conservation of mass to assert that the masses of the fluid in **E** and **E**$'$ are the same, so that,

> *as the density is reciprocally proportional to the volume, the quantity* ρ' *will be related to* ρ *as* $\mathrm{d}x\mathrm{d}y\mathrm{d}z$ *is related to*
>
> $$\mathrm{d}x\mathrm{d}y\mathrm{d}z\left(1 + \mathrm{d}t\frac{\partial u}{\partial x} + \mathrm{d}t\frac{\partial v}{\partial y} + \mathrm{d}t\frac{\partial w}{\partial z}\right);$$
>
> *whence, by carrying out the division, the very remarkable condition which relates from the continuity of the fluid,*
>
> $$\frac{\partial \rho}{\partial t} + u\frac{\partial \rho}{\partial x} + v\frac{\partial \rho}{\partial y} + w\frac{\partial \rho}{\partial z} + \rho\frac{\partial u}{\partial x} + \rho\frac{\partial v}{\partial y} + \rho\frac{\partial w}{\partial z} = 0.$$
>
> *This may be written more simply as*
>
> $$\frac{\partial \rho}{\partial t} + \frac{\partial}{\partial x}(\rho u) + \frac{\partial}{\partial y}(\rho v) + \frac{\partial}{\partial z}(\rho w) = 0$$
>
> *and, for an incompressible fluid, it reduces to*
>
> $$\frac{\partial u}{\partial x} + \frac{\partial v}{\partial y} + \frac{\partial w}{\partial z} = 0.$$

It will be seen from this calculation that Euler treats the volume element **E** not as an atom or monad in the strict sense—as part of a continuum it must of necessity be divisible—but as being of sufficient minuteness to preserve its rectilinear *shape* under infinitesimal flow, yet allowing its *volume* to undergo infinitesimal change. This idea was to become fundamental in continuum mechanics.

Euler also made important contributions to the development of the function concept. As has been pointed out, the notion of functionality or functional relation arose in connection with continuous variation; indeed the term "function" itself, introduced by Leibniz in a manuscript[13] of 1673, was used by him to denote a variable length related in a specified way to a variable point on a curve. In 1718 the scope of the concept was greatly enlarged when John Bernoulli defined a "function

[13]Kline (1972), p. 340.

of a variable magnitude" as a quantity made up in any way of this variable magnitude and constants.[14] In 1730 Bernoulli introduced the distinction between algebraic and transcendental functions,[15] where by the latter he meant integrals of algebraic functions. In 1734 Euler introduced the characteristic function-argument notation $f(x)$. His *Introductio in analysin infinitorum* of 1748 gives unprecedented prominence to the function concept; there he extends Bernoulli's definition of function still further by defining it to be any *analytical* expression defined from variable quantities and constants, where the term "analytical" includes polynomials, power series, and logarithmic and trigonometric expressions. For Euler a *continuous* function meant a function "unbroken" in the sense of being specified by a single analytic formula; hence a *discontinuous* function meant one "broken" in the sense of requiring different analytic expressions in different domains of the independent variable. In Boyer's words, for Euler "functionality became a matter of formal representation, rather than conceptual recognition of a relation."[16] This is true; nevertheless, Euler's formalistic treatment of function freed the concept from its geometric origins and paved the way for the general concept of function which emerged in the middle of the nineteenth century.

3.2 From D'Alembert to Carnot

While Euler treated infinitesimals as formal zeros, that is, as fixed quantities, his contemporary **Jean le Rond d'Alembert** (1717–83) took a different view of the matter. Following Newton's lead, he conceived of infinitesimals or differentials in terms of the limit concept, which he formulated by the assertion that one varying quantity is the limit of another if the second can approach the other more closely than by any given quantity.[17] D'Alembert firmly rejected the idea of infinitesimals as fixed quantities, asserting

> *A quantity is something or nothing: if it is something, it has not yet vanished; if it is nothing, it has literally vanished. The supposition that there is an intermediate state between the two is a chimera.*[18]

D'Alembert saw the idea of limit as supplying the methodological root of the differential calculus:

> *The differentiation of equations consists simply in finding the limits of the ratios of finite differences of two variables included in the equation.*[19]

[14]Art. *Function,* Encyclopedia Britannica, Eleventh Edition, 1910–11.

[15]*Ibid.*

[16]Boyer(1959), p. 243.

[17]*Ibid.*, p. 247.

[18]Quoted in Boyer (1959), pp. 248.

[19]Quoted *ibid.*, pp. 247–8.

For d'Alembert the language of infinitesimals or differentials was just a convenient shorthand for avoiding the cumbrousness of expression required by the use of the limit concept.

Although d'Alembert anticipated the doctrine of limits which would later come to provide the rigorous foundation for analysis long sought by mathematicians, the majority of his contemporaries regarded his formulation of the limit concept as being no less vague, and much less convenient, than the concept of infinitesimal it was intended to supplant. Indeed, of 28 publications on the calculus appearing from 1754 to 1784, just 6 employ the limit concept.[20] The use of the latter was not to become routine until the very end of the nineteenth century.

In 1742 **Colin Maclaurin** (1698–1746) published his *Treatise of Fluxions*. In this work Maclaurin sets out to demonstrate the essential validity of Newton's fluxional theory by rigorous derivation "after the manner of the ancients, from a few unexceptionable principles".[21] His approach involved discarding Leibnizian infinitesimals and differentials, but retaining the fundamental notions of Newton's fluxional theory, in particular instantaneous velocity.

Maclaurin gives a masterly account of the process by which the rigorous procedures of Archimedes and other Greek mathematicians came gradually to be replaced by arguments, convenient but of doubtful logical validity, involving infinities and infinitesimals. He writes:

> *But when the principles and strict methods of the ancients...were so far abandoned, it was difficult for the Geometricians to determine where they should stop. After they had indulged themselves in admitting quantities, of various kinds, that were not assignable, in supposing such things to be done as could not possibly be effected (against the constant practice of the ancients), and had involved themselves in the mazes of infinity; it was not easy for them to avoid perplexity, and sometimes error, or to fix bounds to these liberties once they were introduced. Curves were not only considered as polygons of an infinite number of infinitely little sides, and their differences deduced from the different angles that were supposed to be formed from these sides; but infinites and infinitesimals were admitted of infinite orders...*[22]

From geometry, Maclaurin observes, the infinites and infinitesimals passed into "philosophy", i.e. natural science,

> *carrying with them the obscurity and perplexity that cannot fail to accompany them. An actual division, as well as a divisibility of matter* in infinitum, *is admitted by some. Fluids are imagined consisting of infinitely small particles, which are composed of others infinitely less; and this subdivision is supposed to be continued without end. Vortices are proposed, for solving the phaenomena of nature, of indefinite or infinite degrees, in imitation of the infinitesimals in geometry...*[23]

[20] *Ibid.*,p. 250.

[21] Maclaurin, *Treatise of Fluxions*, Preface, in Ewald (1999) *From Kant to Hilbert*, p. 93.

[22] *Ibid*, pp. 107–8

[23] *Ibid.*, p. 108.

Maclaurin notes that the introduction of infinitesimals into the description of natural phenomena carries with it the constraint that natural processes always occur continuously, so ruling out physical atomism:

> *Nature is confined in her operations to operate by infinitely small steps. Bodies of a perfect hardness are rejected, and the old doctrine of atoms treated as imaginary, because in their actions and collisions they might pass at once from motion to rest, or from rest to motion, in violation of this law. Thus the doctrine of infinites is interwoven with our speculations in geometry and nature.*[24]

Despite the clarity and originality of Maclaurin's treatise, its style of presentation *in modo geometrico* is in truth a backward step. For, as the Continental mathematicians recognized, the analysis of continuous variation had transcended the modes of reasoning in classical geometry and could no longer be adequately accommodated there. Historians of mathematics have noted that the wide influence of Maclaurin's treatise on British mathematicians led them to adopt the mathematical style of Archimedes and to ignore the more fertile methods of analysis emerging on the Continent. As a result, British mathematicians lagged behind their Continental counterparts for nearly a century.

The last efforts of the eighteenth century mathematicians to demystify infinitesimals and banish the persistent doubts concerning the soundness of the calculus both appeared in France in 1797. These were the *Théorie des Fonctions Analytiques* by the great Franco-Italian mathematician **Joseph-Louis Lagrange** (1736–1813) and the *Reflexions sur la Metaphysique du Calcul Infinitesimal* by the mathematician and "organisateur de la victoire" of the French Revolution, **Lazare Carnot** (1753–1823).

The *Théorie des Fonctions Analytiques* embodies an "algebraic" approach to the calculus. Lagrange had long been sceptical concerning infinitesimals but was at the same time less than enamoured of the idea of a limit, considering it metaphysically suspect. Nor did the method of fluxions appeal to him, as it involved the extraneous concept of motion. The treatment of differentials as formal zeros he also regarded as dubious.[25] In seeking a method for obtaining the results of the calculus which avoided these pitfalls, he came up with an idea based on the *Taylor expansion*[26] of a function:

$$f(x+h) = f(x) + hf'(x) + \frac{h^2}{2!}f''(x) + \dots.$$

Here the coefficients $f''(x), f''(x), \dots$ are the first, second, ... *derivatives* of f. These derivatives had originally been determined through the use of fluxions or infinitesimals, but Lagrange proposed to avoid these by *defining* the successive derivatives of

---

[24] *Ibid.*

[25] Boyer (1959), p. 251.

[26] First introduced in 1715 by the English mathematician **Brook Taylor** (1685–1731).

a function to be the coefficients—which Lagrange called the *derived functions*[27] of the given function—in its Taylor expansion. In this way the differential calculus was to be purged of all metaphysical difficulties, in fact becoming no more than a straightforward method for finding the derived functions of a given function.

Carnot's aim, by contrast, was not to banish the infinitesimal but to divest the concept of all trace of vagueness or obscurity. He attempts to achieve this by conceiving of infinitesimals as variable quantities:

> *We will call every quantity, which is considered as continually decreasing (so that it may be made as small as we please, without being at the same time obliged to make those quantities vary the ratio of which it is our object to determine), an Infinitely small Quantity.*[28]
>
> *...You ask me what Infinitesimal quantities mean? I declare to you that I never by that expression mean metaphysical and abstract existences, as this abridged name seems to imply; but real, arbitrary quantities, capable of becoming as small as I wish, without being compelled at the same time to make those quantities vary whose ratio it was my intention to discover.*[29]

But despite their reality, the inherent variability of infinitesimal quantities necessitates that they be discharged at the conclusion of a calculation:

> *...You ask me if my calculation is perfectly exact and rigorous? I reply in the affirmative, as soon as I have arrived at eliminating from it the Infinitesimal quantities spoken of above and have reduced it so as to contain ordinary Algebraic quantities alone.*[30]

But Carnot did not follow d'Alembert in regarding the use of infinitesimals as a convenient shorthand for an underlying use of a limit concept. Rather, Carnot suggests that when infinitesimals are taken as real quantities the efficacy of the calculus is then explained by the compensation of errors, a view defended earlier by Berkeley.[31] He contrasts this with the explanation, resting on the Law of Continuity, associated with Euler's view that infinitesimals are no more than zeros:

> *We may then regard the Infinitesimal Analysis in two points of view: by considering the infinitely small quantities either as real quantities or as absolutely zero. In the former case, Infinitesimal Analysis is nothing more than a calculation of compensation of errors: and in the second it is the art of comparing vanishing quantities together and with others, in order to deduce from these comparisons the ratios, whatever they may be, which exist between the proposed quantities. These quantities, as equal to zero, ought to be overlooked in the calculation when they are found in addition with any real quantity; or when they are subtracted from them: nevertheless they have...ratios very interesting to discover, and such as are determined by the law of continuity, to which the system of auxiliary quantities is subject in its changes. Now in order to readily apprehend this law of continuity, we may easily observe, that we are obliged to consider the quantities in question, at some distance from the term when they vanish altogether, to forestall them from presenting the indefinite ratio of* 0 *to* 0*: but this distance is arbitrary, and has no other object than to enable us to judge the more easily of the ratios which exist between vanishing quantities: these are the*

[27]Hence the term *derivative.* The notation $f\,'(x)$ was also introduced by Lagrange.

[28]Carnot (1832), p. 14.

[29]*Ibid.*, p. 33

[30]*Ibid.*, p. 34.

[31]See below.

ratios which we have in view, whilst we regard the infinitely small quantities as absolutely zero, and not those which have not yet arrived at the term of their annihilation. These last, which have been called *infinitely small, are never intended themselves to enter into the calculation, regarded under the point of view of which we are at present speaking, but are only employed to assist the imagination, and to point out the law of continuity which determines any ratio whatever of the vanishing quantities to which they correspond.*[32]

Carnot's work continued to be influential well into the nineteenth century before it was swept away by the limit concept.

3.3 The Philosophers

Berkeley

Infinitesimals, differentials, evanescent quantities and the like coursed through the veins of the calculus throughout the eighteenth century. Although nebulous—even logically suspect—these concepts provided, *faute de mieux,* the tools for deriving the great wealth of results the calculus had made possible. And while, with the notable exception of Euler, many eighteenth century mathematicians were ill-at-ease with the infinitesimal, they would not risk killing the goose laying such a wealth of golden mathematical eggs. Accordingly they refrained, in the main, from destructive criticism of the ideas underlying the calculus. Philosophers, however, were not fettered by such constraints.

The philosopher **George Berkeley** (1685–1753), noted both for his subjective idealist doctrine of *esse est percipi* and his denial of general ideas, was a persistent critic of the presuppositions underlying the mathematical practice of his day. His most celebrated broadsides were directed at the calculus, but in fact his conflict with the mathematicians went deeper. For his denial of the existence of abstract ideas of any kind went in direct opposition with the abstractionist account of mathematical concepts held by the majority of mathematicians and philosophers of the day. The central tenet of this doctrine, which goes back to Aristotle, is that the mind creates mathematical concepts by *abstraction*, that is, by the mental suppression of extraneous features of perceived objects so as to focus on properties singled out for attention. Berkeley rejected this, asserting that mathematics as a science is ultimately concerned with objects of sense, its admitted generality stemming from the capacity of percepts to serve as signs for all percepts of a similar form.[33]

Berkeley's empiricist philosophy had initially led him to claim that geometry, correctly conceived, can make reference only to actually perceived lines and figures. Thus in his early and unpublished *Philosophical Commentaries* he rejects infinite divisibility and proposes jettisoning classical geometry in favour of a new account of

[32]Carnot (1832), pp. 101–2.

[33]Jesseph (1993), p. 37.

geometry formulated in terms of perceptible minima.[34] By the time he came to write the *Principles of Human Knowledge,* (published in 1710), his radicalism *vis-a-vis* geometry had softened somewhat, but he again attacks infinite divisibility on the grounds that, since to exist is to be perceived, only perceived extension exists, and it is manifest that this is not infinitely divisible:

> *Every particular finite extension which may possibly be an object of our thought is an* idea *existing only in the mind, and consequently each part thereof must be perceived. If, therefore, I cannot perceive innumerable parts in any finite extension that I consider, it is certain that they are not contained in it; but it is evident that I cannot distinguish innumerable parts in any particular line, surface or solid, which I either perceive by sense, or figure to myself in my mind: wherefore I conclude that they are not contained in it. Nothing can be plainer to me than that the extensions I have in view are no other than my own ideas; and it is no less plain that I cannot resolve any one of my ideas into an infinite number of other ideas; that is, they are not infinitely divisible.*[35]

It will be seen that Berkeley, like Epicurus, reads the thesis of infinite divisibility as the assertion that an extended magnitude must contain an actual infinity of parts.[36]

Not surprisingly, Berkeley ridicules those who adhere to the concept of infinitesimal:

> *Of late the speculation about infinites have run so high, and grown to such strange notions, as have occasioned no scruples and disputes among the geometers of the present age. Some there are of great note who, not content with holding that finite lines may be divided into an infinite number of parts, do yet farther maintain that each of those infinitesimals is itself subdivisible into an infinity of other parts or infinitesimals of a second order, and so on* ad infinitum. *These, I say, assert there are infinitesimals of infinitesimals of infinitesimals, etc., without ever coming to an end! so that according to them an inch does not barely contain an infinite number of parts, but an infinity of an infinity of an infinity* ad infinitum *of parts. Others there be who hold all orders of infinitesimals below the first to be nothing at all; thinking it with good reason absurd to imagine there is any positive quantity or part of extension which, though multiplied infinitely, can never equal the smallest given extension. And yet on the other hand it seems no less absurd to think the square, cube, or other power of a positive real root, should itself be nothing at all; which they who hold infinitesimals of the first order, denying all of the subsequent orders, are obliged to maintain.*[37]

Berkeley maintains that the use of infinitesimals in deriving mathematical results is illusory, and that they can be eliminated:

> *If it be said that several theorems undoubtedly true are discovered by methods in which infinitesimals are made use of, which could never have been if their existence included a contradiction in it; I answer that upon a thorough examination it will not be found that in*

---

[34]*Ibid.*, p. 57. The doctrine of perceptible minima is, of course, subject to the same objections that had been raised against previous attempts at analyzing extension in terms of atoms.

[35]Berkeley (1960), §124.

[36]According to Jesseph (1993, p. 67), Berkeley was largely unaware of the tradition of "mathematical atomism" in ancient and medieval philosophy.

[37]Berkeley (1960), §130. It is of interest here to note that the final sentence of this quotation is an explicit rejection of the concept of nilpotent infinitesimal which had been defended by Nieuwentijdt against Leibniz. 250 years later, that concept was to be revived in smooth infinitesimal analysis. See Chap. 10 below.

any instance it is necessary make use of or conceive infinitesimal parts of finite lines, or even quantities less than the minimum sensible; *nay, it will be evident this is never done, it being impossible. And whatever mathematicians may think of fluxions or the differential calculus and the like, a little reflexion will shew them, that in working by those methods, they do not conceive or imagine lines or surfaces less than what are perceivable to sense. They may, indeed, call these little and almost insensible quantities infinitesimals or infinitesimals of infinitesimals, if they please: but at bottom this is all, they being in truth finite, nor does the solution of problems require the supposing any other.*[38]

In his 1721 treatise *De Motu* Berkeley adopts a more tolerant attitude towards infinitesimals, regarding them as useful fictions in somewhat the same way as did Leibniz:

Just as a curve can be considered as consisting of an infinity of right lines, even if in truth it does not consist of them but because this hypothesis is useful in geometry, in the same way circular motion can be regarded as traced and arising from an infinity of rectilinear directions, which supposition is useful in the mechanical philosophy.[39]

In *The Analyst* of 1734 Berkeley launched his most sustained and sophisticated critique of infinitesimals and the whole metaphysics of the calculus. Addressed *To an Infidel Mathematician,*[40] the tract was written with the avowed purpose of defending theology against the scepticism shared by many of the mathematicians and scientists of the day. Berkeley's defense of religion amounts to the claim that the reasoning of mathematicians in respect of the calculus is no less flawed than that of theologians in respect of the mysteries of the divine:

...he who can digest a second or third fluxion, a second or third difference,[41] *need not, methinks, be squeamish about any point in divinity.*[42]

Berkeley's arguments are directed chiefly against the Newtonian fluxional calculus. Typical of his objections is that in attempting to avoid infinitesimals by the employment of such devices as evanescent quantities and prime and ultimate ratios Newton has in fact violated the law of noncontradiction by first subjecting a quantity to an increment and then setting the increment to 0, that is, denying that an increment had ever been present. As for fluxions and evanescent increments themselves, Berkeley has this to say:

And what are these fluxions? The velocities of evanescent increments? And what are these same evanescent increments? They are neither finite quantities nor quantities infinitely small, nor yet nothing. May we not call them the ghosts of departed quantities?[43]

Nor did the Leibnizian method of differentials escape Berkeley's strictures:

---

[38] *Ibid.*, §132.

[39] Berkeley, *De Motu,* §61. In Ewald (1999).

[40] Likely the astronomer Edmund Halley (1656–1742).

[41] By "difference" Berkeley means "differential".

[42] Berkeley, *Analyst,* §7. In Ewald (1999).

[43] *Ibid.*, §35.

> *Instead of flowing quantities and their fluxions,* [the Leibnizians] *consider the variable finite quantities as increasing or diminishing by the continual addition or subduction of infinitely small quantities. Instead of the velocities wherewith increments are generated, they consider the increments or decrements themselves, which they call differences, and which are supposed to be infinitely small. The difference of a line is an infinitely little line; of a plane an infinitely little plane. They suppose finite quantities to consist of parts infinitely little, and curves to be polygons, whereof the sides are infinitely little, which by the angles they make with each other determine the curvity of the line. Now to conceive a quantity infinitely small, that is, infinitely less than any sensible or imaginable quantity, or than any the least finite magnitude is, I confess, above my capacity...*[44]

Berkeley asserts that he is not challenging the conclusions drawn by the analysts, but only the methods by which these conclusions are drawn. His own view is that the methods of the calculus actually work through the introduction of what he calls "contrary errors". In finding the tangent to a curve by means of differentials, for example, increments are first introduced; but these determine the secant, not the tangent. This error is then expunged by ignoring higher differentials, and so

> *by virtue of a twofold mistake you arrive, though not at science, yet at truth.*[45]

This explanation of the validity of the results of the calculus, which became known as the principle of *compensation of errors,* was endorsed by a number of eighteenth century mathematicians, including Euler, Lagrange and, as we have observed, Carnot.

Berkeley's criticisms of the methodology of the calculus, while pointed, were by no means wholly destructive. Indeed, *The Analyst* has been identified as marking "a turning point in the history of mathematical thought in Great Britain."[46] By exposing with such severity the logical inadequacies of the foundations of the calculus, Berkeley provoked an avalanche of responses from mathematicians anxious to clarify the concepts underlying the calculus and so to place its results and methods beyond reasonable doubt.

Hume

The views of the radical empiricist **David Hume** (1711–76) are similar in certain respects to those of Epicurus.[47] For Hume, as for Berkeley, to comprehend or to have an idea of a thing is to have a mental picture of it,[48] a picture not inferior in immediacy to a sense impression. This requirement automatically places beyond comprehension both the infinite and the infinitesimal. Accordingly Hume rejects

---

[44] *Ibid.*, §§5, 6.

[45] *Ibid.*, §22.

[46] Cajori (1919), p. 89.

[47] Furley (1967), Ch. 10.

[48] *Ibid.*, p. 137.

infinite divisibility, even in thought, on the grounds that anything infinitely divisible must contain an infinity of parts, while the mind is incapable of grasping an infinity in its actuality. Thus Hume writes in Part II of his *Treatise of Human Nature:*

> *'Tis universally allow'd, that the capacity of the mind is limited, and can never attain a full and adequate conception of infinity: And tho' it were not allow'd, 'twould be sufficiently evident from the plainest observation and evidence. 'Tis also obvious, that whatever is capable of being divided* in infinitum, *must consist of an infinite number of parts, and that 'tis impossible to set any bounds to the number of parts, without setting bounds at the same time to the division. It requires scarce any induction to conclude from hence, that the* idea, *which we form of any finite quality, is not infinitely divisible, but that by proper distinctions and separations we may run up this idea to inferior ones, which will be perfectly simple and indivisible. In rejecting the infinite capacity of the mind, we suppose it may arrive at an end in the division of its ideas; nor are there any possible means of evading the evidence of this conclusion.*[49]

Hume makes the interesting observation that, even though one may grasp the concept of an arbitrarily small numerical fraction *applied* to an idea, the idea *itself* is not infinitely divisible:

> *When you tell me of the thousandth or the ten thousandth part of a grain of sand, I have a distinct idea of their different proportions; but the images, which I form in my mind to represent the things themselves, are nothing different from each other, nor inferior to that image, by which I represent the grain of sand itself, which is suppos'd so vastly to exceed them. What consists of parts is distinguishable into them, and what is distinguishable is separable. But whatever we may imagine of the thing, the idea of a grain of sand is not distinguishable, nor separable into twenty, much less a thousand, ten thousand, or an infinite number of different ideas.*[50]

Hume concludes that the boundedness of divisibility in imagination leads inevitably to indivisible minima, that is, to atoms:

> *'Tis therefore certain, that the imagination reaches a* minimum, *and may raise up to itself an idea, of which it cannot conceive any sub-division, and which cannot be distinguished without a total annihilation.*[51]

And what holds of the imagination holds equally of the senses:

> *'Tis the same case with the impressions of the senses as with the ideas of the imagination. Put a spot of ink on paper, fix your eye upon that spot, and retire to such a distance, that at last you lose sight of it: 'tis plain, that the moment before it vanish'd the image or impression was perfectly indivisible.*[52]

But while sense perception may present us with apparent minima which reason tells us must be composed of a multitude of parts, Hume urges that the atoms of the imagination are genuinely indivisible, and not so merely as a result of mental limitation:

---

[49]Hume (1962), II, 1.

[50]*Ibid.* But presumably the idea of a grain of sand is separable into the ideas "grain" and "sand".

[51]*Ibid.*

[52]*Ibid.*

> *We may hence discover the error of the common opinion, that the capacity of the mind is limited. . .and that 'tis impossible for the imagination to form an adequate idea, of what goes beyond a certain degree of minuteness. . . . Nothing can be more minute, than some ideas, which we form in the fancy; and images, which appear to the senses; since there are ideas and images perfectly simple and indivisible. The only defect of our senses is, that they give us disproportion'd images of things, and represent as minute and uncompounded what is really great and composed of a vast number of parts. This mistake we are not sensible of: but taking the impressions of those minute objects, which appear to the senses, to be equal or nearly equal to the objects, and finding by reason, that there are other objects vastly more minute, we too hastily conclude, that these are inferior to any idea of our imagination or impression of the senses.*[53]

Hume next contends that what holds of ideas holds equally of the objects represented by them, so that, since ideas are not infinitely divisible, neither are objects:

> *Wherever ideas are adequate representations of objects, the relations, contradictions and agreements of the ideas are all applicable to the objects. . . . But our ideas are adequate representations of the most minute parts of extension; and thro' whatever divisions and subdivisions we may suppose these parts to be arriv'd at, they can never become inferior to some ideas we can form. The plain consequence is, that whatever* appears *impossible and contradictory upon the comparison of these ideas, must be* really *impossible and contradictory, without any farther excuse or evasion.*
>
> *Everything capable of being infinitely divided contains an infinite number of parts; other wise the division would be stopt short by the indivisible parts, which we should immediately arrive at. If therefore any finite extension be infinitely divisible, it can be no contradiction to suppose, that a finite extension contains an infinite number of parts; And* vice versa, *if it be a contradiction to suppose, that a finite extension contains an infinite number of parts, no finite extension can be infinitely divisible. But that this latter supposition is absurd, I easily convince myself by the consideration of my clear ideas. I first take the least idea I can form of a part of extension and being certain that there is nothing more minute than this idea, I conclude, that whatever I discover by its means must be a real quality of extension. I then repeat this idea once, twice, thrice, &c, and find the compound idea of extension, arising from its repetition, always to augment, and become double, triple, quadruple, &c, till at last it swells up to a considerable bulk, greater or smaller, in proportion as I repeat greater or less the same idea. When I stop in the addition of parts, the idea of extension ceases to augment; and were I to carry on this addition* in infinitum, *I clearly perceive, that the idea of extension must also become infinite. Upon the whole I conclude, that the idea of an infinite number of parts is individually the same idea with that of infinite extension; that no finite extension is capable of containing an infinite number of parts; and consequently that no finite extension is infinitely divisible.*[54]

In a nutshell, Hume's argument is the following. No matter how small an actual extension may be, it can always be represented as an idea in the mind. Now suppose that an actual extension were divisible into infinitely many parts. Each of these parts can be represented as an idea, and so each is no smaller than the minimal idea of extension conceivable. But then the idea of the given extension, as the sum of (the ideas of) its parts, would have to be as least as large as the sum of infinitely many

[53] *Ibid.*

[54] *Ibid., II, 2.*

minimal ideas of extension, and so itself infinite. Since the finite mind cannot contain an infinite idea, a contradiction results.

Hume's atomism, to which he refers as "the doctrine of indivisible points", was thus both conceptual and physical. Hume says that in order to be grasped these points or atoms of extension must possess sensible qualities such as colour or solidity:

> *The idea of space is convey'd to the mind by two senses, the sight and touch; nor does anything ever appear extended, that is not either visible or tangible. That compound impression, which represents extension, consists of several lesser impressions, that are indivisible to the eye or feeling, and may be call'd impressions of atoms or corpuscles endow'd with colour and solidity....*[55]

According to Hume the doctrine of sensible points or atoms provides a way of avoiding the infinite divisibility of extension which mathematicians have always considered ineluctable:

> *It has often been maintained in the schools, that extension must be divisible,* in infinitum, *because the system of mathematical points is absurd; and that system is absurd, because a mathematical point is a non-entity, and consequently can never by its conjunction with others form a real existence. This wou'd be perfectly decisive, were there no medium betwixt the infinite divisibility of matter, and the non-entity of mathematical points. But there is evidently a medium, viz. the bestowing of colour or solidity on these points; and the absurdity of both the extremes is a demonstration of the truth and reality of this medium.*[56]

The exact nature of Hume's sensible, or indivisible points remains somewhat unclear. Should they be taken as extended entities like Epicurus's minima or as extensionless mathematical points? Not the former, since Hume denies them "real extension"; but also not the latter since they can be compounded to form extended magnitudes. Whatever his "points" may be, Hume admits that they are "entirely useless" as providing a standard of comparison of the sizes of extended magnitudes, since it is impossible to compute the number of points these contain.

It has been suggested[57] that Hume later came around to the view that sensible points could be taken as minimal parts of extension, a position he decisively rejected in the *Treatise*. This aspect of Hume's doctrine remains puzzling and controversial.

Kant

The opposition between continuity and discreteness plays a significant role in the philosophical thought of **Immanuel Kant** (1724–1804). His mature philosophy, *transcendental idealism*, rests on the division of reality into two realms. The first, the *phenomenal* realm, consists of appearances or objects of possible experience, configured by the forms of sensibility and the epistemic categories. The second, the

[55] *Ibid., II, 3.*

[56] *Ibid., II, 4.*

[57] Furley (1967) p. 142.

noumenal realm, consists of "entities of the understanding to which no objects of experience can ever correspond",[58] that is, things-in-themselves.

Regarded as magnitudes, appearances are spatiotemporally extended and continuous, that is infinitely, or at least limitlessly, divisible. Space and time constitute the underlying order of phenomena, so are ultimately phenomenal themselves, and hence also continuous:

> *The property of magnitudes by which no part of them is the smallest possible, that is, by which no part is simple, is called their continuity. Space and time are* quanta continua, *because no part of them can be given save as enclosed between limits (points or instants), and therefore only in such fashion that this part is again a space or a time. Space therefore consists solely of spaces, time solely of times. Points and instants are only limits, that is, mere positions which limit space and time. But positions always presuppose the intuitions which they limit or are intended to limit; and out of mere positions, viewed as constituents capable of being given prior to space and time, neither space nor time can be constructed. Such magnitudes may also be called* flowing, *since the synthesis of productive imagination involved in their production is a progression in time, and the continuity of time is ordinarily designated by the term flowing or flowing away.*[59]

As objects of knowledge, appearances are continuous *extensive* magnitudes, but as objects of sensation or perception they are, according to Kant, *intensive* magnitudes. By an intensive magnitude Kant means a magnitude possessing a *degree* and so capable of being apprehended by the senses: for example, brightness or temperature. Intensive magnitudes are entirely free of the intuitions of space or time, and "can only be presented as unities". But, like extensive magnitudes, they are continuous:

> *Every sensation... is capable of diminution, so that it can decrease and gradually vanish. Between reality in the field of appearance and negation there is therefore a continuity of many possible intermediate sensations, the difference between any two of which is always smaller than the difference between the given sensation and zero or complete negation.*[60]

Moreover, appearances are always presented to the senses as intensive magnitudes:

> *...the real in the field of appearance always has a magnitude. But since its apprehension by means of mere sensation always takes place in an instant and not through successive synthesis of different sensations, and therefore does not proceed the parts to the whole, the magnitude is to be met only in the apprehension. The real has therefore magnitude, but not extensive magnitude. ... Every reality in the field of appearance has therefore intensive magnitude.*[61]

Kant regards as "remarkable" the facts that

> *of magnitudes in general we can know* a priori *only a single* quality, *namely, that of continuity, and ... in all quality (the real in appearances) we can know nothing save*

[58]Körner (1955), p. 94.

[59]Kant (1964), p. 204.

[60]*Ibid.*, p. 203.

[61]*Ibid.*, p. 203.

> *[in regard to] their intensive* quantity, *namely that they have degree. Everything else has to be left to experience.*[62]

As for the concept of a thing-in-itself, it signifies "only the thought of something in general, in which I abstract from everything that belongs to the form of sensible intuition."[63] Kant seems to have regarded things-in-themselves as discrete entities in the sense of not being divisible to infinity, and hence, like Leibniz's real entities, as being compounded from simples. This may be inferred from his assertion in the *Metaphysical Foundations of Natural Science* of 1786 that a thing-in-itself "must in advance already contain within itself all the parts in their entirety into which it can be divided."[64] So were a thing-in-itself to be infinitely divisible, one could infer that it consists of an infinite multitude of parts. This, however, is impossible "because there is a contradiction involved in thinking of an infinite number as complete, inasmuch as the concept of an infinite number already implies that it can never be wholly complete."[65]

While Kant never deviated from the claim that space and time are divisible without limit, his opinion on the divisibility of matter underwent alteration. In the *Physical Monadology* of 1756, for example, he attempts to establish the compatibility of the *indivisibility* of physical monads or atoms with the infinite divisibility of space itself. Kant's argument is essentially that while substances must be compounded from simple parts and so cannot be infinitely divisible, this does not apply to space because it is not itself a substance but no more than a well-founded phenomenon, a "certain appearance of the external relation of substances".[66] He begins by arguing that bodies must be compounded from monads, or simple parts:

> *Bodies consist of parts, each of which separately has an enduring existence. Since, however, the composition of such parts is nothing but a relation, and hence a determination which is itself contingent, and which can be denied without abrogating the existence of the things having this relation, it is plain that all composition of a body can be abolished, though all the parts which were formerly combined together nonetheless continue to exist. When all composition is abolished, moreover, the parts which are left are not compound at all; and thus they are completely free from plurality of substances, and, consequently, they are simple. All bodies, whatever, therefore, consist of absolutely simple fundamental parts, that is to say, monads.*[67]

The infinite divisibility of space is then established by means of an argument similar to that in the *Port-Royal Logic,* with the consequence that space does not consist of simple parts. In a later commentary Kant remarks that from the fact that bodies are

---

[62]*Ibid.*, p. 208.

[63]*Ibid.*, p. 270.

[64]Kant (1970), p. 53. Cf. *Critique of Pure Reason, Observation on the Second Antinomy:*

> *Though it may be true that when a whole, made up of substances, is thought by the pure understanding alone, we must, prior to all composition, of it, have the simple...*

[65]*Ibid.*, p. 53.

[66]Kant's thus echoes Leibniz – with the exception that Leibniz's monads were not physical.

[67]*Physical Monadology,* Proposition II, in Kant (1992).

composed of monads and yet the space they occupy is infinitely divisible it would be wrong to infer that physical monads are "infinitely small particles of a body":

> *For it is abundantly plain that space, which is entirely free from substantiality and which is the appearance of the external relations of unitary monads, will not be exhausted by division continued to infinity. However, in the case of any compound whatever, where composition is nothing but an accident and in which there are substantial subjects of composition, it would be absurd if it admitted infinite division. For if a compound were to admit infinite division, it would follow that all the fundamental parts whatever of a body would be so constituted that, whether they were combined with a thousand, or ten thousand, or millions of millions—in a word, no matter how many—they would not constitute particles of matter. This would certainly and obviously deprive a compound of all substantiality; it cannot, therefore, apply to the bodies of nature.*[68]

That is, if bodies were infinitely divisible, their fundamental parts, as indivisibles, must perforce be unextended and could not then be recombined to form an extended object. It follows that each body consists of a determinate number of simple elements. Kant goes on to argue that while each such physical monad is not only situated in space, and actually fills the space it occupies, it does not follow from the admitted divisibility of that space that the monad is likewise divisible. In a commentary to this Kant remarks:

> *The line or surface which divides a small space into two parts certainly indicates that one part of the space exists outside the other. But since space is not a substance but a certain appearance of the external relation of substances, it follows that the possibility of dividing the relation of one and the same substance into two parts is not incompatible with the simplicity of, or if you prefer, the unity of that substance. For what exists on each side of the dividing line is not something which can be so separated from the substance that it preserves an existence of its own, apart from the substance itself and in separation from it, which would, of course, be necessary for real division which destroys simplicity. What exists on each side of the dividing line is an action which is exercised on both sides of one and the same substance: in other words, it is a relation, in which the existence of a certain plurality does not amount to tearing the substance itself into parts.*[69]

It is through its indivisibility or simplicity that a physical monad is distinguished from the space it happens to occupy. But this cannot of itself explain the fact that it occupies that *particular part* of space. The explanation, says Kant, is to be found in the relations of the monad with the substances external to it, and so ultimately in the monad's intherent *impenetrability*. This "prevents the monads immediately present to it on each side from drawing closer to each other" and so limits "the degree of proximity by which they are able to approach it". The monad thus "fills the space by the sphere of its activity". Kant identifies this activity as a *force*. The force has two components, one repulsive, preventing penetration by other monads; the other attractive, ensuring that the monad has a determinate form.

As with Leibnizian monads, it is therefore the *activity* of Kant's monads which prevents them from "collapsing" into mere geometrical points. And Kant follows

[68]*Physical Monadology*, Scholium to Prop. IV.

[69]*Ibid.*, Scholium to Prop. V.

Leibniz in describing this activity as a *force*. But Kant's monads, in contradistinction with Leibniz's, are material entities, so that, while Leibniz saw the "force" of his monads as being akin to the impulses of the soul, Kant endowed his monads with the physical forces of attraction and repulsion.

In a startling *volte-face*, Kant came to repudiate his earlier view that matter is not divisible to infinity.[70] In the *Metaphysical Foundations of Natural Science* of 1786 he now insists that, in a space filled with matter, "every part of the space contains repulsive force to counteract on all sides the remaining parts", so that "every part of a space filled by matter is separable from the remaining parts, insofar as they are material substance." From the mathematical divisibility of the space to infinity there now follows the infinite divisibility of matter "into parts each of which is in turn material substance."

The upshot is that, like space and time, matter too must be assigned to the realm of appearance:

> *...space is no property appertaining to anything outside of our senses, but is only the subjective form of our sensibility. Under this form objects of our external senses appear to us, but we do not know them as they are constituted in themselves. We call this appearance matter.*[71]

But, again, the infinite divisibility of matter (or of space, time, or indeed any appearance) does not imply that it consists in *actuality* of infinitely many parts; the infinity involved is Aristotle's potential infinite:

> *That matter consists of infinitely many parts can indeed be thought by reason, even though this thought cannot be constructed and rendered intuitable. For with regard to what is actual only by being given in representation, there is not more given than is met with in the representation, i.e., as far as the progression of the representation reaches. Therefore, one can only say of appearances, whose division goes on to infinity, that there are as many parts of the appearance as we give, i.e., as far as we want to divide. For the parts insofar as they belong to the existence of an appearance exist only in thought, that is, in the division itself. The division indeed goes on to infinity, but it is never given as infinite; and hence it does not follow that the divisible contains within itself an infinite number of parts in themselves, that are outside of our representation, merely because the division goes on to infinity. For it is not the division of the thing but only the division of its representation that can be infinitely continued. Any division of the object (which is itself unknown) can never be completed and hence can never be entirely given. Therefore, any division of the representation proves no actual infinite multitude to be in the object (since such a multitude would be an express contradiction).*[72]

In the *Critique of Pure Reason* (1781) Kant brings a new subtlety (and, it must be said, tortuousity) to the analysis of the opposition between continuity and discreteness. This may be seen in the second of the celebrated Antinomies in that work, which concerns the question of the mereological composition of matter, or extended substance. Is it (*a*) discrete, that is, consists of simple or indivisible parts, or (*b*) continuous, that is, contains parts within parts *ad infinitum*? Although (*a*), which Kant calls the *Thesis* and (*b*) the *Antithesis* would seem to contradict one another,

---

[70]I am grateful to my colleague Lorne Falkenstein for pointing this out to me.

[71]Kant (1970), p. 55.

[72]*Ibid.*, p.54.

Kant offers proofs of both assertions. The resulting contradiction may be resolved, he asserts, by observing that while the antinomy "relates to the division of appearances",[73] the arguments for (*a*) and (*b*) implicitly treat matter or substance as things-in-themselves:

> *For these* [i.e., appearances] *are mere representations; and the parts exist merely in their representation, consequently in the division (i.e., in a possible experience where they are given) and the division reaches only so far as such experience reaches. To assume that an appearance, e.g., that of a body, contains in itself before all experience all the parts which any experience can ever reach is to impute to a mere appearance, which can exist only in experience, an existence previous to experience. In other words, it would mean that mere representations exist before they can be found in our faculty of our representations. Such an assumption is self-contradictory, as also every solution of our misunderstood problem, whether we maintain that bodies in themselves consist of an infinite number of parts or of a finite number of simple parts.*[74]

Kant concludes that both Thesis and Antithesis "presuppose an inadmissible condition" and accordingly "both fall to the ground, inasmuch as the condition, under which alone either of them can be maintained, itself falls."

Kant identifies the inadmissible condition as the implicit taking of matter as a thing-in-itself, which in turn leads to the mistake of taking the division of matter into parts to subsist independently of the act of dividing. In that case, the Thesis implies that the sequence of divisions is finite; the Antithesis, that it is infinite. These cannot be both be true of the *completed* (or at least completable) sequence of divisions which would result from taking matter or substance as a thing-in-itself.[75] Now since the truth of both assertions has been shown to follow from that assumption, *it* must be false, that is, matter and extended substance are appearances only. And for appearances, Kant maintains, divisions into parts are not completable in experience, with the result that such divisions can be considered neither finite nor infinite:

> *We must therefore say that the number of parts in a given appearance is in itself neither finite nor infinite. For an appearance is not something existing in itself, and its parts given in and through the regress of the decomposing synthesis, a regress which is never given in its absolute completeness, either as finite or infinite.*[76]

It follows that, for appearances, both Thesis and Antithesis are *false*.

Later in the *Critique* Kant enlarges on the issue of divisibility:

> *If we divide a whole which is given in intuition, we proceed from something conditioned to the conditions of its possibility. The division of the parts. . .is a regress in the series of these conditions. The absolute totality of this series would be given only if the regress could reach* simple *parts. But if all the parts in a continuously progressing decomposition are themselves again divisible, the division, that is, the regress from the conditioned to its conditions,* proceeds in infinitum. *For the conditions (the parts) are themselves contained in the conditioned, and since this is given complete in an intuition that is enclosed between limits,*

[73]Kant (1977), p. 83.

[74]*Ibid.*

[75]As already observed, Kant would probably maintain the truth of the Thesis in that event.

[76]Kant (1964), p. 448.

> *the parts are all one and all given together with the conditioned. The regress may not, therefore be entitled merely a regress* in indefinitum. . . . *Nevertheless we are not entitled to say of a whole which is divisible to infinity that* it is made up of infinitely many parts. *For although all parts are contained in the intuition of the whole,* the whole division *is not so contained, but consists only in the continuous decomposition, that is, the regress itself, whereby the series first becomes actual. Since this regress is infinite, all the members or parts at which it arrives are contained in the whole, viewed as an* aggregate. *But the whole* series of the division *is not so contained, for it is a successive infinite and never* whole, *and cannot, therefore, exhibit an infinite multiplicity, or any combination of an infinite multiplicity in a whole.*[77]

What Kant seems to be saying here is that, while each part generated by a sequence of divisions of an intuited whole is given with the whole, the sequence's incompletability prevents *it* from forming a whole; *a fortiori* no such sequence can be claimed to be actually infinite.

More on Kant's Antinomies can be found in Appendix D.

Hegel

The concepts of continuity and discreteness, albeit in a unorthodox and esoteric form, play an important role in the philosophy of **G. W. F. Hegel** (1770–1831). Hegel saw continuity and discreteness as being locked in an indissoluble dialectical relationship—a "unity of opposites". Continuity and discreteness are the "moments", that is, the defining or constituting attributes, of the category of Quantity; the latter is itself a "simple unity of Discreteness and Continuity".[78]

In Hegel's conception continuity is, as it was for Parmenides, first and foremost a form of unity or identity; in the *Science of Logic* (1812–16) continuity is characterized as

> *simple and self-identical self-relation, interrupted by no limit or exclusion; not however, an immediate unity, but a unity of the Ones which are for themselves. The externality of plurality is still here contained, but as something undifferentiated and uninterrupted. In continuity, plurality is posited as it is in itself; each of the many is what the others are, each is equal to the other, and hence plurality is simple and undifferentiated equality.*[79]

Continuity thus still entails plurality, but as "something undifferentiated and uninterrupted." In continuity, Hegel says

> *plurality is posited as it is itself; each of the many is what the others are, each is equal to the other, and hence [the] plurality is simple and undifferentiated equality.*[80]

---

[77] *Ibid.*, p. 459.

[78] Hegel (1961), p. 204.

[79] *Ibid.*, p. 200.

[80] *Ibid.*

He remarks that imagination "easily changes Continuity into Combination, that is, into an external relation of the Ones to one another"[81]. But on the other hand, "continuity is not external but peculiar to [the One] and founded in its essence". Atomism, says Hegel, "remains entangled" in this "externality of continuity". By contrast, mathematics

> *rejects a metaphysic which should be content to allow time to consist of points of time, space in general (or as a first step, the line) of points in space, the plane, of lines, and the whole of space, of planes; it allows no validity to such discontinuous Ones. And although it determines, for instance, the magnitude of a plane as consisting of the sum of an infinity of lines, yet this discreteness is taken only as a momentary image; and the infinite plurality of lines implies, since the space which they are meant to constitute is after all limited, that their discreteness has already been transcended.*[82]

It is to this fact, Hegel says, that we must attribute "the conflict or Antinomy of the infinite divisibility of Space, of Time, of Matter, and so on.": here he is referring to Kant's Second Antinomy in the *Critique of Pure Reason.* For Hegel,

> *This antinomy consists solely in the necessity of asserting Discreteness as much as Continuity. The one-sided assertion of Discreteness gives an infinite or absolute division (and thus something indivisible) for principle; and the one-sided assertion of Continuity, infinite divisibility.*[83]

Hegel finds Kant's analysis of the antinomy wanting in that an absolute separation is made, inadmissibly, between continuity and discreteness. He writes:

> *Looked at from the point of view of mere discreteness, substance, matter, space and time, and so on, are absolutely divided, and the One is their principle. From the point of view of continuity, this One is merely suspended: division remains divisibility, the possibility of dividing remains possibility, without ever actually reaching the atom. Now . . . still continuity contains the moment of the atom, since continuity exists simply as the possibility of division; just as accomplished division, or discreteness, cancels all distinctions between the Ones—for each simple One is what every other is, —and for that very reason contains their equality and therefore their continuity. Each of the two opposed sides contains the other in itself, and neither can be thought of without the other; and thus it follows that, taken alone, neither determination has truth, but only their unity. This is the true dialectic consideration of them, and the true result.*[84]

Hegel next embarks on a discussion of continuous and discrete magnitude. Having observed that Quantity embodies both continuity and discreteness, continuous magnitude is identified as Quantity "posited only in one of its determinations, namely, continuity."[85] Now "continuity is one of the moments of Quantity which requires the other moment, discreteness, to complete it", so that

---

[81] *Ibid.*

[82] *Ibid.*, pp. 202–3.

[83] *Ibid.*, p. 204.

[84] *Ibid.*, p. 211.

[85] *Ibid.*, p. 213.

> *continuity is only coherent and homogeneous unity as unity of discrete elements, and posited thus, it is no longer mere moment but complete Quantity: this is Continuous Magnitude.*[86]

Quantity is identified by Hegel as "externality in itself", and Continuous Magnitude is "this externality as propagating itself without negation, as a context which remains at one with itself." By contrast Discrete Magnitude is "this externality as non-continuous or interrupted". If continuity is identity, then discreteness is distinguishability. But while Discrete Magnitude is multiplicity, this multiplicity is not that of "the multitude of atoms and the void", for

> *Discrete Magnitude is Quantity; and for that very reason* [its] *discreteness is continuous. The continuity in discreteness consists in the fact that the Ones are equal to one another, or have the same unity. Discrete magnitude, then, is the externality of much One posited as the same, and not of the many Ones in general; it is posited as the Many of one unity.*[87]

For Leibniz, the Many in One was manifested in continuous extension, but Hegel saw it in discrete magnitude.

There is an interesting attempt by Bertrand Russell, in *The Principles of Mathematics,* to elucidate the Hegelian conception of continuity and discreteness:

> *The word* continuity *has borne among philosophers, especially since the time of Hegel, a meaning totally unlike that given to it by Cantor. Thus Hegel says: "Quantity, as we saw, has two sources: the exclusive unit, and the identification or equalization of these units. When we look, therefore, at its immediate relation to self, or at the characteristic of selfsameness made explicit by abstraction, quantity is* Continuous *magnitude; but when we look at the One implied in it, it is* Discrete *magnitude." When we remember that quantity and magnitude, in Hegel, both mean "cardinal number", we may conjecture that this assertion amounts to the following: "Many terms, considered as having a cardinal number, must all be members of one class; in so far as they are merely an instance of the class-concept, they are indistinguishable from one another, and in this aspect the whole that they compose is called* continuous*; but in order to their manyness, they must be* different *instances of the class-concept, and in this respect the whole that they compose is called* discrete.[88]

If this is right, then Hegel's conception of *discrete* magnitude may be seen as corresponding to Cantor's famous definition of *set*:

> *By a "set" we mean any collection M into a whole of definite distinct objects m... of our perception or thought.*[89]

And Hegel's conception of *continuous* magnitude would further correspond to Cantor's notion of *power* or *cardinal number:*

> *By the* power *or* cardinal number *of a set M (which consists of distinct, conceptually separate elements m, m', ... and is to this extent determined and limited), I understand the general concept or generic character (universal) which one obtains by abstracting from the elements of the set, as well as from all connections which the elements may have (be it between*

---

[86] *Ibid.*, p. 214.

[87] *Ibid.*, p. 214.

[88] Russell (1964), p. 346.

[89] Quoted in Dauben (1979), p. 170.

> *themselves or other objects), but in particular from the order in which they occur, and by reflecting only upon that which is common to all sets which are equivalent to M.*[90]

It is delightfully dialectical that Hegel seems to have identified as continuous what Cantor held to be discrete.

The *Science of Logic* contains an extensive discussion of the ideas underlying the calculus. Like Berkeley, d'Alembert and Lagrange, Hegel was critical of the use mathematicians had made of infinitesimals and differentials. But far from rejecting the infinitesimal, Hegel was concerned to assign it a proper location within his philosophical scheme, whose reigning principle was the division of reality into the triad of Being, Nothing, and Becoming. For Cavalieri infinitesimals possessed Being, and for Euler they were Nothing, but for Hegel they fell under the category of Becoming. He writes:

> *In an equation where x and y are posited primarily as determined through a ratio of powers, x and y as such are still meant to denote Quanta*[91]*: now this meaning is entirely lost in the so-called infinitesimal differences. dx and dy are no longer Quanta and are not supposed to signify such; they have a significance only in their relation, a meaning merely as moments. They no longer are Something (Something being taken as Quantum), nor are they finite differences; but they are also not Nothing or the indeterminate nil. Apart from their relation they are pure nil; but they are meant to be taken as moments of the relation, as determinations of the differential coefficient* $\frac{dx}{dy}$.[92]

> *Now when the mathematics of the infinite* [i.e., the infinitesimal] *still maintained that these quantitative determinations were vanishing magnitudes, that is, magnitudes which no longer are any Quantum but also not nothing, it seemed abundantly clear that such an intermediate state, as it was called, between Being and Nothing did not exist. ... The unity of Being and Nothing is indeed not a* state; *for a state would be a determination of Being and Nothing such as might have been reached by these moments only contingently, as it were through disease or external influence, and through erroneous thinking; but, on the contrary, this mean and unity, this vanishing and, equally, Becoming is, in fact, their only truth.*[93]

In Hegel's subsequent review of how the infinitesimal has been conceived by mathematicians of the past, those who regarded infinitesimals as fixed quantities receive short shrift, while those who saw infinitesimals in terms of the limit concept (which in Hegel's eyes fell under the appropriate category of Becoming) are praised. Thus, for example, Newton is praised for his explanation of fluxions not in terms of indivisibles, but in terms of "vanishing *divisibilia*", and, further, "not [in terms of] sums and ratios of determinate parts, but [in terms of] the limits (*limites*) of the sums and ratios."[94] Newton's conception of generative or variable magnitudes also receives Hegel's endorsement. He quotes Newton to the effect that

---

[90] Quoted *ibid.*, p. 221.

[91] By *Quantum* Hegel means determinate Quantity, that is, Quantity of a definite size.

[92] Hegel (1961), p. 269.

[93] *Ibid.*, pp. 269–70.

[94] *Ibid.*, p. 271.

> *"[Any finite magnitude] is considered as variable in its incessant motion and flow of increase and decrease, and so its momentary augmentation [I give] the name of* Moments. *These, however, must not be taken as particles of determinate magnitude (particulae finitae). They are not moments themselves, but magnitudes produced by moments; the generative principles or beginnings of finite magnitudes must here be understood.*

And then comments:

> *—An internal distinction is here made in Quantum; it is taken first as product or Determinate Being, and next in its Becoming, as its beginning and principle, that is , as it is in its concept or (what is the same thing) in its qualitative determination. In the latter the qualitative differences, the infinitesimal incrementa or decrementa, are moments only; and it is only in what has been generated that we have Quantum or that which has passed over into the indifference of determinate existence and into externality.*[95]

The use of fixed infinitesimals, on the other hand, Hegel deplores:

> *The idea of infinitely small quantities (latent also in increment and decrement) is far inferior to the mode of conception* [just] *indicated. The idea supposes them to be of such a nature that they may be neglected in relation to finite magnitudes; and not only that, but also their higher orders relative to the lower order, and the products of several relative to one.—With Leibniz this demand to neglect (which previous inventors of methods referring to this kind of magnitude also bring into play) becomes more strikingly prominent. It is this chiefly which gives an appearance of inexactitude and express incorrectness, the price of convenience, to this calculus in the course of its operation.*[96]

Nor does Euler's view of infinitesimals as formal zeros fare much better:

> *In this regard Euler's idea especially must be cited. On the basis of Newton's general definition, he insists that the differential calculus considers the ratios of incrementa of a magnitude, while the infinitesimal difference as such is to be regarded wholly as nil.—It will be clear from the above how this is to be understood: the infinitesimal difference is nil only quantitatively, it is not a qualitative nil, but, as nil of quantum, it is pure moment of a ratio only. There is no magnitudinal difference; but for that reason it is, in a manner, wrong to express as incrementa or decrementa as differences those moments which are called infinitely small magnitudes. . . . the difficulty is self-evident when it is said that for themselves the incrementa are each nil, and that only their ratios are being considered; for a nil is altogether without determinateness. Thus this image, although it reaches the negative aspect of Quantum and expressly asserts it, yet does not simultaneously seize this negative in its positive meaning of qualitative determinations of quantity, which, if torn away from the ratio and treated as Quanta, would each be but a nil.*[97]

Hegel goes on to discuss some of the methods that mathematicians have employed to resolve the conceptual difficulties caused by the use of infinitesimals. He pays particular attention to Lagrange's attempt to eliminate infinitesimals from the calculus through the use of Taylor expansions. Hegel considers that the Taylor expansion of a function "must not only be regarded as a sum, but as qualitative moments of a conceptual whole."[98] That being the case, he says, the basic calculus

---

[95] *Ibid.*, p. 274.

[96] *Ibid.*

[97] *Ibid.*, p. 275–6.

[98] *Ibid.*, p. 280.

procedure of omitting from the Taylor series terms with higher powers has "a significance wholly different from that which belongs to their omission on the ground of relative smallness"[99] He continues:

> *And here the general assertion can be made that the whole difficulty of the principle would be removed, if the qualitative meaning of the principle were indicated and the operation made dependent on it, in place of the formalism which identifies the determination of the differential with the problem which gives it its name, the distinction generally between a function and its variation after its variable magnitude has received an increment. In this sense it is clear that the first term of the series resulting from the development of* $(x + dx)^n$ *quite exhausts the differentia of* x^n*. Thus the neglect of the other terms is not due to their relative smallness;—and no inexactitude, no mistake or error is here assumed which is supposed to be compensated or rectified by another error .... A ratio, and not a sum, is here in question; and, therefore, with the first term the differential is fully found....*[100]

Hegel (correctly) regards the differential coefficient $\frac{dy}{dx}$ as the limit of a *ratio* in which the term *dy*, in particular, is not to be "taken as difference or increment in the sense that it is only the numerical difference of the Quantum from the Quantum of another ordinate."[101] If differentials *are*, incorrectly, construed in this way, the matter is "obscured", for then:

> *Limit has not here the meaning of ratio: it counts only as the ultimate value which another and similar magnitude steadily approaches in such a manner, that the difference between them may be as small as desired, and the ultimate relation a relation of equality. Thus the infinitesimal difference is the ghost of a difference between one Quantum and another, and, when it is thus imagined, the qualitative nature, according to which dx is related essentially as a determination of ratio not to x but to dy, is in the background. ...—In this kind of determination, geometers are chiefly at pains to make intelligible the approximation of a magnitude to its limit, clinging to this aspect of the difference between Quantum and Quantum, where it is no difference and yet still is a difference. But in any case Approximation is a category which in itself means and makes intelligible nothing; dx has already passed through approximation, it is neither near nor is it nearer; and "infinitely near" itself means the negation of nearness and approximation.*[102]

Hegel observes that this incorrect construal of differentials amounts to considering "the incrementa or infinitesimal differences ... only from the side of the Quantum that vanishes in them, and as the limit of this: they are, then, taken as unrelated moments.".[103] And from this, he says, "the inadmissible idea would follow, that it would be permissible to equate in the ultimate ratio abscissae and ordinates, or else sines and cosines, tangents, and versed sines,—anything, in fact." That is to say, the illegitimate equating of entities of differing *types*. Interestingly, Hegel does not see this inadmissible procedure at work when infinitesimal portions of curves are taken to be straight lines. He writes:

---

[99] *Ibid.*, p. 280–1. Hegel regards as highly dubious the procedure of omitting terms in a sum because of their "relative smallness".

[100] *Ibid.*, p. 281–2.

[101] *Ibid.*, p. 287.

[102] *Ibid.*

[103] *Ibid.*

> *This idea seems to be operating when a curve is treated as a tangent; for the curve too is incommensurable with the straight line, and its element of a different quality from the element of the straight line. And it seems even more irrational and less permissible than the confusion of abscissae and ordinates, versed sine and cosine, and so forth, when* (quadrata rotundis) *a part—though infinitely small—of a curve is taken for part of a tangent and thus is treated as a straight line.—However this treatment differs essentially from the confusion we have just denounced; and this is its justification,—in the triangle,*[104] *which has for its sides the element of a curve and the elements of its abscissae and ordinates, the relation is the same as though this element of the curve were the element of a straight line—the tangent: the angles, which constitute the essential relation (that is, that which remains in these elements after abstraction made from the finite magnitudes belonging to them), are the same.*[105]

He concludes:

> *We can also express ourselves in this matter as follows:— straight lines, as being infinitely small, have passed over into curves, and the ratio which subsists between them in their infinity is a ratio of curves. The straight line according to its definition is the shortest distance between two points, and therefore its difference from the curve is based upon the determination of amount, upon the smaller number of distinguishable steps on this route,—and this is a quantitative determination. But when it is taken as intensive magnitude,*[106] *or infinite moment, or element, this determination vanishes with it, and with it vanishes the difference from the curve, which is based solely on a difference in Quantum.—Taken as infinitesimal, therefore, straight line and curve have no quantitative relation, and hence (on the basis of the accepted definition) no qualitative difference relatively to each other: the latter now passes into the former.*[107]

Hegel is often regarded a philosopher who did not take mathematics very seriously. The fact that he devoted a substantial portion of *The Science of Logic* to the infinitesimal calculus speaks to the contrary.[108]

---

[104] I.e., the differential triangle.

[105] *Ibid.*, p. 287–8.

[106] Hegel distinguishes between *extensive* and *intensive* magnitude. When a magnitude is regarded as a multiplicity, it is extensive; regarded as a unity, it is intensive.

[107] *Ibid.*, p. 288.

[108] Hegel's disciple Karl Marx was also preoccupied by the infinitesimal calculus. See Marx (1983).

Chapter 4
The Reduction of the Continuous to the Discrete in the Nineteenth and Early Twentieth Centuries

The rapid development of mathematical analysis in the eighteenth century had not concealed the fact that its underlying concepts not only lacked rigorous definition but were even (e.g. in the case of differentials and infinitesimals) of doubtful logical character. The lack of precision in the notion of continuous function—still vaguely understood as one which could be represented by a formula and whose associated curve could be smoothly drawn—had led to doubts concerning the validity of a number of procedures in which that concept figured. For example when Lagrange had formulated his method for "algebraizing" the calculus he had implicitly assumed that every continuous function could be expressed as an infinite series by means of Taylor's theorem. Early in the nineteenth century this and other assumptions began to be questioned, thereby initiating an inquiry into what was meant by a function in general and by a continuous function in particular.

4.1 Bolzano and Cauchy

A pioneer in the matter of clarifying the concept of continuous function was the Bohemian priest, philosopher and mathematician **Bernard Bolzano** (1781–1848). In his *Reine analytischer Beweis* of 1817 he defines a (real-valued) function f to be continuous at a point x if the difference $f(x + \omega) - f(x)$ can be made smaller than any preselected quantity once we are permitted to take ω as small as we please. This is essentially the same as the definition of continuity in terms of the limit concept given a little later by Cauchy. Using this definition Bolzano goes on to show that both the difference and the composition of any two continuous functions is continuous. Bolzano also formulated a definition of the derivative of a function free of the notion of infinitesimal, which later became standard:

> *I have no need then, of so restrictive a hypothesis in this matter as the one so often considered necessary, to wit: that the quantities to be calculated can become* infinitely

J. L. Bell, *The Continuous, the Discrete and the Infinitesimal in Philosophy and Mathematics*, The Western Ontario Series in Philosophy of Science 82,
https://doi.org/10.1007/978-3-030-18707-1_4

small. ... *I ask one thing only: that these quantities, when they are* variable, *and not independently variable but* dependently *variable upon one or more other quantities, should possess a* derivative, *what Lagrange calls "une fonction dérivée"—if not for every value of their* determining *variable, then at least for all the values to which the process is to be validly applied. In other words: when x designates one of the independent variables, and y = f(x) designates a variable dependent on it, then, if our calculation is to give a correct result for all values of x between x = a and x = b, the mode of dependence of y upon x must be such that for all values of x between a and b the quotient*

$$\frac{\Delta y}{\Delta x} = \frac{f(x + \Delta x) - f(x)}{\Delta x}$$

(which arises from the division of the increase in y by the increase in x) can be brought as close as we wish to some constant, or to some quantity f'(x) depending solely on x, by taking Δx *sufficiently small; and subsequently, on our making* Δx *smaller still, either remains as close thereto or comes closer still.*[1]

Sometime before 1830 Bolzano invented a process for constructing continuous but nowhere differentiable functions.[2] In this he anticipated Weierstrass's better known construction by some 30 years.

Like Galileo before him, Bolzano considered that actual infinity could come into existence through aggregation. In particular he held that a continuum was to be regarded as an infinite aggregate of points, and, like Ockham and Leibniz, he thought that density alone was sufficient for a set of points to make up a continuum:

If we try to form a clear idea of what we call a 'continuous extension' *or* 'continuum', *we are forced to declare that a continuum is present when, and only when, we have an aggregate of simple entities (instants or points or substances) so arranged that each individual member of the aggregate has, at each individual and sufficiently small distance from itself, at least one other member of the aggregate for a neighbour.*

...one final question might be posed: how are we to interpret the assertion of those mathematicians who declare both that extension can never be generated by the mere accumulation of points however numerous, and also that it can never be resolved into simple points? Strictly speaking, we should on the one hand *certainly teach that extension is never produced by a finite set of points, and produced by an infinite set only when, but always when, the...oft-mentioned condition is fulfilled—namely that each point of the set has, at each sufficiently small distance, a neighbour also belonging to the set; and* on the other hand *we should admit that not* every partition *of a spatial object works down to its simple parts: in no case one whose subsets are finite in number, and not even all that subdivide to infinity, say by successive halvings. Nevertheless, we must still insist that every continuum can be made up in the last analysis of points and points alone. Once the two [apparent opposites] are properly understood, they are perfectly consistent with one another.*[3]

Bolzano admitted the existence of infinitesimal quantities as the reciprocals of infinitely great quantities:

---

[1]Bolzano (1950), p. 123.

[2]*Ibid.*, p. 30.

[3]*Ibid.* pp. 129–131.

> *Now that the possibility of* calculating with the infinitely great *has been vindicated. . ., we assert the like for the* infinitely small. *For if* N_0 *is infinitely great,* $\frac{1}{N_0}$ *necessarily represents an infinitely small quantity, and we shall have no reason for denying objective reference to such an idea, at any rate in the* general *theory of quantity.*[4]

Bolzano repudiated Euler's treatment of differentials as formal zeros in expressions such as dy/dx. He suggested instead that in determining the derivative of a function, increments Δx, Δy,... be *finally set* to zero. He writes:

> *Once an equation between x and y is given it is usually a very easy and well-known matter to find [the] derivate of y. If for example*
>
> $$y^3 = ax^2 + a^3,$$
>
> *then we should have for ever* Δx *other than zero,*
>
> $$(y + \Delta y)^3 = a(x + \Delta x)^2 + a^3,$$

whence by the known rules

$$\frac{\Delta y}{\Delta x} = \frac{2ax + a\Delta x}{3y^2 + 3y\Delta y + \Delta y^2}$$

> *and the* derived function *of y, or in Lagrange's notation y', would be discovered to be*
>
> $$\frac{2ax}{3y^2},$$
>
> *a function obtained from the expression for*
>
> $$\frac{\Delta y}{\Delta x}$$
>
> *by first suitably developing it, namely into a fraction whose numerator and denominators separate the terms multiplying* Δx and Δy *from those which do not, and then putting* Δx *and* Δy *equal to zero in the expression*
>
> $$\frac{2ax + a\Delta x}{3y^2 + 3y\Delta y + \Delta y^2}$$
>
> *thus arrived at.*[5]

Bolzano then remarks that it is perfectly in order to symbolize the derivative by $\frac{dy}{dx}$, provided that two things are understood, namely:

> *(i) that all the* Δx *and* Δy *(or, if you like, the dx or dy written in their stead) which occur in the development of* $\Delta y/\Delta x$ *are to be regarded and treated as* mere zeros; *and (ii) that the symbol dy/dx shall not be regarded as the* quotient *of dy by dx, but expressly and exclusively as a* symbol *for the derivate of y with respect to x.*[6]

---

[4] *Ibid.*, p. 109.

[5] *Ibid.*, p. 124. Bolzano's calculation here is formally equivalent to the procedure adopted in smooth infinitesimal analysis in which squares and higher powers of infinitesimals (but not the infinitesimals themselves) are set to zero: see Chapter 10 below.

[6] *Ibid.*, p. 125.

Expressions involving differentials are construed by Bolzano as having a purely formal significance. In particular differentials themselves are never to be regarded as "the symbols of actual quantities, but always as equivalents to zero". An equation in which differentials figure is to be considered as

> *nothing but a compound symbol so constituted that (i) if we carry out only such changes as algebra allows with the symbols of actual quantities (in this case, therefore, also divisions by dx and the like) and (ii) if we finally succeed in getting rid of the symbols dx, dy and so forth on both sides of the equation, then no false result will ever be the outcome.*[7]

Thus for Bolzano differentials have the status of "ideal elements", purely formal entities such as points and lines at infinity in projective geometry, or (as Bolzano himself mentions) imaginary numbers, whose use will never lead to false assertions concerning "real" quantities.

Given Bolzano's construal of infinitesimals as pure symbols, it was natural that he should repudiate as chimerical the idea of actual infinitesimal geometric entities such as infinitesimal lines or areas:

> Infinitely small distances *have been assumed hardly less often than infinitely large ones, especially when it seemed necessary to treat as straight (or plane) those lines (or surfaces) of which no portion was both extended and really straight (or plane) on the plea for example of more easily determining their actual length or the magnitude of their curvature.... People even went so far as to invent fictitious distances supposed to be measured by infinitely small quantities of the second order, of the third order, or even of still higher orders.*
>
> *Now if this process seldom led to false results, particularly in geometry, the only circumstance we can thank for it was...that in order to apply to* determinable *spatial extensions, variable quantities must be so constituted that, with the exception at most of* isolated individual values, *their first and second and all subsequent* derivates *exist. For if they do exist, then what was being asserted of the 'infinitely small' lines and surfaces and volumes can, as a general rule, quite rightly be asserted of all lines and surfaces and volumes which, while always remaining finite, nevertheless can be taken as small as we please, or as we express it, infinitely decreased. The former assertions, mistakenly applied to 'infinitely small distances' were thus really true of the 'variable quantities'.*
>
> *It can thus be understood, however, that such methods of describing the situation were bound to produce, and appear to prove, much that was paradoxical and even quite false. How scandalous it sounded, for example, when they said that every curve and surface was nothing but an assemblage of infinitely many straight lines and plane surfaces, which only needed to be considered in infinite multitude; and aggravated even this by adding the supposition of infinitely small lines and surfaces which were themselves curved...*[8]
>
> *In reality, however, infinitely small arcs are just as non-existent as infinitely small chords, and the statements which mathematicians make about their so-called 'infinitely small arcs and chords' are statements which they only prove for arcs and chords that we can take as small as we please.*[9]

---

[7] *Ibid.*, p. 127.

[8] *Ibid.*, pp. 139–40.

[9] *Ibid.*, pp. 172–73.

Although Bolzano anticipated the form that the rigorous formulation of the concepts of the calculus would assume, his work was largely ignored in his lifetime. The cornerstone for the rigorous development of the calculus was supplied by the ideas—essentially similar to Bolzano's—of the great French mathematician **Augustin-Louis Cauchy** (1789–1857). Cauchy's approach is presented in the three treatises *Cours d'anaylse de l'École Polytechnique* (1821), *Résumé des leçons sur le calcul infinitésimal* (1823), and *Leçons sur le calcul différentiel* (1829). In Cauchy's work, as in Bolzano's, a central role is played by a purely arithmetical concept of limit freed of all geometric and temporal intuition. In the *Cours d'analyse* Cauchy defines the limit concept in the following way:

When the successive values attributed to a variable approach indefinitely a fixed value so as to end by differing from it as little as one wishes, this last is called the limit of the others.[10]

In the *Cours* Cauchy also formulates the condition for a sequence of real numbers to converge to a limit, and states his familiar criterion for convergence,[11] namely, that a sequence $<s_n>$ convergent if and only if $s_{n+r} - s_n$ can be made less in absolute value than any preassigned quantity for all r and sufficiently large n. Cauchy proves that this is necessary for convergence, but as to sufficiency of the condition merely remarks "when the various conditions are fulfilled, the convergence of the series is assured."[12] In making this latter assertion he is implicitly appealing to geometric intuition, since he makes no attempt to define real numbers, observing only that irrational numbers are to be regarded as the limits of sequences of rational numbers.

Cauchy chose to characterize the continuity of functions in terms of a rigorized notion of infinitesimal, which he defines in the *Cours d'analyse* as "a variable quantity [whose value] decreases indefinitely in such a way as to converge to the limit 0." Here is his definition of continuity:

Let f(x) be a function of the variable x, and suppose that, for each value of x intermediate between two given limits [bounds], this function constantly assumes a finite and unique value. If, beginning with a value of x contained between these two limits, one assigns to the variable x an infinitely small increment α, the function itself will take on as increment the difference f(x + α) − f(x), which will depend at the same time on the new variable α and on the value of x. This granted, the function f(x) will be, between the two limits assigned to the variable x, a continuous *function of the variable if, for each value of x intermediate between these two limits, the numerical value of the difference f(x + α) − f(x) decreases indefinitely with that of α. In other words,* the function *f*(*x*) will remain continuous with respect to *x* between the given limits if, between these limits, an infinitely small increment of the variable always produces an infinitely small increment of the function itself.

We also say that the function f(x) is a continuous function of x in the neighbourhood of a particular value assigned to the variable x, as long as it [the function] is continuous between these two limits of x, no matter how close together, which enclose the value in question.[13]

---

[10]Quoted in Kline (1972), p. 951.

[11]This had been previously given by Bolzano.

[12]Kline (1972), p. 963.

[13]Quoted *ibid.*, pp. 951–2.

Cauchy's definition of continuity of $f(x)$ in the neighbourhood of a value a amounts to the condition, in modern notation, that $\lim_{x \to a} f(x) = f(a)$.

In the *Résumé des leçons* Cauchy defines the derivative $f'(x)$ of a function $f(x)$ in a manner essentially identical to that of Bolzano. He then defines the differentials dy and dx in terms of the derivative by taking dx to be any finite quantity h and the corresponding differential dy of $y = f(x)$ to be $hf'(x)$. In defining the differentials dy and dx in such a way that their quotient is precisely $f'(x)$, Cauchy decisively reversed Leibniz's definition of the derivative of a function as the quotient of differentials.

The work of Cauchy (as well as that of Bolzano) represents a crucial stage in the renunciation by mathematicians—adumbrated in the work of d'Alembert—of (fixed) infinitesimals and the intuitive ideas of continuity and motion. Certain mathematicians of the day, such as Poisson and Cournot, who regarded the limit concept as no more than a circuitous substitute for the use of infinitesimally small magnitudes—which in any case (they claimed) had a real existence—felt that Cauchy's reforms had been carried too far.[14] But traces of the traditional ideas did in fact remain in Cauchy's formulations, as evidenced by his use of such expressions as "variable quantities", "infinitesimal quantities", "approach indefinitely", "as little as one wishes" and the like.[15]

4.2 Riemann

If analysts strove to eliminate the infinitesimal from the foundations of their discipline, the same cannot be said of the geometers, especially the differential geometers. Differential geometry[16] had first emerged in the seventeenth century through the injection of the methods of the calculus into coordinate geometry. While Euclidean or projective geometry is concerned with the *global* properties of geometric objects, the focus of differential geometry is the *local* or *infinitesimal* properties of such objects, that is, those arising in the immediate neighbourhoods of points, and which may vary from point to point. The language of differentials or infinitesimals is natural to differential geometry, and it was freely employed by those mathematicians, such as the Bernoullis, Clairaut, Euler, and Gauss, in their investigations into the subject.[17]

---

[14]Boyer (1959), p. 283.

[15]Boyer (1959), p. 284. Fisher (1978) argues that here and there in his work Cauchy did "argue directly with infinitely small quantities treated as actual infinitesimals." More recently Bair et al. (to appear) have shown that in a number of places Cauchy "used [infinitesimals] neither as variable quantities nor as sequences but rather as *numbers*".

[16]The term "differential geometry" was introduced in 1894 by the Italian mathematician *Luigi Bianchi* (1856–1928).

[17]Cauchy too made important contributions to differential geometry, but he was much more circumspect than his fellows in the use of infinitesimals.

The infinitesimal plays an important role in the revolutionary extension of differential geometry conceived by the great German mathematician **Bernhard Riemann** (1826–66).[18] In 1854[19] Riemann introduced the idea of an intrinsic geometry for an arbitrary "space" which he termed a *multiply extended manifold.* Riemann conceived of a manifold as being the domain over which varies what he terms a *multiply extended magnitude*. Such a magnitude M is called *n-fold extended,* and the associated manifold *n-dimensional,* if n quantities—called *coordinates*—need to be specified in order to fix the value of M. For example, the position of a rigid body is a six-fold extended magnitude because three quantities are required to specify its location and another three to specify its orientation in space. Similarly, the fact that pure musical tones are determined by giving intensity and pitch show these to be two-fold extended magnitudes. In both of these cases the associated manifold is *continuous* in so far as each magnitude is capable of varying continuously with no "gaps". By contrast, Riemann termed *discrete* a manifold whose associated magnitude jumps discontinuously from one value to another, such as, for example, the number of leaves on the branches of a tree. Of discrete manifolds Riemann remarks:

> *Concepts whose modes of determination form a discrete manifold are so numerous, that for things arbitrarily given there can always be found a concept...under which they are comprehended, and mathematicians have been able therefore in the doctrine of discrete quantities to set out without scruple from the postulate that given things are to be considered as being all of one kind. On the other hand, there are in common life only such infrequent occasions to form concepts whose modes of determination form a continuous manifold, that the positions of objects of sense, and the colours, are probably the only simple notions whose modes of determination form a continuous manifold. More frequent occasion for the birth and development of these notions is first found in higher mathematics.*[20]

The size of parts of discrete manifolds can be compared, says Riemann, by straightforward counting, and the matter ends there. In the case of continuous manifolds, on the other hand, such comparisons must be made by *measurement.* Measurement, however, involves superposition, and so requires the positing of some magnitude—not a pure number—independent of its place in the manifold. Moreover, in a continuous manifold, as we pass from one element to another in a necessarily continuous manner, the series of intermediate terms passed through itself forms a one-dimensional manifold. If this whole manifold is now induced to pass over into another, each of its elements passes through a one-dimensional manifold, so generating a two-dimensional manifold. Iterating this procedure yields

---

[18]In the words of Hermann Weyl:

> *The principle of gaining knowledge of the external world from the behaviour of its infinitesimal parts is the mainspring of the theory of knowledge in infinitesimal physics as in Riemann's geometry ,and, indeed, the mainspring of all the eminent work of Riemann* (1950, p. 92).

[19]*On the Hypotheses which Lie at the Foundations of Geometry*, published 1868, translated in *A Source Book in Mathematics*, Smith (1959), pp. 411–25.

[20]*Ibid.*, pp. 412–13.

n-dimensional manifolds for an arbitrary integer n. Inversely, a manifold of n dimensions can be analyzed into one of one dimension and one of $n - 1$ dimensions. Repeating this process finally resolves the position of an element into n magnitudes.

Riemann thinks of a continuous manifold as a generalization of the three-dimensional space of experience and refers to the coordinates of the associated continuous magnitudes as *points*. He was convinced that our acquaintance with physical space arises only *locally*, that is, through the experience of phenomena arising in our immediate neighbourhood. Thus it was natural for him to look to differential geometry to provide a suitable language in which to develop his conceptions. In particular, the *distance* between two points in a manifold is defined in the first instance only between points which are at *infinitesimal* distance from one another. This distance is calculated according to a natural generalization of the distance formula in Euclidean space. In n-dimensional Euclidean space, the distance ε between two points P and Q with coordinates $(x_1,\ldots, x_n)$ and $(x_1 + \varepsilon_1,\ldots, x_n + \varepsilon_n)$ is given by

$$\varepsilon^2 = \varepsilon_1^2 + \ldots + \varepsilon_n^2. \quad (4.1)$$

In an n-dimensional manifold, the distance between the points P and Q—assuming that the quantities ε_i are infinitesimally small—is given by Riemann as the following generalization of (1):

$$\varepsilon^2 = \Sigma g_{ij}\varepsilon_i\varepsilon_j,$$

where the g_{ij} are functions of the coordinates $x_1,\ldots, x_n$, $g_{ij} = g_{ji}$ and the sum on the right side, taken over all i, j such that $1 \leq i, j \leq n$, is always positive. The array of functions g_{ij} is called the *metric* of the manifold. In allowing the g_{ij} to be functions of the coordinates Riemann allows for the possibility that the nature of the manifold or "space" may vary from point to point, just as the curvature of a surface may so vary.

Riemann concludes his discussion with the following words, the last sentence of which proved to be prophetic:

> *While in a discrete manifold the principle of metric relations is implicit in the notion of this manifold, it must come from somewhere else in the case of a continuous manifold. Either then the actual things forming the groundwork of a space must constitute a discrete manifold, or else the basis of metric relations must be sought for outside that actuality, in colligating forces that operate on it. A decision on these questions can only be found by starting from the structure of phenomena that has been confirmed in experience hitherto...and by modifying the structure gradually under the compulsion of facts which it cannot explain...This path leads out into the domain of another science, into the realm of physics.*[21]

Riemann is saying, in other words, that if physical space is a continuous manifold, then its geometry cannot be derived *a priori*—as claimed, famously, by Kant—but

[21] *Ibid.*, pp. 424–25.

can only be determined by experience. In particular, and again in opposition to Kant, who held that the axioms of Euclidean geometry were necessarily and exactly true of our conception of space, these axioms may have no more than approximate truth.

The prophetic nature of Riemann's final sentence was realized in 1916 when his geometry—*Riemannian geometry*—was used as the basis for a landmark development in physics, Einstein's celebrated *General Theory of Relativity.* In Einstein's theory, the geometry of space is determined by the gravitational influence of the matter contained in it, thus perfectly realizing Riemann's contention that this geometry must come from "somewhere else", to wit, from physics.

4.3 Weierstrass and Dedekind

Meanwhile the German mathematician **Karl Weierstrass** (1815–97) was completing the expulsion of spatiotemporal intuition, and the infinitesimal, from the foundations of analysis. To instill complete logical rigour Weierstrass proposed to establish mathematical analysis on the basis of number alone, to "arithmetize"[22] it—in effect, to replace the continuous by the discrete. "Arithmetization" may be seen as a form of mathematical atomism.[23] In pursuit of this goal Weierstrass had first to formulate a rigorous "arithmetical" definition of real number. He did this by defining a (positive) real number to be a countable set of positive rational numbers for which the sum of any finite subset always remains below some preassigned bound, and then specifying the conditions under which two such "real numbers" are to be considered equal, or strictly less than one another.

Weierstrass was concerned to purge the foundations of analysis of all traces of the intuition of continuous motion—in a word, to replace the variable by the fixed. For Weierstrass a variable x was simply a symbol designating an arbitrary member of a given set of numbers, and a continuous variable one whose corresponding set S has the property that any interval around any member x of S contains members of S other than x.[24] Weierstrass also formulated the familiar (ε, δ) definition of continuous function:[25] a function $f(x)$ is continuous at a if for any $\varepsilon > 0$ there is $\delta > 0$ such that

[22]According to Hobson (1907, p. 22), "the term 'arithmetization' is used to denote the movement which has resulted in placing analysis on a basis free from the idea of measurable quantity, the fractional, negative, and irrational numbers being so defined that they depend ultimately upon the conception of integral number."

[23]It would perhaps be too much of a conceptual stretch to regard arithmetization as a kind of neo-Pythagoreanism.

[24]Boyer (1959), p. 286. This property—now known as *density in itself*—later came to be regarded as too weak to characterize a continuum; see the remarks on Cantor below.

[25]The concept of function had by this time been greatly broadened: in 1837 Dirichlet suggested that a variable y should be regarded as a function of the independent variable x if a rule exists according to which, whenever a numerical value of x is given, a unique value of y is determined. (This idea was later to evolve into the set-theoretic definition of function as a set of ordered pairs.) Dirichlet's

$|f(x) - f(a)| < \varepsilon$ for all x with $|x - a| < \delta$.[26] He also proved his famous *approximation theorem for continuous functions*: any continuous function defined on a closed interval of real numbers can be uniformly approximated to by a sequence of polynomials.

Following Weierstrass's efforts, another attack on the problem of formulating rigorous definitions of continuity and the real numbers was mounted by **Richard Dedekind** (1831–1916). We learn from the introductory remarls to *Continuity and Irrational Numbers* (1872), that Dedekind was stimulated to embark on his investigations by his belief that, in presenting the differential calculus, "geometric intuition", while "exceedingly useful from a didactic standpoint", can "make no clam to being scientific". Accordingly he made "the fixed resolve to keep meditating on the question till I should find a purely arithmetic and perfectly rigorous foundation for the principles of infinitesimal analysis". He goes on to observe

> *The statement is so frequently made that that the differential calculus deals with continuous magnitude, and yet an explanation of this continuity is nowhere given; even the most rigorous expositions of the differential calculus do not base their proofs upon continuity but, with more or less consciousness of the fact, they either appeal to geometric notions or those suggested by geometry, or upon theorems which are never established in a purely arithmetic manner.*

Dedekind considered the concept of number in general to be a part of *logic,* and not, as Kant had claimed, dependent on spatio-temporal intuition. In the Preface to his later book *The Nature and Meaning of Numbers* (1888) he says

> *In speaking of arithmetic (algebra, analysis) as a part of ligic I mean to imply that I consider the number-concept entirely independent of the notions or intuitions of space and time, that I consider it an immediate result from the laws of thought.*

In his investigations of continuity Dedekind focussed attention on the question: exactly what is it that distinguishes a continuous domain from a discontinuous one? He seems to have been the first to recognize that the property of density, possessed by the ordered set of rational numbers, is insufficient to guarantee continuity. In *Continuity and Irrational Numbers* he remarks that when the rational numbers are associated to points on a straight line, "there are infinitely many points [on the line] to which no rational number corresponds"[27] so that the rational numbers manifest "a gappiness, incompleteness, discontinuity", in contrast with the straight line's "absence of gaps, completeness, continuity."[28] He goes on:

---

definition of function as a correspondence from which all traces of continuity had been purged, made necessary Weirstrass's independent definition of continuous function.

[26]The notion of *uniform continuity* for functions was later introduced (in 1870) by Heine: a real valued function f is uniformly continuous if for any $\varepsilon > 0$ there is $\delta > 0$ such that $|f(x) - f(y)| < \varepsilon$ for all x and y in the domain of f with $|x - y| < \delta$. In 1872 Heine proved the important theorem that any continuous real-valued function defined on a closed bounded interval of real numbers is uniformly continuous.

[27]Ewald (1999), p. 770.

[28]*Ibid.*, p. 771.

> *In what then does this continuity consist? Everything must depend on the answer to this question, and only through it shall we obtain a scientific basis for the investigations of* all *continuous domains. By vague remarks upon the unbroken connection in the smallest parts obviously nothing is gained; the problem is to indicate a precise characteristic of continuity that can serve as the basis for valid deductions. For a long time I pondered over this in vain, but finally I found what I was seeking. This discovery will, perhaps, be differently estimated by different people; but I believe the majority will find its content quite trivial. It consists of the following. In the preceding Section attention was called to the fact that every point p of the straight line produces a separation of the same into two portions such that every point of one portion lies to the left of every point of the other. I find the essence of continuity in the converse, i.e., in the following principle:*
>
> *'If all points of the straight line fall into two classes such that every point of the first class lies to the left of every point of the second class, then there exists one and only one point which produces this division of all points into two classes, this severing of the straight line into two portions.'*[29]

Dedekind regards this principle as being essentially indemonstrable; he ascribes to it, rather, the status of an axiom "by which we attribute to the line its continuity, by which we think continuity into the line."[30] It is not, Dedekind stresses, necessary for *space* to be continuous in this sense, for "many of its properties would remain the same even if it were discontinuous."[31] And in any case, he goes on,

> *if we knew for certain that space were discontinuous there would be nothing to prevent us . . . from filling up its gaps in thought and thus making it continuous; this filling up would consist in a creation of new point-individuals and would have to be carried out in accordance with the above principle.*[32]

The filling-up of gaps in the rational numbers through the "creation of new point-individuals" is the key idea underlying Dedekind's construction of the domain of real numbers. He first defines a *cut* to be a partition (A_1, A_2) of the rational numbers such that every member of A_1 is less than every member of A_2.[33] After noting that each rational number corresponds, in an evident way, to a cut, he observes that infinitely many cuts fail to be engendered by rational numbers. The discontinuity or incompleteness of the domain of rational numbers consists precisely in this latter fact. That being the case, he continues,

> *whenever we have a cut* (A_1, A_2) *produced by no rational number, we* create *a new number, an* irrational *number* α, *which we regard as completely defined by this cut* (A_1, A_2); *we shall say that the number* α *corresponds to this cut, or that it produces this cut. From now on, therefore, to every definite cut there corresponds a definite rational or irrational number, and we regard two numbers as* different *or* unequal *if and only if they correspond to essentially different cuts.*[34]

---

[29] *Ibid.*, p. 771.

[30] *Ibid.*, p. 772.

[31] *Ibid.*

[32] *Ibid.*

[33] *Ibid.*

[34] *Ibid.*, p. 773.

It is to be noted that Dedekind does not identify irrational numbers with cuts; rather, each irrational number is newly "created" by a mental act and remains quite distinct from its associated cut.

Dedekind goes on to show how the domain of cuts, and thereby the associated domain of real numbers, can be ordered in such a way as to possess the property of *continuity*, viz.

> *if the system* R *of all real numbers divides into two classes* A_1, A_2 *such that every number* a_1 *of the class* A_1 *is less than every number* a_2 *of the class* A_2, *then there exists one and only one number* α *by which this separation is produced.*[35]

Dedekind notes that this property of continuity is actually equivalent to two principles basic to the theory of limits; these he states as:

> *If a magnitude grows continually but not beyond all limits it approaches a limiting value*

and

> *if in the variation of a magnitude* x *we can, for every given positive magnitude* δ, *assign a corresponding interval within which* x *changes by less than* δ, *then* x *approaches a limiting value.*[36]

Dedekind's definition of real numbers as cuts was to become basic to the set-theoretic analysis of the continuum.

4.4 Cantor

The most visionary "arithmetizer" of all was **Georg Cantor** (1845–1918). Cantor's analysis of the continuum in terms of infinite point sets led to his theory of transfinite numbers and to the eventual freeing of the concept of set from its geometric origins as a collection of points, so paving the way for the emergence of the concept of general abstract set central to today's mathematics.[37]

---

[35] *Ibid.*, p. 776.

[36] *Ibid.*, p. 778.

[37] The following observations in 1900 of A. Schoenflies, one of set theory's earliest contributors, are pertinent in this connection. He saw the emergence of set theory, and the eventual discretization of mathematics, as issuing primarily from the effort to clarify the function concept, and only secondarily as the result of the struggle to tame the continuum:

> *The development of set theory had its source in the effort to produce clear analyses of two fundamental mathematical concepts, namely the concepts of argument and of function. Both concepts have undergone quite essential changes through the course of years. The concept of argument, specifically that of independent variable, originally coincided with [the] no further defined, naive concept of the geometric continuum; today it is common everywhere, to allow as domain of arguments any chosen value-set or point-set, which one can make up out of the continuum by rules defined in any way at all. Even more decisive is the change which has befallen the notion of function. This change may be tied internally to Fourier's theorem, that a so-called arbitrary function can be represented by a trigonometric series;*

At about the same time that Dedekind published his researches into the nature of the continuous, Cantor formulated his theory of the real numbers. This was presented in the first section of a paper of 1872 on trigonometric series.[38] Like Weierstrass and Dedekind, Cantor aimed to formulate an adequate definition of the irrational numbers which avoided the presupposition of their prior existence, and he follows them in basing his definition on the rational numbers. Following Cauchy, Cantor calls a sequence $a_1, a_2, \ldots, a_n, \ldots$ of rational numbers a *fundamental sequence* if there exists an integer N such that, for any positive rational ε, $|a_{n+m} - a_n| < \varepsilon$ for all m and all $n > N$. Any sequence $<a_n>$ satisfying this condition is said to have a *definite limit* b. Dedekind had taken irrational numbers to be "mental objects" associated with cuts, so, analogously, Cantor regards these definite limits as nothing more than *formal symbols* associated with fundamental sequences.[39] The domain B of such symbols may be considered an enlargement of the domain A of rational numbers, since each rational number r may be identified with the formal symbol associated with the fundamental sequence $r, r, \ldots, r, \ldots$. Order relations and arithmetical operations are then defined on B: for example, given three such symbols b, b', b'' associated with the fundamental sequences $<a_n>, <a'_n>, <a''_n>$, the inequality $b < b'$ is taken to signify that, for some $\varepsilon > 0$ and N, $a_n - a'_n > \varepsilon$ for all $n > N$, while the equality $b + b' = b''$ is taken to express the relation $\lim(a_n + a'_n - a''_n) = 0$.

Having imposed an arithmetical structure on the domain B, Cantor is emboldened to refer to its elements as (real) *numbers*. Nevertheless, he still insists that these "numbers" have no existence except as representatives of fundamental sequences: in his theory

> *the numbers (above all lacking general objectivity in themselves) appear only as components of theorems which have objectivity, for example, the theorem that the corresponding sequence has the number as limit.*[40]

Cantor next considers how real numbers are to be associated with points on the linear continuum. If a given point on the line lies at a distance from the origin O bearing a rational relation to the point at unit distance from that origin, then it can be represented by an element of A. Otherwise, it can be approached by a sequence $a_1, a_2, \ldots, a_n, \ldots$ of points each of which corresponds to an element of A. Moreover, the sequence $<a_n>$ can be taken to be a fundamental sequence; Cantor writes:

> *externally it finds expression in the definition which goes back to Dirichlet, which treats the general concept of function, to put it briefly, as equivalent to an arbitrary table . . . It was left to Cantor to find the concepts which proved proper for a methodical investigation, and which made it possible to force infinite sets under the dominion of mathematical formulas and laws* . . (Quoted in McLarty (1988), p. 83.)

[38] *Über die Ausdehnung eines Satzes aus der Theorie der trigonometrischen Reihen*, Mathematische Annalen 5, 123–132.

[39] Dauben (1979), p. 38.

[40] Quoted *ibid.*, p. 39.

The distance of the point to be determined from the point O (the origin) is equal to b, where b is the number corresponding to the sequence.[41]

In this way Cantor shows that each point on the line corresponds to a definite element of *B*. Conversely, each element of *B* should determine a definite point on the line. Realizing that the intuitive nature of the linear continuum precludes a rigorous proof of this property, Cantor simply assumes it as an axiom, just as Dedekind had done in regard to his principle of continuity:

Also conversely, to every number there corresponds a definite point of the line, whose coordinate is equal to that number.[42]

For Cantor, who began as a number-theorist, and throughout his career cleaved to the discrete, it was numbers, rather than geometric points, that possessed objective significance. Indeed, the isomorphism between the discrete numerical domain *B* and the linear continuum was regarded by Cantor essentially as a device for facilitating the manipulation of numbers.

Cantor's arithmetization of the continuum had another important consequence. It had long been recognized that the sets of points of any pair of line segments, even if one of them is infinite in length, can be placed in one-one correspondence. This fact was taken to show that such sets of points have no well-defined "size". But Cantor's identification of the set of points on a linear continuum with a domain of numbers enabled the *sizes* of point sets to be compared in a definite way, using the well-grounded idea of *one-one correspondence* between sets of numbers. Thus in a letter to Dedekind written in November 1873 Cantor notes that the totality of natural numbers can be put into one-one correspondence with the totality of positive rational numbers, and, more generally, with the totality of finite sequences of natural numbers. It follows that these totalities have the same "size"; they are all *denumerable.* Cantor now raises the question of whether the natural numbers can be placed in one-one correspondence with the totality of all positive *real* numbers.[43] He quickly answers his own question in the negative. In letters to Dedekind written during December 1873. Cantor shows that, for any sequence of real numbers, one can define numbers in every interval that are not in the sequence. It follows in particular that the whole set of real numbers is nondenumerable. Another important consequence concerns the existence of *transcendental numbers*, that is, numbers which are not algebraic in the sense of being the root of an algebraic equation with rational coefficients. In 1844 Liouville had established the transcendentality of any number of the form

$$\frac{a_1}{10} + \frac{a_2}{10^{2!}} + \frac{a_3}{10^{3!}} + \dots$$

[41]Quoted *ibid.*, p. 40.

[42]Quoted *ibid.*, p. 40.

[43]Ewald (1999), p. 844.

where the a_i are arbitrary integers from 0 to 9.[44] In his reply to Cantor's letter of November 1873, Dedekind had observed that the set of algebraic numbers is denumerable; it followed from the nondenumerability of the real numbers that *there must be many*[45] *transcendental numbers.*

By this time Cantor had come to regard nondenumerability as a necessary condition for the continuity of a point set, for in a paper of 1874 he asserts:

> *Moreover, the theorem...represents the reason why aggregates of real numbers which constitute a so-called continuum (say the totality of real numbers which are* $\geq\geq 0$ *and* $\leq$ *1), cannot be uniquely correlated with the aggregate (ν); thus I found the clear difference between a so-called continuum and an aggregate like the entirety of all real algebraic numbers.*[46]

Cantor next became concerned with the question of whether the points of spaces of different dimensions—for instance a line and a plane—can be put into one-one correspondence. In a letter to Dedekind of January 1874 he remarks:

> *It still seems to me at the moment that the answer to this question is very difficult—although here too one is so impelled to say* no *that one would like to hold the proof to be almost superfluous.*[47]

Nevertheless, 3 years later Cantor, in a dramatic *volte-face*, established the existence of such correspondences between spaces of different dimensions. He showed, in fact, that (the points of) a space of any dimension whatsoever can be put into one-one correspondence with (the points of) a line. This result so startled him that, in a letter to Dedekind of June 1877 he was moved to exclaim: *Je le vois, mais je ne le crois pas.*[48]

Cantor's discovery caused him to question the adequacy of the customary definition of the dimension of a continuum. For it had always been assumed that the determination of a point in a an n-dimensional continuous manifold requires n independent coordinates, but now Cantor had shown that, in principle at least, the job could be done with just a *single* coordinate. For Cantor this fact was sufficient to justify the claim that

> *... all philosophical or mathematical deductions that use that erroneous presupposition are inadmissible. Rather the difference that obtains between structures of* different *dimension-number must be sought in quite other terms than in the number of independent coordinates—the number that was hitherto held to be characteristic.*[49]

In his reply to Cantor, Dedekind conceded the correctness of Cantor's result, but balked at Cantor's radical inferences therefrom. Dedekind maintained that the

---

[44]Kline (1972), p. 981.

[45]In fact nondenumerably many, although Cantor did not make this fact explicit until later.

[46]Quoted in Dauben (1979), p. 53.

[47]Ewald (1999), p. 850.

[48]"I see it, but I don't believe it." *Ibid.*, p. 860.

[49]*Ibid.*

dimension-number of a continuous manifold was its "first and most important invariant",[50] and emphasized the issue of continuity:

> *For all authors have clearly made the tacit, completely natural presupposition that in a new determination of the points of a continuous manifold by new coordinates, these coordinates should also (in general) be* continuous *functions of the old coordinates, so that whatever appears as continuously connected under the first set of coordinates remains continuously connected under the second.*[51]

Dedekind also noted the extreme discontinuity of the correspondence Cantor had set up between higher dimensional spaces and the line:

> *... it seems to me that in your present proof the* initial *correspondence between the points of the ρ-interval (whose coordinates are all irrational) and the points of the unit interval (also with irrational coordinates) is, in a certain sense (smallness of the alteration), as continuous as possible; but to fill up the gaps, you are compelled to admit, a frightful, dizzying discontinuity in the correspondence, which dissolves everything to atoms, so that every continuously connected part of the one domain appears in its image as thoroughly decomposed and discontinuous.*[52]

Dedekind avows his belief that no one-one correspondence between spaces of different dimensions can be continuous:

> *If it is possible to establish a reciprocal, one-to-one, and complete correspondence between the points of a continuous manifold A* of *a dimensions and the points of a continuous manifold B of b dimensions, then this* correspondence itself, *if a and b are* unequal, *is necessarily* utterly discontinuous.[53]

In his reply to Dedekind of July 1877 Cantor clarifies his remarks concerning the dimension of a continuous manifold:

> *...I unintentionally gave the appearance of wishing by my proof to oppose altogether the concept of a ρ-fold extended continuous manifold, whereas all my efforts have rather been intended to clarify it and to put it on the correct footing. When I said: "Now it seems to me that all philosophical and mathematical deductions which use that erroneous presupposition—" I meant by this presupposition* not *"the determinateness of the dimension-number" but rather the determinateness of the independent coordinates, whose number is assumed by certain authors to be in all circumstances equal to the number of dimensions. But if one takes the concept of coordinate* generally, *with no presuppositions about the nature of the intermediate functions, then the number of independent, one-to-one, complete coordinates, as I showed, can be set to any number.*[54]

But he agrees with Dedekind that if "we require that the correspondence be continuous, then only structures with the same number of dimensions can be related to each other one-to-one."[55] In that case, an invariant *can* be found in the number of

[50]*Ibid.*, p. 863.

[51]*Ibid.*

[52]*Ibid.*, pp. 863–4.

[53]*Ibid.*, p. 863.

[54]*Ibid.*, p. 864.

[55]*Ibid.*.

independent coordinates, "which ought to lead to a definition of the dimension-number of a continuous structure."[56] The problem is to correlate that dimension-number, a perfectly definite mathematical object, with something as elusive as an arbitrary continuous correspondence. Cantor writes:

> *However, I do not yet know how difficult this path (to the concept of dimension-number) will prove, because I do not know whether one is able to limit the concept of* continuous correspondence in general. *But everything in this direction seems to me to depend on the possibility of such a limiting.*
>
> *I believe I see a further difficulty in the fact that this path will probably fail if the structure ceases to be thoroughly continuous; but even in this case one wants to have something corresponding to the dimension-number—all the more so, given how difficult it is to prove that the manifolds that occur in nature are thoroughly continuous.*[57]

In rendering the continuous discrete, and thereby admitting *arbitrary* correspondences "of [a] frightful, dizzying discontinuity" between geometric objects "dissolved to atoms", Cantor grasps at the same time that he has rendered the intuitive concept of spatial dimension a hostage to fortune.[58]

In 1878 Cantor published a fuller account[59] of his ideas. Here Cantor explicitly introduces the concept of the *power* [60] of a set of points: two sets are said to be of equal power if there exists a one-one correspondence between them. Cantor presents demonstrations of the denumerability of the rationals and the algebraic numbers, remarking that "the sequence of positive whole numbers constitutes. . .the least of all powers which occur among infinite aggregates."[61]

The central theme of Cantor's 1878 paper is the study of the powers of continuous n-dimensional spaces. He raises the issue of invariance of dimension and its connection with continuity:

> *Apart from making the* assumption, *most are silent about how it follows from the course of this research that the correspondence between the elements of the space and the system of values $x_1, x_2, \ldots, x_n$ is a* continuous *one, so that any infinitely small change of the system $x_1, x_2, \ldots, x_n$ corresponds to an infinitely small change of the corresponding element, and conversely, to every infinitely small change of the element a similar change in the coordinates corresponds. It may be left undecided whether these assumptions are to be considered as sufficient, or whether they are to be extended by more specialized conditions in order to*

---

[56] *Ibid.*

[57] *Ibid.*

[58] At the same time Cantor's recognition that the problem of defining dimension depends on the possibility of restricting the correspondences between the structures concerned is indicative of his great prescience as a mathematician. For the idea of taking as primary data *correspondences* between mathematical structures, as opposed to the structures themselves, was, through category theory, to prove seminal in the mathematics of the twentieth century.

[59] *Ein Beitrag zur Mannigfaltigkeitslehre,* Journal für die reine und angewandte Mathematik 84, 242–58.

[60] *Mächtigkeit.* Cantor mentions that the origin of this concept lies in the work of the Swiss geometer Jakob Steiner, who used it in his study of invariants of conic sections.

[61] Quoted in Dauben (1979), p. 59.

> *consider the intended conceptual construction of n-dimensional continuous spaces as one ensured against any contradictions, sound in itself.*[62]

The remarkable result obtained when one no longer insists on continuity in the correspondence between the spatial elements and the system of coordinates is described by Cantor in the following terms:

> *As our research will show, it is even possible to determine uniquely and completely the elements of an n-dimensional continuous space by a single real coordinate t. If no assumptions are made about the kind of correspondence, it then follows that the number of independent, continuous, real coordinates which are used for the unique and complete determination of the elements of an n-dimensional continuous space can be brought to any arbitrary number, and thus is* not *to be regarded as a unique feature of the space.*[63]

Cantor shows how this result can be deduced from the existence of a one-one correspondence between the set of reals and the set of irrationals, and then, by means of an involved argument, constructs such a correspondence.

Cantor seems to have become convinced by this time that the essential nature of a continuum was fully reflected in the properties of sets of points—a conviction which was later to give birth to abstract set theory. In particular a continuum's key properties, Cantor believed, resided in the range of powers of its subsets of points. Since the power of a continuum of any number of dimensions is the same as that of a linear continuum, the essential properties of arbitrary continua were thereby reduced to those of a line. In his investigations of the linear continuum Cantor had found its infinite subsets to possess just two powers, that of the natural numbers and that of the linear continuum itself. This led him to the conviction that these were the only possible powers of such subsets—a thesis later to be enunciated as the famous *continuum hypothesis.*

But the problem of establishing the invariance of dimension of spaces under continuous correspondences remained a pressing issue. Soon after the publication of Cantor's 1878 paper, a number of mathematicians, for example Lüroth, Thomae, Jürgens and Netto attempted proofs, but all of these suffered from shortcomings which did not escape notice.[64] In 1879 Cantor himself published a proof which seems to have passed muster at the time, but which also contained flaws that were not detected for another 20 years.[65]

Satisfied that he had resolved the question of invariance of dimension, Cantor returned to his investigation of the properties of subsets of the linear continuum. The results of his labours are presented in six masterly papers published during 1879–84, *Über unendliche lineare Punktmannigfaltigkeiten* ("On infinite, linear point manifolds"). Remarkable in their richness of ideas, these papers contain the first accounts of Cantor's revolutionary theory of infinite sets and its application to the

---

[62]*Ibid.*, p. 60.

[63]*Ibid.*, p. 60.

[64]Dauben (1979), pp. 70–72

[65]*Ibid.*, pp. 72–76. The matter was only placed beyond doubt in 1911 when Brouwer showed definitively that the dimension of a Euclidean space is a topological invariant.

classification of subsets of the linear continuum. In the third and fifth of these are to be found Cantor's observations on the nature of the continuum.

In the third article, that of 1882, which is concerned with multidimensional spaces, Cantor applies his result on the nondenumerability of the continuum to prove the startling result that continuous motion is possible in discontinuous spaces. To be precise, he shows that, if M is any countable dense subset of the Euclidean plane R^2, (for example the set of points with both coordinates algebraic real numbers), then any pair of points of the discontinuous space $A = R^2 - M$ can be joined by a continuous arc lying entirely within A.[66] In fact, Cantor claims even more:

> *After all, with the same resources, it would be possible to connect the points...by a continuously running line given by a* unique analytic rule *and completely contained within the domain* A.[67]

Cantor points out that the belief in the continuity of space is traditionally based on the evidence of continuous motion, but now it has been shown that continuous motion is possible even in discontinuous spaces. That being the case, the presumed continuity of space is no more than a hypothesis. Indeed, it cannot necessarily be assumed that physical space contains every point given by three real number coordinates. This assumption, he urges,

> *...must be regarded as a free act of our constructive mental activity. The hypothesis of the continuity of space is therefore nothing but the assumption, arbitrary in itself, of the complete, one-to-one correspondence between the 3-dimensional purely arithmetic continuum (x, y, z) and the space underlying the world of phenomena.*[68]

These facts, so much at variance with received views, confirmed for Cantor once again that geometric intuition was a poor guide to the understanding of the continuum. For such understanding to be attained reliance must instead be placed on arithmetical analysis.

In the fifth paper in the series, the *Grundlagen* of 1883, is to be found a forthright declaration of Cantor's philosophical principles, which leads on to an extensive discussion of the concept of the continuum. Cantor distinguishes between the *intrasubjective* or *immanent* reality and *transsubjective* or *transient* reality of concepts or ideas. The first type of reality, he says, is ascribable to a concept which may be regarded

> *as actual in so far as, on the basis of definitions, [it] is well distinguished from other parts of our thought and stand[s] to them in determinate relationships, and thus modifies the substance of our mind in a determinate way.*[69]

---

[66]In modern terminology, spaces like A are *arcwise connected.*

[67]Quoted in Dauben (1979), p. 86.

[68]*Ibid.*

[69]Ewald (1999), p. 895. Cantor refers specifically here to the concept of the natural numbers.

The second type, transient or transsubjective reality, is ascribable to a concept when it can, or must, be taken

> *as an expression or copy of the events and relationships in the external world which confronts the intellect.*[70]

Cantor now presents the principal tenet of his philosophy, to wit, that the two sorts of reality he has identified invariably occur together,

> *in the sense that a concept designated in the first respect as existent always also possesses, in certain, even infinitely many ways, a transient reality.*[71]

Cantor's thesis is tantamount to the principle that correct thinking is, in its essence, a reflection of the order of Nature.[72] In a footnote Cantor places his thesis in the context of the history of philosophy. He claims that "it agrees essentially both with the principles of the Platonic system and with an essential tendency of the Spinozistic system" and that it can be found also in Leibniz's philosophy. But philosophy since that time has come, in Cantor's eyes, to deviate from this cardinal principle:

> *Only since the growth of modern empiricism, sensualism, and scepticism, as well as of the Kantian criticism that grows out of them, have people believed that the source of knowledge and certainty is to be found in the senses or in the so-called pure form of intuition of the world of appearances, and that they must confine themselves to these. But in my opinion these elements do not furnish us with any secure knowledge. For this can be obtained only from concepts and ideas that are stimulated by external experience and are essentially formed by inner induction and deduction as something that, as it were, was already in us and is merely awakened and brought to consciousness.*[73]

This linkage between the immanent and transient reality of mathematical concepts—the fact that correct mathematical thinking reflects objective reality—has, in Cantor's view, the important consequence that

> *mathematics, in the development of its ideas has* only *to take account of the* immanent *reality of its concepts and has* absolutely no *obligation to examine their* transient *reality.*[74]

It follows that

> *mathematics in its development is entirely free*[75] *and is only bound in the self-evident respect that its concepts must both be consistent with each other and also stand in exact relationships, ordered by definitions, to those concepts which have previously been introduced and*

---

[70]*Loc. cit.*

[71]*Loc. cit.*

[72]Or, perhaps, vice-versa.

[73]*Ibid.*, p. 918.

[74]*Ibid.*, p. 896.

[75]Cantor goes on to assert, in a famous pronouncement, "the *essence* of *mathematics* lies precisely in its *freedom.*".

> *are already at hand and established.*[76] *In particular, in the introduction of new numbers it is only obliged to give definitions of them which will bestow such a determinacy and, in certain circumstances, such a relationship to older numbers that they can in any given instance be precisely distinguished.*[77]

For these reasons Cantor bestows his blessing on rational, irrational, and complex numbers, which "one must regard as being every bit as existent as the finite positive integers"; even Kummer's introduction of "ideal" numbers into number theory meets with his approval. But not infinitesimal numbers, as we shall see.

Cantor begins his examination of the continuum with a tart summary of the controversies that have traditionally surrounded the notion:

> *The concept of the 'continuum' has not only played an important role everywhere in the development of the sciences but has also evoked the greatest differences of opinion and event vehement quarrels. This lies perhaps in the fact that, because the exact and complete definition of the concept has not been bequeathed to the dissentients, the underlying idea has taken on different meanings; but it must also be (and this seems to me the most probable) that the idea of the continuum had not been thought out by the Greeks (who may have been the first to conceive it) with the clarity and completeness which would have been required to exclude the possibility of different opinions among their posterity. Thus we see that Leucippus, Democritus, and Aristotle consider the continuum as a composite which consists* ex partibus sine fine divisilibus,[78] *but Epicurus and Lucretius construct it out of their atoms considered as finite things. Out of this a great quarrel arose among the philosophers, of whom some followed Aristotle, others Epicurus; still others, in order to remain aloof from this quarrel, declared with Thomas Aquinas that the continuum consisted neither of infinitely many nor of a finite number of parts, but of* absolutely no *parts. This last opinion seems to me to contain less an explanation of the facts than a tacit confession that one has not got to the bottom of the matter and prefers to get genteely out of its way. Here we see the* medieval-scholastic origin *of a point of view which we still find represented today, in which the continuum is thought to be an unanalysable concept, or, as others express themselves, a pure* a priori *intuition which is scarcely susceptible to a determination through concepts. Every arithmetical attempt at*

---

[76]Here Cantor inserts a characteristic footnote, in which he puts forward a theory of the formation of concepts reminiscent of Plato's doctrine of anamnesis:

> *The procedure in the correct formulation of concepts is in my opinion everywhere the same. One posits a thing with properties that at the outset is nothing other than a name or a sign A, and then in an orderly fashion gives it different, or even infinitely many, intelligible predicates whose meaning is known on the basis of ideas already at hand, and which may not contradict one another. In this way one determines the connection of A to the concepts that are already at hand, in particular to related concepts. If one has reached the end of this process, then one has met all the preconditions for awakening the concept A which slumbered inside us, and it comes into being accompanied by the intrasubjective reality which is all that can be demanded of a concept; to determine its transient meaning is then a matter for metaphysics.* (Ewald 1999, p. 918}

[77]Ewald (1999), p. 896.

[78]"from parts divisible without end". It may strike one as odd to find the atomists Leucippus and Democritus here bracketed with Aristotle as upholders of divisionism; presumably Cantor is implicitly distinguishing the formers' *material* atomism from their probable, or at least possible, assent to *theoretical* divisionism. Cf. Heath (1981), vol. I, p. 181, where the idea that Democritus may have upheld geometric indivisibles is dismissed on the grounds that "Democritus was too good a mathematician to have anything to do with such a theory."

> *determination of this* mysterium *is looked on as a forbidden encroachment and repulsed with due vigour. Timid natures thereby get the impression that with the 'continuum' it is not a matter of a* mathematically logical concept *but rather of* religious dogma.[79]

It is not Cantor's intention to "conjure up these controversial questions again". Rather, he is concerned to "develop the concept of the continuum as soberly and briefly as possible, and only with regard to the *mathematical* theory of sets". This opens the way, he believes, to the formulation of an exact concept of the continuum—nothing less than a demystification of the *mysterium*. Cantor points out that the idea of the continuum has heretofore merely been presupposed by mathematicians concerned with the analysis of continuous functions and the like and has "not been subjected to any more thorough inspection."

Cantor next repudiates any use of temporal intuition in an exact determination of the continuum:

> *...I must explain that in my opinion to bring in* the concept of time *or* the *intuition of time in discussing the much more fundamental and more general concept of the continuum is* not *the correct way to proceed;* time *is in my opinion a representation, and its clear explanation presupposes the concept of continuity upon which it depends and without whose assistance it cannot be conceived either objectively (as a substance) or subjectively (as the form of an* a priori *intuition), but is nothing other than a* helping and linking concept, *through which one ascertains the relation between various different motions that occur in nature and are perceived by us. Such a thing as* objective *or* absolute time *never occurs in nature, and therefore* time *cannot be regarded as the measure of motion; far rather motion as the measure of* time—*were it not that* time, *even in the modest role of a* subjective necessary a priori *form of intuition, has not been able to produce any fruitful, incontestable success, although since Kant the time for this has not been lacking.*[80]

These strictures apply, *pari passu*, to spatial intuition:

> *It is likewise my conviction that with the so-called* form of intuition *of* space *one cannot even begin to acquire knowledge of the* continuum. *For only with the help of a conceptually already* completed *continuum do* space *and the structure thought into it receive that content with which they can become the object, not merely of aesthetic contemplation or philosophical cleverness or imprecise comparisons, but of sober and exact mathematical investigations.*[81]

Cantor now embarks on the formulation of a precise arithmetical definition of a continuum. Making reference to the definition of real number he has already provided (i.e., in terms of fundamental sequences), he introduces the n-dimensional arithmetical space G_n as the set of all n-tuples of real numbers $(x_1 | x_2 | \ldots | x_n)$, calling each such an *arithmetical point* of G_n. The distance between two such points is given by

$$\sqrt{(x_1' - x_1)^2 + (x_2' - x_2)^2 + \ldots (x_n' - x_n)^2}.$$

[79]Ewald (1999), p. 903.

[80]*Ibid.*, p. 904.

[81]*Ibid.*

Cantor defines an *arithmetical* point-set in G_n to be any "aggregate of points of the points of the space G_n that is given in a lawlike way".

After remarking that he has previously shown that all spaces G_n have the same power as the set of real numbers in the interval (0, 1), and reiterating his conviction that any infinite point sets has either the power of the set of natural numbers or that of (0, 1),[82] Cantor turns to the definition of the general concept of a continuum within G_n. For this he employs the concept of *derivative* or *derived set* of a point set introduced in his 1872 paper on trigonometric series. Cantor had defined the derived set of a point set P to be the set of *limit points* of P, where a limit point of P is a point of P with infinitely many points of P arbitrarily close to it. A point set is called *perfect* if it coincides with its derived set.[83] Cantor observes that this condition does not suffice to characterize a continuum, since perfect sets can be constructed in the linear continuum which are dense in no interval, however small: as an example of such a set he offers the set[84] consisting of all real numbers in (0, 1) whose ternary expansion does not contain a "1".

Accordingly, an additional condition is needed to define a continuum. Cantor supplies this by introducing the concept of a *connected* set. A point set T is connected in Cantor's sense if for any pair of its points t, t' and any arbitrarily small number ε there is a finite sequence of points $t_1, t_2, \ldots, t_n$ of T for which the distances $\overline{tt_1}, \overline{t_1t_2}, \overline{t_2t_3}, \ldots, \overline{t_nt'}$ are all less than ε. Cantor now observes:

> *...all the geometric point-continua known to us fall under this concept of* connected *point-set, as it easy to see; I believe that in these* two *predicates 'perfect" and 'connected' I have discovered the necessary and* sufficient *properties of a point-continuum. I therefore define a point-continuum inside G_n as a* perfect-connected set. *Here 'perfect' and 'connected' are not merely words but completely general predicates of the* continuum; *they have been conceptually characterized in the sharpest way by the foregoing definitions.*[85]

Cantor points out the shortcomings of previous definitions of continuum such as those of Bolzano and Dedekind, and in a note dilates on the merits of his own definition:

> *Observe that this definition of a continuum is free from every reference to that which is called the* dimension *of a continuous structure; the definition includes also continua that are composed of connected pieces of different dimensions, such as lines, surfaces, solids, etc....I know very well that the word 'continuum' has previously* not *had a precise meaning in mathematics; so my definition will be judged by some as too* narrow, *by others as too* broad. *I trust that that I have succeeded in finding a proper* mean *between the two.*
>
> *In my opinion, a* continuum *can* only *be a* perfect *and* connected *structure. So, for example, a straight line segment lacking one or both of its end-points, or a disc whose boundary is excluded, are not* complete *continua; I call such point-sets* semi-continua.[86]

---

[82]This, *Cantor's continuum hypothesis*, is actually stated in terms of the transfinite ordinal numbers introduced in previous sections of the *Grundlagen.*

[83]In the terminology of general topology, a set is perfect if it is closed and has no isolated points.

[84]This set later became known as the *Cantor ternary set* or the *Cantor discontinuum.*

[85]Ewald (1999), p. 906.

[86]*Ibid,* p. 919.

It will be seen that Cantor has advanced beyond his predecessors in formulating what is in essence a *topological* definition of continuum, one that, while still dependent on metric notions, does not involve an order relation.[87] It is interesting to compare Cantor's definition with the definition of continuum in modern general topology. In a well-known textbook[88] on the subject we find a *continuum* defined as a compact connected subset of a topological space. Now within any *bounded* region of Euclidean space it can be shown that Cantor's continua coincide with continua in the sense of the modern definition. While Cantor lacked the definition of compactness, his requirement that continua be "complete" (which led to his rejecting as continua such noncompact sets as open intervals or discs) is not far away from the idea.

Cantor's analysis of infinite point sets had led him to introduce transfinite numbers,[89] and he had come to accept their objective existence as being beyond doubt. But throughout his mathematical career he maintained an unwavering, even dogmatic opposition to infinitesimals, attacking the efforts of mathematicians such as du Bois-Reymond and Veronese[90] to formulate rigorous theories of actual infinitesimals. As far as Cantor was concerned, the infinitesimal was beyond the realm of the possible; infiinitesimals were no more than "castles in the air, or rather just nonsense", to be classed "with circular squares and square circles".[91] His abhorrence of infinitesimals went so deep as to move him to outright vilification, branding them as "Cholera-bacilli of mathematics."[92]

Cantor believed that the theory of transfinite numbers could be employed to explode the concept of infinitesimal once and for all. Cantor's specific aim was to refute all attempts at introducing infinitesimals through the abandoning of the *Archimedean principle*—i.e. the assertion that for any positive real numbers $a < b$, there is a sufficiently large natural number n such that $na > b$. (Domains in which this principle fails to hold are called nonarchimedean). In a paper of 1887, Cantor attempted to demonstrate that the Archimedean property was a necessary consequence of the "concept of linear quantity" and "certain theorems of transfinite number theory", so that the linear continuum could contain no infinitesimals. He concludes that "the so-called Archimedean axiom is not an axiom at all, but a theorem which follows with logical necessity from the concept of linear quantity.".[93] Cantor's argument relied on the claim that the product of a positive infinitesimal,

[87]Cantor later turned to the problem of characterizing the linear continuum as an ordered set. His solution was published in 1895 in the *Mathematische* Annalen. For a modern presentation, see §3 of Ch. 6 of Kuratowski-Mostowski (1968).

[88]Hocking and Young (1961). See also Chap. 6 below.

[89]For an account, see Hallett (1984).

[90]Other proponents of infinitesimals of the time include Johannes Thomae and Otto Stolz: see Fisher (1981). For du Bois-Reymond and Veronese see Chap. 5 below.

[91]Fisher (1981), p. 118.

[92]Dauben (1979), p. 233.

[93]Fisher (1981), p. 118.

should such exist, with one of his transfinite numbers could never be finite.[94] But since a proof of this claim was not supplied, Cantor's alleged demonstration that infinitesimals are impossible must be regarded as inconclusive.[95]

Cantor's rejection of infinitesimals stemmed from his conviction that his own theory of transfinite ordinal and cardinal numbers exhausted the realm of the numerable, so that no further generalization of the concept of number, in particular any which embraced infinitesimals, was admissible.[96] For Cantor, transfinite

---

[94]In this connection Fraenkel observes (1976, p. 123) that Cantor's cardinals and ordinals themselves constitute nonarchimedean domains for the simple reason that, when α is finite and β is transfinite, then $n\alpha < \beta$ for any n. While admitting the legitimacy of the sort of nonarchimedean domains put forward by mathematicians such as Veronese and du Bois-Reymond, Fraenkel distinguishes between Cantor's from the other nonarchimedean domains through the fact that in the former any cardinal or ordinal can be reached by the repeated addition of unity (if necessary, transfinitely), while in the latter such a procedure is not even definable. Fraenkel concludes: "This contrast explains in what sense other non-Archimedean domains contain 'relatively infinite' magnitudes while the transfiniteness of cardinals and ordinals is an 'absolute' one."

It is worth quoting Fraenkel's full endorsement of Cantor's rejection of infinitesimal numbers:

> *Cantor, when undertaking a "continuation of the series of real integers beyond the infinite" and showing the usefulness of this generalization of the process of counting, refused to consider infinitely small magnitude beyond the "potential infinite" of analysis based on the concept of limit. The 'infinitesimals' of analysis, as is well known, refer to an infinite* process *and not to a constant positive value which, if greater than zero, could nit be infinitely small. . .*
>
> *Opposing this attitude, some schools of philosophers. . . and later sporadic mathematicians proposed resuming the vague attempts of most 17th and 18th century mathematicians to base calculus on infinitely small magnitudes, the so-called infinitesimals or differentials. After the introduction of transfinite numbers by Cantor such attitudes pretended to be justified by set theory because there ought to be reciprocals (inverse ratios) to the transfinite numbers, namely the ostensible infinitesimals of various degrees representing the ratios of finite to transfinite numbers.*
>
> *These views have been thoroughly rejected by Cantor and by the mathematical world in general. The reason for this uniformity was not dogmatism, which is a rare feature in mathematics and then almost invariably fought off; nobody has pleaded more ardently than Cantor himself that liberty of thought was the essence of mathematics and that prejudices had a very short life. The argument was not even that the admission of infinitesimals was self-contradictory, but just that it was* sterile and useless. . .*This uselessness contrasts strikingly with the success of the transfinite numbers regarding both their applications and their task of generalizing finite counting and ordering.* (Fraenkel 1976, pp. 121–2.)

But it has to be said (as remarked by Fisher 1981) that concerning infinitesimals Cantor did display a dogmatic attitude and did argue, in effect, that the admission of infinitesimals was self-contradictory. While Cantor's intolerant attitude towards the infinitesimal is not, strictly speaking, inconsistent with the "freedom" of mathematics in his sense, it does seem to reflect his deep-seated conviction (reported in Dauben 1979, pp. 288–91) that his transfinite set theory was the product of "divine inspiration", so that anything in conflict with it must be anathematized.

[95]Fisher (1981), p. 118. Even so, Cantor's argument (in amended form) seems to have convinced a number of influential mathematicians, including Peano and Russell, of the untenability of infinitesimals.

[96]Dauben (1979), p. 235.

numbers were grounded in *transient reality*, while infinitesimals and similar chimeras could not be accorded such a status.[97] Recall Cantor's assertion:

> *In particular, in the introduction of new numbers it is only obliged to give definitions of them which will bestow such a determinacy and, in certain circumstances, such a relationship to older numbers that they can in any given instance be precisely distinguished.*

Accordingly Cantor could not grant the infinitesimal an immanent reality which was compatible with "older" numbers—among which he of course included his transfinite numbers— for had he done so he would perforce have had, in accordance with his own principles, to grant the infinitesimal transient reality. This seems to be the reason for Cantor's determination to demonstrate the inconsistency of the infinitesimal with his concept of transfinite number.[98]

4.5 Russell

Bertrand Russell (1872–1970) began his philosophical career as a Hegelian, but he soon abandoned Hegel in favour of the logical approach to philosophy espoused by mathematicians such as Cantor, Frege and Peano. In 1903 Russell published his great work *The Principles of Mathematics*. In writing this work Russell's principal objective was to demonstrate the *logicist* thesis that "pure mathematics deals exclusively with concepts definable in terms of a very small number of fundamental logical concepts, and that all its propositions are deducible from a very small number of fundamental logical principles". In particular the concept of *continuity* comes under close scrutiny. The work's Part V—a kind of paean to Weierstrass and Cantor—is devoted to an analysis of the idea of continuity, and its relation to the infinite and the infinitesimal.

Before getting down to a full analysis of these topics, which "[have] been generally considered the fundamental problem[s] of mathematical philosophy", Russell remarks that, thanks to the labours of modern mathematicians, the whole problem has been completely transformed:

> *Since the time of Newton and Leibniz, the nature of infinity and continuity had been sought in discussions of the so-called Infinitesimal Calculus. But it has been shown that this Calculus is not, as a matter of fact, in any way concerned with the infinitesimal, and that a large and most important branch of mathematics is logically prior to it. It was formerly supposed—and herein lay the real strength of Kant's mathematical philosophy—that continuity had an essential reference to space and time, and that the calculus (as the word* fluxion *suggests) in some way presupposed motion or at least change. In this way, the philosophy of space and time was prior to that of continuity, the Transcendental Aesthetic preceded the Transcendental Dialectic, and the antinomies (at least the mathematical ones) were essentially spatio-temporal. All this has been changed by modern mathematics. What is called the*

[97] *Ibid.*, p. 236.

[98] A related point is made by Fraenkel (1976), p. 123.

> *arithmetization of mathematics has shown that all the problems presented, in this respect, by space and time, are already present in pure arithmetic.*[99]

While the theory of infinity has two forms, "cardinal and ordinal, of which the former springs from the logical theory of number", the new theory of the continuous that Russell champions so enthusiastically is "purely ordinal". Indeed, he goes on,

> *we shall find it possible to give a general definition of continuity, in which no appeal is made to the mass of unanalyzed prejudice which Kantians call "intuition"; and ... we shall find that no other continuity is involved in space and time. And we shall find that, by a strict adherence to the doctrine of limits, it is possible to dispense entirely with the infinitesimal, even in the definition of continuity and the foundations of the calculus.*[100]

Russell's presentation of the theory of real numbers in the *Principles* begins with characteristic *brio:*

> *The philosopher may be surprised, after all that has already been said concerning numbers, to find that he is only now to learn about* real *numbers; and his surprise will be turned to horror when he learns that real is opposed* to *rational. But he will be relieved to learn that real numbers are not numbers at all, but something quite different.*[101]

Real numbers, according to Russell, are nothing more than certain *classes* of rational numbers. By way of illustration, he observes "the class of rationals less than ½ is a real number, associated with, but obviously not identical with, the rational number ½." Russell remarks of this theory:

> *[It] is not, so far as I know, explicitly advocated by any other author, though Peano suggests it and Cantor comes very near to it.*[102]

Russell proposes that real numbers are to be what he calls *segments* of rational numbers, where

> *a segment of rationals may be defined as a class of rationals which is not null, nor yet coextensive with the rationals themselves ... and which is identical with the class of rational less than a (variable) term of itself, i.e. with the class of rationals x such that there is a rational y of the said class such that x is less than y.*[103]

That is, a subset S of the set $\mathbb{Q}$ of rational numbers is a segment provided that $\varnothing \neq S \neq \mathbb{Q}$ and, for any $x \in \mathbb{Q}$, $x \in S$ if and only if $\exists y \in S.\ x < y$. It is curious that Russell does not mention Dedekind in connection with this definition (although Dedekind's construction of the irrational numbers is outlined a few pages further on). For the definition of real numbers in terms of Russell's "segments" is, from a purely formal

---

[99]Russell (1964), p. 259. Earlier (in Ch. XXIII) Russell presents an argument to show that, while continuity, infinity and the infinitesimal have been traditionally associated with the category of Quantity, in fact they are more properly regarded as being ordinal and arithmetical in nature.

[100]*Ibid.*, p. 260.

[101]*Ibid.*, p. 270.

[102]*Ibid.*

[103]*Ibid.*, p. 271.

standpoint, the same as that given by Dedekind in terms of his "cuts".[104] But Russell seems to have regarded the *content* of his definition as differing in a significant way from that of Dedekind. In Russell's view, Dedekind merely *postulates* the existence of an irrational number corresponding to each of his "cuts". Dedekind does not actually construct or otherwise prove the existence of such numbers, in particular he does not claim that the cut *itself* is the number. In Russell's eyes, his own definition had the merit of *explicitly identifying* the object (number) in question as a class without resorting to abstraction or "intuition." It is interesting to note that this point, so important to the philosopher Russell, was not ascribed so great a significance by practicing mathematicians, who saw Russell's definition as no more than a "less abstract" version of Dedekind's.[105]

After reviewing Cantor's definition of real numbers, Russell next proceeds to a discussion of Cantor's definitions of continuity. He begins with a light-hearted dig at Hegel:

> *The notion of continuity has been treated by philosophers, as a rule, as though it were incapable of analysis. They have said many things about it, including the Hegelian dictum that everything discrete is continuous and vice-versa. This remark, as being an exemplification of Hegel's usual habit of combining opposites, has been tamely repeated by all his followers. But as to what they meant by continuity and discreteness, they have preserved a discreet and continuous silence; only one thing was evident, that whatever they did mean could not be relevant to mathematics, or to the philosophy of space and time.*[106]

Russell contrasts the "continuity" of the rational numbers—the fact that between any two there is another[107]—with "that other kind of continuity, which was seen to belong to space." This latter form of continuity, Russell says,

> *was treated, as Cantor remarks, as a kind of religious dogma, and was exempted from that conceptual analysis which is requisite to its comprehension. Indeed it was often held to show, especially by philosophers, that any subject-matter possessing it was not validly analyzable into elements. Cantor has shown that this view is mistaken, by a precise definition of the kind of continuity which must belong to space. This definition, if it is to be explanatory of space, must, as he rightly says, be effected without any appeal to space.*[108]

In his *Introduction to Mathematical Philosophy* of 1919[109] Russell enlarges wittily on the contrast between the "arithmetical" characterization of continuity and the intuitive notion:

[104]Indeed, in Hobson (1957), we find Russell's definition of real number (introduced, with due acknowledgment, as Russell's "form" of the definition) included in the section of the book entitled "The Dedekind Theory of Irrational Numbers" (*loc. cit.*, pp. 23–27). Hobson refers to Russell's "segments" as "lower segments"; and in fact Russell's real numbers are the same as those lower segments of Dedekind cuts which possess no largest member.

[105]Hobson (1957), p. 24.

[106]Russell (1964), p. 287.

[107]Russell calls this property "compactness"; it is now usually referred to as "density".

[108]Russell (1964), p. 288.

[109]Written while Russell was in prison for his pacifist activities during the First World War.

> *The definitions of continuity ... of Dedekind and Cantor do not correspond very closely to the vague idea which is associated with the notion in the mind of the man in the street or of the philosopher. They conceive continuity rather as absence of separateness, the sort of general obliteration of distinctions which characterises a thick fog. A fog gives an impression of vastness without definite multiplicity or division. It is this sort of thing that the metaphysician means by "continuity", declaring it, very truly, to be characteristic of his mental life and that of children and animals.*
>
> *The general idea of vaguely indicated by the word "continuity" when so employed, or by the word "flux" is one which is certainly quite different from that which we have been defining* [i.e., the arithmetical one]. *Take, for example, the series of real numbers. Each is what it is, quite definitely and uncompromisingly; it does not pass over by imperceptible degrees into another; it is a hard, separate unit, and its distance from every other unit is finite, though it can be made less than any given finite amount assigned in advance. The question of the relation between the kind of continuity existing among the real numbers and the kind exhibited, e.g., by what we see at a given time, is a difficult and intricate one. It is not to be maintained that the two kinds are simply identical, but it may, I think, be very well maintained that the mathematical conception ... gives the abstract logical scheme to which it must be possible to bring empirical material by suitable manipulation, if that material is to be called "continuous" in any precisely definable sense.*[110]

Returning now to the *Principles*, Russell considers continuity to be a purely *ordinal* notion, and accordingly Cantor's later definition of continuity in purely order-theoretic terms is superior, in Russell's eyes, to the earlier one which involves metric considerations.[111]

Following a brief examination of the theory of infinite cardinals and ordinals,[112] Russell turns to the calculus and the infinitesimal. It is Russell's contention that, despite its traditional denomination as the "Infinitesimal" Calculus, "there is no allusion to, or implication of, the infinitesimal in any part of this branch of mathematics." Russell's unbounded enthusiasm for the actual infinite was accompanied by a certain hostility to the infinitesimal, although it ran less deep than Cantor's. Russell begins his discussion of the calculus with a withering account of Leibniz's muddled use of infinitesimals:[113]

---

[110]Russell (1995) p. 105.

[111]With the later subsumption of both order and metric under general topology, Cantor's two definitions of continuity, when referred to ordered and metric topological spaces, respectively, become equivalent.

[112]This too receives a sparkling introduction which is worth quoting:

> *The mathematical theory of infinity may almost be said to begin with Cantor. The Infinitesimal Calculus, though it cannot wholly dispense with infinity, has as few dealings with it as possible, and contrives to hide it away before facing the world. Cantor has abandoned this cowardly policy and has brought the skeleton out of its cupboard. He has been emboldened in this course by denying that it is a skeleton. Indeed, like many other skeletons, it was wholly dependent on its cupboard, and vanished in the light of day.* (*Ibid.*, p. 304).

[113]While Russell's disdain for the infinitesimal may have tempered somewhat his estimation of Leibniz, he never wavered in his regard for the latter as "one of the supreme intellects of all time" (1945, p. 581).

The philosophical theory of the Calculus has been, ever since the subject was first invented, in a somewhat disgraceful condition. Leibniz himself—who, one would have supposed, should have been competent to give a correct account of his own invention—had ideas, upon this topic, which can only be described as extremely crude. He appears to have held that, if metaphysical subtleties are left aside, the calculus is only approximate, but is justified practically by the fact that the errors to which it gives rise are less than those of observation. When he was thinking of Dynamics, his belief in the actual infinitesimal hindered him from discovering that the calculus rests on the doctrine of limits and made him regard his dx and dy as neither zero, nor finite, nor mathematical fictions, but as really representing the units to which, in his philosophy, infinite division was supposed to lead. And in his mathematical expositions of the subject, he avoided giving careful proofs, contenting himself with the enumeration of rules. At other times, it is true, he definitely rejects infinitesimals as philosophically valid; but he failed to show how, without the use of infinitesimals, the results obtained by means of the Calculus could be exact, and not approximate.[114]

Newton, however, fares rather better in Russell's account (as he did in Hegel's); Russell continues:

In this respect, Newton is preferable to Leibniz: his Lemmas give the true foundation of the Calculus in the doctrine of limits, and, assuming the continuity of space and time in Cantor's sense, they give valid proofs of its rules so far as spatio-temporal magnitudes are concerned. But Newton was, of course, entirely ignorant of the fact that his Lemmas depend upon the modern theory of continuity; moreover, the appeal to time and change, which appears in the word fluxion, and to space, which appears in the Lemmas, was wholly unnecessary, and served merely to hide the fact that no definition of continuity had been given.[115]

But Russell is not quite finished with his censure of Leibniz and the infinitesimalists, for he continues:

Whether Leibniz avoided this error, seems highly doubtful: it is at any rate certain that, in his first published account of the Calculus, he defined the differential coefficient by means of the tangent to a curve. And by his emphasis on the infinitesimal, he gave a wrong direction to speculation as to the Calculus, which misled all mathematicians before Weierstrass (with the exception, perhaps, of De Morgan), and all philosophers down to the present day.[116]

Russell proceeds to show, in painstaking detail, that the definitions of continuity, differentiability and integrability of a function in terms of limits involve no reference to the infinitesimal whatsoever, so establishing his claim that the Calculus can be washed entirely free of the notion. He remarks:

Until recent times, it was universally believed that continuity, the derivative, and the definite integral all involved actual infinitesimals, i.e., that even if the definitions of these notions could be formally freed from explicit mention of the infinitesimal, yet, where the definitions applied, the actual infinitesimal must always be found. This belief is now generally abandoned. The definitions which have been given in previous chapters do not in any way imply the infinitesimal, and this notion appears to have become mathematically useless.[117]

---

[114]Russell (1964)., p. 325.

[115]*Ibid.*, p. 325–6.

[116]*Ibid.*, p. 326.

[117]*Ibid.*, p. 331.

But what of the concept of the infinitesimal on its own account? It is to this issue that Russell next turns his attention.

Russell begins by observing that the infinitesimal has, in general, lacked precise definition:

> *It has been regarded as a number or magnitude which, though not zero, is less than any finite number or magnitude. It has been the dx or dy of the Calculus, the time during which a ball thrown vertically upwards is at rest during the highest point of its course, the distance between a point on a line and the next point, etc., etc.*[118]

None of these amount to anything like a precise definition in view of the facts, first, that the differential is not a quantity, nor $\frac{dy}{dx}$ a fraction; secondly, a properly developed theory of motion shows that there is no time during which a ball is at rest at its highest point; and lastly, the idea of the distance between consecutive points "presupposes that there are consecutive points—a view which there is every reason to deny."

Russell suggests that the sole precise definition of infinitesimal makes of it a purely *relative* notion, "correlative to something arbitrarily assumed to be finite." This relative notion is obtained by denying the Archimedean principle that any pair of numbers or comparable magnitudes P, Q are relatively finite in the sense that, if P be the lesser, then there is a finite integer n such than $nP > Q$. In that case P may be defined to be infinitesimal with respect to Q, and Q infinite with respect to P, if $nP < Q$ for any integer n. As far as magnitudes are concerned, Russell says, the only way of defining the infinitesimal and, indeed, the infinite, is through the use (or denial) of the Archimedean principle, for

> *Of a magnitude not numerically measurable, there is nothing to be said except that it is greater than some of its kind, and less than others; but from such propositions infinity cannot be obtained. Even if there be a magnitude greater than all others of its kind, there is no reason for regarding it as infinite. Finitude and infinity are essentially numerical notions, and it is only by relation to numbers that these terms can be applied to other entities.*[119]

Russell admits that instances of the (relative) infinitesimal can be found, even some of significance. One example he offers arises in connection with his introduction, earlier on in the *Principles*, of the concept of *magnitude of divisibility*. Russell proposes this concept as a way of overcoming one difficulty posed by Cantor's reduction of the continuous to the discrete, namely the consequence that any two continua, however different they may be in metric size, are always the same size numerically in the sense of being composed of the same (transfinite) number of points (or "terms"). As Russell says, "there must be . . . some other respect in which [say] two periods of 12 h are equal, while a period of 1 h and another of 23 h are unequal."[120] Russell's suggestion is then to introduce, corresponding to each aggregate, a magnitude called its *magnitude of divisibility*. Roughly speaking, an

---

[118] *Ibid.*

[119] *Ibid.*, p. 332.

[120] *Ibid.*, p. 151.

aggregate's magnitude of divisibility represents the "number" of parts into which it can be divided with respect to a given determination of the meaning of "part". For example, if the aggregates in question are finite sets, and "parts" are singletons, then the corresponding magnitudes of divisibility are natural numbers; if the aggregates are infinite sets and "parts" are again singletons, then the corresponding magnitudes of divisibility are transfinite cardinals. On the other hand, if the aggregates are intervals on a line and the "parts" are intervals of unit length, then the corresponding magnitudes of divisibility are the nonnegative real numbers.

Now if divisibility can be regarded as a magnitude in the sense above, then "it is plain", says Russell, "that the divisibility of any whole containing a finite number of simple parts is infinitesimal as compared with one containing an infinite number. The number of parts being taken as the measure, every infinite whole will be greater than n times every finite whole, whatever finite number n may be."[121]

Russell offers a number of further examples of infinitesimals in the same spirit: any line is infinitesimal with respect to an area, an area with respect to a volume, and a bounded volume with respect to the whole of space. On the other hand, the real numbers have been provided with an unequivocal definition as segments of rationals, and this fact "renders the non-existence of infinitesimals [among the real numbers] demonstrable." Russell's conclusion is

> *if it were possible, in any sense to speak of infinitesimal numbers, it would have to be in some radically new sense.*[122]

Strangely, Russell (whose first published work was devoted to the foundations of geometry) fails to mention the geometer Veronese's attempts at introducing infinitesimals. But he touches on du Bois-Reymond's orders of infinity and infinitesimality of functions. This Russell does with reluctance, given that "on this question the greatest authorities are divided." But in the end Russell sides with Cantor in deciding that "these infinitesimals are mathematical fictions."

In sum, Russell concludes,

> *the infinitesimal... is a restricted and mathematically very unimportant conception, of which continuity and infinity are alike independent.*[123]

So much for the infinitesimal as a mathematical concept. Russell next turns to the philosophical import of the notion. Again, a playful introduction:

> *We have concluded our summary review of what mathematics has to say concerning the continuous, the infinite, and the infinitesimal. And here, if no previous philosophers had treated of these topics, we might leave the discussion, and apply our doctrines to space and time. For I hold the paradoxical opinion that what can be mathematically demonstrated is true. As, however, almost all philosophers disagree with this opinion, and as many have written elaborate arguments in favour of views different from those above expounded, it will*

---

[121] *Ibid.*, pp. 332–3. This example is essentially the same as that mentioned above by Fraenkel.

[122] *Loc. cit*, p. 335.

[123] *Loc. cit*, p. 337.

> *be necessary to examine controversially the principal types of opposing theories and to defend, as far as possible, the points in which I differ from standard writers.*[124]

As his source for these "opposing theories", Russell focuses on Hermann Cohen's 1883 neo-Kantian work, *Das Prinzip der Infinitesimal-Methode und Seine Geschichte* ("The Principle of the Method of Infinitesimals and its History").[125] In this work Cohen develops the view that the infinitesimal is essentially intensive magnitude, possessing the capacity of acting as a kind of generating element of the real as it is presented to the mind. For Cohen, mathematical infinitesimals are entities which, while real, "cannot be directly intuited as insular or discrete elements of being".[126] As such, the infinitesimal mirrors "the relation of thinking and intuition which is to characterize all of modern science." Cohen argues that, "far from being limited only to mathematical or scientific knowledge, the same process at the heart of the infinitesimal lies at the heart of all forms of perception."[127] As an index of change, the infinitesimal "allows a proper understanding of change in the world."[128] Cohen ascribed equal importance to the idea of continuity, which, he says, "is the general basis of consciousness." The wider context in which the mind is constrained to place each particular presented to it is necessarily a "continuous plenum."[129]

---

[124] *Loc. cit*, p. 338.

[125] For an illuminating discussion of this work and its historical context see Moynahan (2003).

[126] *Op. cit.*, p. 44.

[127] *Ibid.*, p. 45.

[128] *Ibid.*

[129] In this connection it is worth mentioning that no less a figure than Leo Tolstoy ascribed similar importance to the continuous and the infinitesimal. In Book 3, Part 3 of *War and Peace* we read:

> *To elicit the laws of history we must leave aside kings, ministers, and generals, and select for study the homogeneous, infinitesimal elements which influence the masses. No one can say how far it is possible for a man to advance in this way to an understanding of the laws of history; but it is obvious that this is the only path to that end...*
>
> *It is impossible for the human intellect to grasp the idea of absolute continuity of motion. Laws of motion only become comprehensible to man when he can examine arbitrarily selected units of that motion. But at the same time it is this arbitrary division of continuous motion into discontinuous units which gives rise to a large proportion of human error.*
>
> *A new branch of mathematics* [i.e., the calculus], *having attained the art of reckoning with infinitesimals, can now yield solutions in ... complex problems of motion which before seemed insoluble. ... This new branch of mathematics ... by admitting the conception, when dealing with problems of motion, of the infinitely small and thus conforming to the chief condition of motion (absolute continuity), corrects the inevitable error which the human intellect cannot but make if it considers separate units of motion instead of continuous motion.*
>
> *In the investigation of the laws of historical movement precisely the same principle operates. The maarch of humanity, springing as it does from an infinite multitude of individual wills, is continuous. The discovery of the laws of this continuous movement is the aim of history.*

It happens that Cohen's work was the subject of a review by Frege in 1885.[130] Here is how Frege sums up the basic idea of Cohen's treatise:

> *Cohen brings reality into a peculiar connection with the differential by going back, it seems, to the anticipations of perception whose principle according to Kant is this: 'In all appearances, the real that is an object of sensation has intensive magnitude, that is, a degree.' Now the differential is an intensive magnitude. If, e.g., x is a distance on the straight line, then dx, its differential or infinitesimal increment, is not to be thought of as an extensive magnitude or as itself a distance; this would lead to contradictions, and it is precisely because mathematicians wanted to let the differential pass throughout as an extensive magnitude that they got entangled in the well-known difficulties. These difficulties can be removed, not by logic, but by the critique of knowledge, which is the term the author uses for 'theory of knowledge', because it shows that an infinitesimal number is an intensive magnitude which, as such, has a power of realization: 'It does not merely represent the* unit of reality*; but also* realizes as such; *it confers reality upon Being in Quality'.* [Again], *'If the* differential *constitutes* reality *as a constitutive condition of thought, then the* integral *designates the* real as object.' *The dx is therefore to be conceived of as, say, an intensive magnitude concentrated at the end point of x, comparable to an electrical charge, or as a power to increase the distance, like for example the last bud on a bough in which we can recognize a striving for growth. These pictures occurred to me as I was reading the book; they are not to be attributed to the author himself. ...*
>
> *The contrast between extensive and intensive goes back to the contrast between intuition and thought, since the quality which corresponds to the intensive magnitude is a category of thought. The extensive magnitude of intuition is thus opposed to the intensive one of sensation. The two sources of knowledge must always be combined if the knowledge is to be objective. Reality, as a means of thought, is able to come in where intuition alone fails, for the latter has the character of ideality. If the infinitesimal is to be fit to do justice to the requirements of reality, it must be withdrawn from intuition, provided that reality is to mean a condition of experience* on the part of thought. *Accordingly, continuity is also separated from intuition and assigned to thought.*[131]

> *Only by assuming an infinitesimally small unit of observation—a differential of history (that is, the common tendencies of men)—and arriving at the art of integration (finding the sum of the infinitesimals) can we hope to discover the laws of history.*

[130]An English translation of Frege's review may be found in Frege (1984), pp. 108–12.

[131]*Loc. cit.*, pp. 109–10. In 1906 Cohen's student Ernst Cassirer summed up Cohen's outlook more sympathetically:

> *The basic idea of Cohen's work can be stated quite briefly: if we want to achieve a true scientific grounding of logic, we should not begin from any sort of completed existence. What naive intuition takes as its obvious and secure possession, this is for logic a real problem; what it assumes as directly 'given', this is what must be critically analyzed and taken apart in its crucial conditions for thought. We should not begin with any objective Being, no matter of what sort and no matter [in] what relation we place ourselves to it: for every "being" is in the first place a product and a result which the operation of thought and its systematic unity has as a presupposition. A foundational conceptual setting of this sort, an intellectual tradition in which we can first speak of "reality" in a scientific sense, is found by Cohen in the idea of the infinitesimal as it is detailed and fixed in modern mathematics.* (Quoted in Moynahan 2003, p. 42.)

Frege, not surprisingly, is very critical of all this, remarking that "I do not find that the infinitesimal has an intimate connection with reality." Above all, he says, "What I miss everywhere is proofs."

Russell, for his part, is equally critical of Cohen's claims for the infinitesimal. A first objection is that Cohen unquestioningly treats differentials as actual entities, a view which Russell regards as having been exploded by the theory of limits:

> *But when we turn to such works as Cohen's, we find the dx and dy treated as separate entities, as the intensively real elements of which the continuum is composed .... The view that the Calculus requires infinitesimals is apparently not thought open to question; at any rate, no arguments whatever are brought up to support it. This view is certainly assumed as self-evident by most philosophers who discuss the Calculus.*[132]

But Russell's principal objection is that approaches to the problem of the infinitesimal such as Cohen's identify the problem as possessing an epistemological, rather than a purely logical character, and so "depends upon the pure intuitions as well as the categories." Russell rejects "this Kantian opinion [which] is wholly opposed to the philosophy which underlies the present work." Cohen's claim that intensive magnitude is infinitesimal extensive magnitude is also rejected, on the grounds that

> *the latter must always be smaller than finite extensive magnitudes and must therefore be of the same kind with them; while intensive magnitudes seem never in any sense smaller than any extensive magnitudes.*[133]

Russell quotes Cohen's summary of his own theory:

> *That I may be able to posit an element* in and for itself, *is the* desideratum, *to which corresponds the* instrument of thought *reality. This instrument of thought must first be set up, in order to be able to enter into that combination with intuition, with the* consciousness of being given, *which is completed in the* principle of intensive magnitude. *This presupposition of intensive reality is latent in all principles and must be made independent.* This presupposition is the meaning of reality and the secret of the concept of the differential.[134]

Russell rejects most of this, but remarks that,

> *What we can agree to, and what, I believe, confusedly underlies the above statement, is, that every continuum must consist of elements or terms; but these, as we have seen, will not fulfill the function of the dx or dy which occur in old-fashioned accounts of the Calculus. Nor can we agree that "this finite" (i.e. that which is the object of physical science* [to quote Cohen] *"can be thought of as a sum of those infinitesimal intensive realities, as a* definite integral." *The definite integral is not a sum of elements of a continuum, although there are such elements: for example, the length of a curve, as obtained by integration, is not the sum of its points, but strictly and only the limit of the lengths of inscribed polygons... There is no such thing as an infinitesimal stretch; if there were, it would not be an element of a continuum.*[135]

---

[132] Russell (1964), p. 339.

[133] *Ibid.*, p. 344.

[134] *Ibid.*

[135] *Ibid.*, p.345.

In sum, says Russell,

> *infinitesimals as explaining continuity must be regarded as unnecessary, erroneous, and self-contradictory.*[136]

Having dismissed the philosophical claims of the infinitesimal, Russell finally turns his attention to the philosophical difficulties posed by the continuum. It is made clear that attention is to be confined to the *arithmetical*—that is to Cantor's—continuum, and that the continuum as it is presented to "intuition" is to be excluded from consideration. Following this decree Russell remarks:

> *It has always been held to be an open question whether the continuum is composed of elements; and even when it has been allowed to contain elements, it has been often alleged to be not* composed *of these. This latter view was maintained by even by so stout a supporter of elements in everything as Leibniz. But all these views are only possible in regard to such continua as space and time. The arithmetical continuum is an object selected by definition, and known to embodied in at least one instance, namely the segments of the rational numbers. I shall* [later] *maintain ... that spaces afford other instances of the arithmetical continuum. The chief reason for the elaborate and paradoxical theories of space and time and their continuity, which have been constructed by philosophers, has been the supposed contradictions in a continuum composed of elements. The thesis of the present chapter is, that Cantor's continuum is free from contradictions. This thesis, as is evident, must be firmly established, before we can allow that spatial-temporal continuity is of Cantor's kind. In this argument I shall assume as proved ... that the continuity to be discussed does not involve the admission of actual infinitesimals.*[137]

Russell demonstrates the freedom from contradiction of Cantor's continuum through a new analysis of Zeno's paradoxes. He translates each paradox into arithmetical language and then shows that the resulting assertions are not paradoxical at all. He introduces his demonstration with another dramatic paragraph:

> *In this capricious world, nothing is more capricious than posthumous fame. One of the most notable victims of posterity's lack of judgment is the Eleatic Zeno. Having invented four arguments, all immeasurably subtle and profound, the grossness of subsequent philosophers pronounced him to be a mere ingenious juggler, and his arguments to be one and all sophisms. After two thousand years of continual refutation, these sophisms were reinstated, and made the foundation of a mathematical renaissance, by a German professor, who probably never dreamed of a connection between himself and Zeno. Weierstrass, by strictly banishing all infinitesimals, has at last shown that we live in an unchanging world, and that the arrow, at every moment of its flight, is truly at rest. The only point where Zeno probably erred was in inferring (if he did infer) that, because there is no change, therefore the world must be in the same state at one time as at another. This consequence by no means follows, and in this point the German professor is more constructive than the ingenious Greek. Weierstrass, being able to embody his opinions in mathematics, where familiarity with truth eliminates the vulgar prejudices of common sense, has been able to give to his propositions the respectable air of platitudes; and if his result is less delightful to the lover of reason than Zeno's bold defiance, it is at any rate more calculated to appease the mass of academic mankind.*[138]

---

[136] *Ibid.* Russell would have been greatly surprised to learn that Cohen's conception of infinitesimals as intensive magnitudes can in fact be given a precise mathematical sense. See Chap. 10 below.

[137] *Ibid.*, p. 347.

[138] *Ibid.*, pp. 347–8.

Here we consider Russell's resolutions of two of the paradoxes, that of Dichotomy and that of the Arrow. The Dichotomy is stated by Russell as "There is no motion, for what moves must reach the middle of its course before it reaches the end." That is to say, Russell continues, "whatever motion we assume to have taken place, this presupposes another motion, and so on *ad infinitum.* Hence there is an endless regress in the mere idea of assigned motion." To state this argument in arithmetical form, Russell considers the class, or set, of real numbers between 0 and 1. This class, he says, is an infinite whole,

> *whose parts are logically prior to it: for it has parts, and it cannot subsist if any of the parts are lacking. Thus the numbers from 0 to 1 presuppose those from 0 to ½, these presuppose those from 0 to ¼, and so on.*[139]

So it would seem to follow that

> *there is an infinite regress in the notion of any infinite whole; but without infinite wholes, real numbers cannot be defined, and arithmetical continuity, which applies to an infinite series, breaks down.*[140]

Russell refutes this argument by observing that a class of real numbers, being given *intensionally* as the class of terms satisfying a given predicate (rather than *extensionally* by enumeration of its members) is not logically posterior to its parts. In particular, the class of all real numbers between 0 and 1 forms

> *a definite class, whose meaning is known as soon as we know what is meant by* real number, 0, 1, *and* between.[141]

It follows that "the particular members of the class, and the smaller classes contained in it, are not logically prior to the class." The infinite regress, which "consists merely in the fact that every segment of real numbers has parts which are again segments" is rendered harmless by observing that these parts are not logically prior to it. Russell concludes that "the solution of the difficulty lies in the theory of denoting and the intensional definition of a class." So much for the Dichotomy.

Of Zeno's arrow puzzle—"If everything is in rest or in motion in a space equal to itself, and if what moves is always in the instant, the arrow in its flight is immovable"—, Russell remarks:

> *This has usually been thought so monstrous a paradox as scarcely to deserve serious discussion. To my mind, I must confess, it seems a very plain statement of a very elementary fact, and its neglect has, I think, caused the quagmire in which the philosophy of change has long been immersed.*[142]

---

[139] *Ibid.*, p. 348.

[140] *Ibid.*

[141] *Ibid.*, p. 350.

[142] *Ibid.*, p. 350.

Russell's dissolution of the paradox is to divest it of all reference to change, so revealing it to be

> *a very important and very widely applicable platitude, namely: "Every possible value of a variable is a constant."*[143]

Russell's claim here is that the variable position of the arrow is in essence a variable in the mathematical sense; and since a mathematical variable is (according to Russell) just a symbol denoting an arbitrary constant, the arrow's flight is, in the eyes of pure logic, just a conjunction of assertions of the form "the arrow was at a certain place at a certain time". Each of these correlated places and times is a constant, and the arrow is at rest at all of them. "This simple logical fact", says Russell, "seems to constitute the essence of Zeno's contention that the arrow is always at rest."

Russell contends that in addition to its purely logical character, Zeno's argument says something fundamental about continua. In the case of motion, it is the denial that there is such a thing as a *state* of motion; once "change" is eradicated, motion is in fact nothing more than the occupation of different places at different times. In the case of a continuous variable, the thrust of Zeno's argument may be taken to be the denial of the existence of actual infinitesimals. "For," says Russell,

> *infinitesimals are an attempt to extend to the* values *of a variable the variability which belongs to it alone. When once it is firmly realized that all the values of a variable are constants, it becomes easy to see, by taking* any *two such values, that their difference is always finite, and hence there are no infinitesimal differences. If x be a variable which may take all values between 0 and 1, then, taking any two of these values, we see that their difference is finite, although x is a continuous variable... This static conception of the variable is due to the mathematicians, and its absence in Zeno's day led him to suppose that continuous change was impossible without a state of change, which involves infinitesimals and the contradiction of a body being where it is not.*[144]

From these analyses Russell infers that

> *Zeno's arguments, though they prove a very great deal, do not prove that the continuum, as we have become acquainted with it, contains any contradiction whatever. Since his day the attacks on the continuum, have not, so far as I know, been conducted with any new or pore powerful weapons.*

Russell again raises his hat to Cantor:

> *The notion to which Cantor gives the name of* continuum *may, of course, be called by any other name in or out of the dictionary, and it is open to every one to assert that he himself means something quite different by the continuum. But these verbal questions are purely frivolous. Cantor's merit lies, not in meaning what other people mean, but in telling us what he means himself—an almost unique merit, where continuity is concerned. He has defined, accurately and generally, a purely ordinal notion, free, as we now see, from contradictions,*

[143]*Ibid.*, p. 351.

[144]*Ibid.*, p. 351–2.

> *and sufficient for all Analysis, Geometry, and Dynamics. ... The salient points in the definition of the continuum are (1) the connection with the doctrine of limits, (2) the denial of infinitesimal segments. These two points being borne in mind, the whole philosophy of the subject becomes illuminated.*[145]

One final conclusion is drawn:

> *The denial of infinitesimal segments resolves an antinomy which has long been an open scandal, I mean the antinomy that the continuum both does and does not consist of elements. We see now that both may be said, though in different senses. Every continuum is a series consisting of terms, and the terms, if not indivisible, at any rate are not divisible into new terms of the continuum. In this sense there are elements. But if we take consecutive terms together with their asymmetrical relations as constituting what may be called ... an* ordinal *element, then, in this sense, our continuum has no elements. If we take a stretch to be essentially serial, so that it must consist of at least two terms, then there are no elementary stretches; and if our continuum be one in which there is distance, then like wise there are no elementary distances. But in neither of these cases is there the slightest logical ground for elements. The demand for consecutive terms springs ... from an illegitimate use of mathematical induction. And as regards distance, small distances are no simpler than large ones, but all .. are alike simple. And large distances do not presuppose small ones .... Thus the infinite regress from greater to smaller distances or stretches is of the harmless kind, and the lack of elements need not cause any logical inconvenience. Hence the antinomy is resolved, and the continuum, so far at least as I am able to discover, is wholly free from contradictions.*[146]

As we have seen, Russell's analysis of the continuum rests chiefly on denying the existence of infinitesimals. He correctly identifies infinitesimals as "an attempt to extend to the values of a variable the variability which belongs to it alone", and he is right in his assertion that if the values of a variable must always be constants, infinitesimals as "variable values" cannot exist. But he could not have foreseen that, some decades after writing this passage (yet still, remarkably, within his own lifetime), developments in mathematics would enable variation to be reincorporated into the subject in such a way as to allow the admission of the infinitesimal in essentially the sense that he repudiates with such *élan*. Another development that Russell could not have anticipated is that the rigorous reintroduction of the infinitesimal in this sense would not require abandoning the law of noncontradiction, as he seems, with some justice, to have thought: the requisite logical adjustment turned out to be the dropping of the law of excluded middle.[147] Russell would also have been greatly surprised—perhaps even chagrined—to learn that the conception of infinitesimals as intensive magnitudes can in fact be given a precise mathematical sense.[148]

---

[145] *Ibid.*, p. 353.

[146] *Ibid.*, pp. 353–4.

[147] See Chap. 10 below.

[148] See Chap. 10 below.

4.6 Hobson's Choice

Published in 1907, *The Theory of Functions of a Real Variable*, by the prominent English mathematician **E. W. Hobson** (1856–1933), was the first systematic exposition in English of the new analysis. In this work, which was to prove very influential, Hobson makes a number of interesting observations concerning the process by which the continuum of intuition had, in the course of the nineteenth century, come to be replaced by the arithmetical continuum:

> *Before the development of analysis was made to rest upon a purely arithmetical basis, it was usually considered that the field of operations was the continuum given by our intuition of extensive magnitude especially of spatial or temporal magnitude, and of the motion of bodies through space.*
>
> *The intuitive idea of continuous motion implies that, in order that a body may pass from one position A too another position B, it must pass through every intermediate position in its path. An attempt to answer the question, what is meant by* every *intermediate position, reveals the essential difficulties of this question, and gives rise to a demand for an exact theoretical treatment of continuous magnitude.*
>
> *The implication in the idea of continuous magnitude shews that, between A and B, other positions A', B' exist, which the body must occupy at definite times; that between A', B', other such positions exist, and so on. The intuitive notion of the continuum, and that of continuous motion, negate the idea that such a process of subdivision can be conceived of as having a definite termination. The view is prevalent that the intuitional notions of continuity and of continuous motion are fundamental and* sui generis; *and that they are incapable of being exhaustively described by a scheme of specification of positions. Nevertheless, the aspect of the continuum as a field of possible positions is the one which is accessible to Arithmetic Analysis, and with which alone Mathematical Analysis is concerned. That property of the intuitional continuum, which may be described as unlimited divisibility, is the only one that is immediately available for use in Mathematical thought,; and this property is not sufficient for the purposes in view, until it has been supplemented by a system of axioms and definitions which shall suffice to provide a complete and exact description of the possible positions of points and other geometrical objects which can be determined in space. Such a scheme constitutes an abstract theory of spatial magnitude .*[149]

Hobson defends the arithmetic continuum as the necessary outcome of an exact theory of measurable quantity:

> *The term arithmetic continuum is used to denote the aggregate of real numbers, because it is held that the system of numbers of this aggregate is adequate for the complete analytical representation of what is known as continuous magnitude. The theory of the arithmetic continuum has been criticized on the ground that it is an attempt to find the continuum within the domain of number, whereas number is essentially discrete. Such an objection presupposes the existence of some independent conception of the continuum, with which that of the aggregate of real numbers can be compared. At the time when the theory of the arithmetic continuum was developed, the only conception of the continuum which was extant was that of the continuum as given by intuition; but this, as we shall shew, is too vague a conception to be fitted for an object of exact mathematical thought, until its character as a pure intuitional datum has been clarified by exact definitions and axioms. The discussions*

[149]Hobson (1957), pp. 54–5.

connected with arithmetization have led to the construction of abstract theories of measurable quantity; and these all involve the use of some system of arithmetic, as providing the necessary language for the description of the relations of magnitudes and quantities. It would thus appear to be highly probable that, whatever abstract conception of the intuitional continuum of quantity and magnitude may be developed, a parallel conception of the arithmetic continuum, though not necessarily identical with the one which we have discussed, will be required. To any such scheme of numbers, the same objection might be raised as has been referred to above; but if the objection were a valid one, the complete representation of continuous magnitudes would, under any theory of such magnitudes, be impossible. It is clear that only in connection with an exact abstract theory of magnitude that any question as to the adequacy of the continuum of real numbers for the measurement of magnitudes can arise. For actual measurement of physical, or of spatial, or temporal magnitudes, the rational numbers are sufficient; such measurement being essentially of an approximate character only, the degree of error depending on the accuracy of the instruments employed.[150]

Hobson admits the possibility of constructing arithmetic continua with essentially different properties, perhaps even containing infinitesimals:

The disputable idea that the theory here explained [i.e. that of Dedekind] *necessarily implies that a continuum is to be regarded as a set of points, which are elements not possessing magnitude, has frequently been a stumbling-block in the way of the acceptance of the view of the spatial continuum which has been indicated above. It has been held that, if space is to be regarded as made up of elements, these elements must themselves possess spatial character; and this view has given rise to various theories of infinitesimals or of indivisibles, as components of spatial magnitudes. The most modern and complete theory of this kind has been developed by Veronese*[151] *and is based on a denial of the Principle of Archimedes . . .*[152]

But in the end archimedean systems are to be preferred on the grounds of simplicity:

The validity of Veronese's system has been criticized by Cantor and others on the ground that the definitions contained in it, relating to equality and inequality, lead to contradiction; it is however unnecessary for our purpose to enter into the controversy on this point. The straight line of geometry is an ideal object of which any properties whatever may be postulated, provided that they satisfy the conditions, (1), that they form a valid scheme, i.e. one that does not lead to contradiction, and (2), that the object defined is such that it is not in contradiction with empirical straightness and linearity. There is no a priori *objection to the existence of two or more such adequate conceptual systems, each self-consistent even if they be incompatible with one another; but of such rival schemes the simplest will naturally be chosen for actual use. Assuming then the possibility of setting up a valid non-Archimedean system for the straight line, still the simpler system, in which the system of Archimedes is assumed, is to be preferred, because it gives a simpler conception of the nature of the straight line and is adequate for the purposes for which it was devised.*[153]

As Hobson's work shows, by the beginning of the twentieth century, mathematical analysis had come to be placed on a set-theoretic foundation, supplanting the older methods of analysis based on infinitesimals and the intuitive continuum. In

[150] Hobson (1957), pp. 53–4.

[151] See Chap. 5 below.

[152] Hobson (1957), pp. 57–8.

[153] *Ibid.*, p. 58.

geometry, by contrast, the process of set-theorization was considerably less rapid. The geometer Sophus Lie, for example, was "untouched by it in the 1890s".[154] Hermann Weyl's and Tullio Levi-Civita's work in the 1920s in mathematics and physics avoids the use of set-theoretic methods, making extensive use of infinitesimals, even though both "believed that ε–δ style foundations were better in principle."[155] Nor is there much trace of set theory in the work of the geometer Élie Cartan, who was active in the 1930s. In fact set theory did not come to dominate geometry until the mid-twentieth century.

[154] McLarty (1988), p. 87.
[155] *Ibid.*

Chapter 5
Dissenting Voices: Divergent Conceptions of the Continuum in the Nineteenth and Early Twentieth Centuries

Despite the great success of Weierstrass, Dedekind and Cantor in constructing the continuum from arithmetical materials, a number of thinkers of the late nineteenth and early twentieth centuries remained opposed, in varying degrees, to the idea of explicating the continuum concept entirely in discrete terms. These include the mathematicians *du Bois-Reymond, Veronese, Poincaré, Brouwer* and *Weyl,* and the philosophers *Brentano* and *Peirce*.

5.1 Du Bois-Reymond

Paul du Bois-Reymond (1831–1889), against whose theory of infinities and infinitesimals Cantor fought so hard, was a prominent mathematician of the later nineteenth century who made significant contributions to real analysis, differential equations, mathematical physics and the foundations of mathematics. While accepting many of the methods of the Dedekind-Cantor school, and indeed embracing the idea of the actual infinite, he rejected its associated philosophy of the continuum on the grounds that it was committed to the reduction of the continuous to the discrete. In 1882 he writes:

> *The conception of space as static and unchanging can never generate the notion of a sharply defined, uniform line from a series of points however dense, for, after all, points are devoid of size, and hence no matter how dense a series of points may be, it can never become an interval, which must always be regarded as the sum of intervals between points.*[1]

Du Bois-Reymond took a somewhat mystical view of the continuum, asserting that its true nature, being beyond the limits of human cognition, would forever elude

[1]Quoted in Ehrlich (1994), p. x

J. L. Bell, *The Continuous, the Discrete and the Infinitesimal in Philosophy and Mathematics*, The Western Ontario Series in Philosophy of Science 82,
https://doi.org/10.1007/978-3-030-18707-1_5

the understanding of mathematicians.[2] Nevertheless, this did not prevent him from developing his own theory of the mathematical continuum, a *continuum of functions*, during the 1870s and 80s. This was introduced in an article of 1870–1 as the *calculus of infinities*. Here du Bois-Reymond considers "functions ordered according to the limit of their quotients"[3] .The orderings of functions, in du Bois-Reymond's notation,

$$f(x) \succ \varphi(x), f(x) \approx \varphi(x), f(x) \prec \varphi(x),$$

are defined respectively by

$$\lim_{x\to\infty} f(x)/\varphi(x) = \infty,\ \lim_{x\to\infty} f(x)/\varphi(x) \text{ is finite and } \neq 0,\ \lim_{x\to\infty} f(x)/\varphi(x) = 0.$$

Thus, for example, $e^x \succ x \succ \log(x)$, $x^p \succ x$ for any $p > 1$, while $cx^r \approx x^r$ for any c and r. When $f(x) \succ \varphi(x)$, $f(x)$ is said to have an "infinity greater than $\varphi(x)$"; when $f(x) \prec \varphi(x)$, $f(x)$ may be thought of (although du Bois-Reymond does not say this explicitly) as being infinitesimal in comparison with $\varphi(x)$. Du Bois-Reymond considers sequences of functions linearly ordered under $\prec$ or $\succ$. Such "scales of infinity"[4] can be caused to become arbitrarily complex by the continued interpolation of new such sequences between terms. Du Bois-Reymond draws an analogy with the ordered set of real numbers:

> *Just as between two functions as close with respect to their infinities as one may want, one can imagine an infinity of others forming a kind of passage from the first function to the second, one can compare the sequence F [a scale of infinity] to the sequence of real numbers, in which one can also pass from one number to a number very little different from it by an infinity of other ones.*[5]

While du Bois-Reymond uses the term "infinities" in connection with his classification of functions, he does not at this point speak of infinite numbers or actual infinities. But in an article of 1875 he drops his reservations on the matter, and boldly begins by asserting:

> *I decided to publish this continuation of my research on functions becoming infinite in German after I overcame my aversion to using the word 'infinite* (unendlich)' *as a substantive, like the French their 'infini'. I even flatter myself that, by this 'infinite* (unendlich)', *I have enriched our mathematical vocabulary in a noteworthy way.*[6]

---

[2]This echoes his older brother Emile's famous *ignorabimus* decaration in 1880 concerning "world riddles," certain of which, such as the ultimate nature of matter and force, and the origin of sensations, would always elude explanation.

[3]Fisher(1981), p. 102.

[4]The term is Hardy's: see Hardy (1910).

[5]Quoted in Fisher (1981), p. 104.

[6]Quoted *ibid.*, p. 105.

He goes on to say:

> *In earlier articles I have distinguished the different infinities of functions by their different magnitudes so that they form a domain of quantities (the infinitary) with the stipulation that the infinity of $\varphi(x)$ is to be regarded as larger than that of $\psi(x)$ or equal to itaccording as the quotient $\varphi(x)/\psi(x)$ is infinite or finite. Thus in the infinitary domain of quantities the quotient enters in place of the difference in the ordinary domain of numbers. Between the two domains there are many analogies... I can add further that the most complete symmetry exists between functions becoming zero and becoming infinity, in such a way that everywhere the positive numbers correspond in the most striking way to becoming infinity, the negative numbers to becoming zero, zero to remaining finite. Instead of numbers as fixed signs in the domain of numbers, one has in the infinitary domain of quantities an unlimited number of simple functions; the exponential functions, the powers, the logarithmic functions, that likewise form fixed points of comparisons, and between whose arbitrarily close infinities a limitless number of infinities different from each other can be inserted.*[7]

In a paper of 1877 du Bois-Reymond compares his system of "infinities" and that of "ordinary" numbers. He introduces the concept of "numerical continuity", an idea which he suggests underlies the introduction of irrational numbers. To illustrate the idea, du Bois-Reymond offers as a metaphor the distribution of the stars on a great circle in the sky.[8] The readily identified brighter stars he compares to rational numbers with small numerators and denominators. Use of telescopes reveals the presence of new stars in any region, however small, but patches of darkness are always found between them. And then

> *our imagination, or speculation, peoples this as it were asymptotically uniform nothingness which always remains, with matter whose radiation or our observation can no longer make accessible. In our thought, we may believe there is no end, and we admit no empty spot in the sky.*[9]

This is analogous to the generation of rational and irrational numbers:

> *Thus through more precise consideration the rational numbers always approach more closely to one another, yet in our minds gaps are always left between them, which mathematical speculation then fills with the irrationals.*[10]

According to du Bois-Reymond this is essentially the way in which "numerical continuity" has arisen. He sees mathematical intuition as assigning equal authenticity to geometric and numerical quantity, but the attainment of complete equality between the two can only be attained through the use of the limit concept in introducing the irrationals. And the insertion of the irrationals between the rationals is an extension of the primitive concept of number to an equally primitive, but more comprehensive, concept of continuous quantity. That being the case, the comparison between numerical and geometric quantities may conceal further subtleties.[11]

---

[7] *Ibid.*, p. 106.
[8] *Ibid.*, pp. 107–8.
[9] Quoted *ibid.*, p. 107
[10] Quoted *ibid.*, p. 107.
[11] *ibid.*, p. 108.

One such subtlety is brought to light in connection with continuous families of curves. When these are allowed to increase with growing rapidity their approximative behaviour is quite different from that associated with ordinary spatial continuity. Du Bois-Reymond writes:

> *If we think of two different quickly increasing functions, then all the transitions from one to the other are spatially conceivable and present in our minds. We cannot conceive anywhere a gap between two curves increasing to infinity or in the neighbourhood of one such curve, which could not be filled with curves; on the contrary, each curve is accompanied by curves which proceed arbitrarily close to it, to infinity.*[12]

Now, unlike the points on a line segment, the curves which run between two such curves do not form a "simple infinity", that is, they do not depend on just a single parameter. Du Bois-Reymond shows that this infinity is "unlimited" in the sense that it is not n-fold for any finite n.[13] He continues:

> *just as in the ordinary domain of quantities we can only express quantities numerically exactly by means of rational numbers, since the other numbers are not actual numbers but only limits of such numbers*[14]*: so we can only express infinities with well-defined functions, of which we only have at our disposal up to now those belonging to the family of logarithms, powers, exponential functions.*[15]

Du Bois-Reymond next notes the difference between the approximative behaviour of real numbers and that of "infinities" associated with functions.[16] While one can approximate a number, say ½, by many sequences in such a way that any number, however close to ½, will fall between two members of any such sequence, the situation is quite different for the functions associated with infinities. For example, consider the sequence of functions

$$x^{1/2}, x^{2/3}, \ldots, x^{p/p+1}, \ldots.$$

The exponents of the members of this sequence approach 1, but it is not hard to establish the existence of functions whose infinities fall between *all* of the infinities of the members of this sequence and the function x to which the sequence converges in an appropriate sense. For example, $x^{\log\log x/\log\log x+1}$ is readily shown to be such a function.

Du Bois-Reymond next proceeds to demonstrate the generality of this phenomenon:

> *One cannot approximate a given infinity $\lambda(x)$ with any sequence of functions $\varphi_p(x)$, $p = 1, 2, \ldots$ in such a way that one could not always specify a function $\psi(x)$ which satisfies for arbitrarily large values of p $\lambda(x) \gtrless \psi(x) \gtrless \varphi_p(x)$.*[17]

---

[12]Quoted *ibid.*, p. 108.

[13]*Ibid.*, p. 108.

[14]This view of irrational numbers is evidently in direct opposition to Cantor's.

[15]Quoted *ibid.*, pp. 108–9.

[16]*Ibid.*, p.109.

[17]Quoted *ibid.*, p. 109.

He continues:

> *Now the fact that we can with no conceivable sequence of functions approach without limit a given infinity, certainly has something strange about it. For it would be ... completely counter to our intuition to suppose that there is necessarily a gap, for example around the line $y = x$. We can always fill this gap in our thoughts with curves which accompany the line $y = x$ to infinity.*[18]

However, he finds here

> *no irreconcilable conflict of the results of different forms of thought, but only one of the idea of a perhaps not very familiar but still not inaccessible spatial behaviour.*[19]

For du Bois-Reymond this only indicates the presence of "a gap in the analogy between ordinary and infinitary quantities", the manifestation of "a behaviour peculiar to the infinitary domain".[20]

In his book *Die allgemeine Functionentheorie* of 1882 du Bois-Reymond presents his views on the nature and existence of infinitesimals. He begins by stating that in the analysis of "continuous mathematical quantities", one begins with a "geometric quantity" and tries to relate other quantities to it.[21] So the finite decimals are assigned correlates on a segment, that is, "points". This correlation between finite decimals and points is then extended to infinite decimals by a limit process. But the totality of such points can never form a complete segment, since

> *points are just dimensionless, and therefore an arbitrarily dense sequence of points can never become a distance.*[22]

Here we see once again a rejection of the idea that the continuous is reducible to the discrete. Consequently, a geometric segment must contain something other than finite and infinite decimals. These "others", according to du Bois-Reymond, are infinitesimal segments: there are infinitely many of these in any line segment, however short.

Du Bois-Reymond provides just a few rules of calculation for infinitesimal segments, reminiscent of those used by 17th and 18th mathematicians. To wit:

> *A finite number of infinitely small segments joined to one another do not form a finite segment, but again an infinitely small segment ... no upper bound can be specified either for the finite or for the infinitely small.*
>
> *I say two finite segments are equal when there is no* finite *difference between them ... Two finite quantities whose difference is infinitely small are equal to one another ... A finite quantity does not change if an infinitely small quantity is added to it or taken away from it.*[23]

[18]Quoted *ibid.*, p. 109.
[19]Quoted *ibid.*, p. 109.
[20]*Ibid.*, p. 110.
[21]*Ibid.*,p. 114.
[22]Quoted *ibid.*, p. 114.
[23]Quoted *ibid.*, p. 115.

While we may be incapable of forming a mental image of the *relation* of the infinitesimal to the finite, according to du Bois-Reymond we can visualize the infinitesimal *in itself*, and when we do we find that it behaves just like the finite:

The infinitely small is a mathematical quantity and has all its properties in common with the finite.[24]

The admission of the infinitesimal in relation to the *finite* opens the way to the infinitesimal in relation to the *infinitesimal* (so entailing, reciprocally, the presence of the infinitely large):

In this way, there arises a series of types of quantities, whose successive relation always is that a finite number of quantities of one kind never yields a quantity of the preceding kinds.[25]

Such quantities accordingly form a nonarchimedean domain. Moreover,

If within one and the same of these types of quantities, the properties of ordinary mathematical quantities hold, hence the same types of calculation as in the finite, then the comparison of the different types of quantities with each other is the object of the so-called infinitary calculus. *This calculus reckons with the relations of the infinitely large or infinitely small from type to type, and these types show connections with each other that do not fall under the ordinary concept of equality. The passages of one type into another do not show, for example, the continuity of change of mathematical quantities, although no jump changes result.*[26]

Du Bois-Reymond concludes his musings on the infinitesimal with the observation that there is an imbalance between belief in the infinitely large and belief in the infinitely small. A majority of educated people, he says, will admit an "infinite" (i.e., actual infinite) in space and time, and not just an "unboundedly large" (i.e., potential infinite). But only with difficulty will they accept the infinitely small, despite the fact that it has the same "right to existence" as the infinitely large.[27] In sum,

A belief in the infinitely small does not triumph easily. Yet when one thinks boldly and freely, the initial mistrust will soon mellow into a pleasant certainty.[28] ... *Were the sight of the starry sky lacking to mankind; had the race arisen and developed troglodytically in enclosed spaces; had its scholars, instead of wandering through the distant places of the universe telescopically, only looked for the smallest constituents of form and so were used in their thoughts to advancing into the boundless in the direction of the unmeasurably small: who would doubt that then the infinitely small would take the same place in our system of concepts that the infinitely large does now? Moreover, hasn't the attempt in mechanics to go back down to the smallest active elements long ago introduced into science the atom, the embodiment of the infinitely small? And don't as always skilful attempts to make it*

---

[24]Quoted *ibid.*, p. 115. Du Bois-Reymond took a dim view of the conception of infinitesimals as being ordinary magnitudes continually in a state of flux towards zero, remarking sarcastically

As long as the book is closed there is perfect repose, but as soon as I open it there commences a race of all the magnitudes which are provided with the letter d towards the zero limit. (Quoted in Ehrlich (1994), pp. 9–10.)

[25]Quoted *ibid.*, p. 115.

[26]Quoted *ibid.*, p. 115.

[27]*Ibid.*, p. 116.

[28]*Ibid.*, p. 116.

> *superfluous for physics face with certainty the same fate as Lagrange's battle against the differential?*[29]

Du Bois-Reymond was, indeed, a doughty champion of the infinitesimal.

5.2 Veronese

While du Bois-Reymond's conception of the infinite and infinitesimal derived from his work as an analyst, that of the second of Cantor's critical targets, **Giuseppe Veronese** (1854–1917) originated in geometry. An outstanding member of the Italian school of geometry in the last quarter of the nineteenth century, Veronese in 1891 published his exhaustive work on the foundations of geometry, whose title in approximate English translation reads: *Foundations of geometry of several dimensions and several kinds of linear unit, presented in elementary form*. In this work Veronese develops n-dimensional projective geometry, including non-Euclidean geometries, in a synthetic and unified way from first principles. Controversially, he also introduces "non-Archimedean" geometries containing both infinitesimal and infinitely large segments. On publication this work attracted the scathing criticism not only of Cantor, but also of Peano and Killing. Yet Hilbert later called it "profound" and incorporated some of Veronese's ideas into his own later *Grundlagen der Geometrie*.

As a geometer Veronese naturally took an essentially geometric view of the continuum. He begins his *Foundations* with a complaint about the use of real numbers as the basis of geometry. Spatial intuition, he says, is what furnishes us with the basal geometric objects and their inherent properties, so that the proper procedure in geometry is a synthetic one,

> *which always treats figures as figures, works directly with the elements of the figures and separates and unites them so that each truth and each step of a proof is accompanied as far as possible by intuition.*[30]

In answer to the question "What is the *continuum*?" Veronese writes, in striking contrast with Cantor:

> *This is a word whose meaning we understand without any mathematical definition, since we* intuit *the continuum in its simplest form as the common characteristic of many concrete things, such as, for example, to give some of the simplest, the* time *and the* place occupied *in the external neighbourhood of the object sketched here, or by a plumb line, if one takes no account of its physical properties and its thickness (in the empirical sense). Noting the particulars of this intuitive continuum, we should approach an abstract definition of the continuum in which intuition or perceived representation of it doesn't enter any more as a necessary part, in such a way that, conversely, this definition can serve abstractly, with complete logical rigour, for the deduction of other properties of this intuitive continuum.*

---

[29]Quoted *ibid.*, p. 116.

[30]Quoted in Fisher (1994), p. 135.

That one can give this mathematically abstract definition, we shall see later. On the other hand, if the definition of the continuum is not merely nominal and we want it instead to conform to the intuitive one, it must clearly arise from investigating the intuitive one, even if later the abstract definition, conforming to mathematically possible principles, contains this continuum as a special case.[31]

Veronese continues by considering a rectilinear continuum *L*, which, "within certain limits of obsevation", is seen to be divided into a sequence of consecutive identical parts *a, b, c, d, etc.*, placed from left to right:

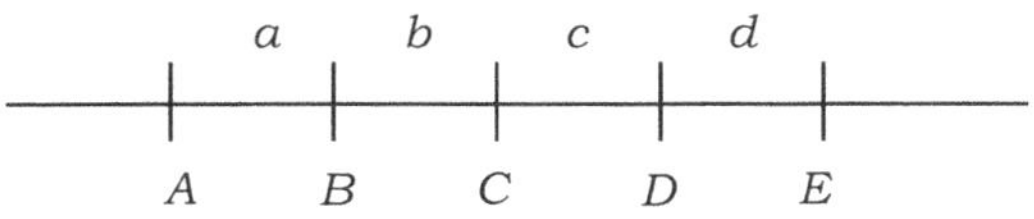

He continues:

We see further that we can experimentally (that is, with a bounded natural sequence [i.e., a finite set] *of decompositions) as well as abstractly (that is, according to any mathematically possible hypothesis or operation which doesn't contradict the results of experience) arrive at a part which is not further decomposable into parts* (indivisible) *of which the continuum is composed (as an instant is for time).*

It is then experience itself which moves us to look for the indivisible in such a way that we cannot obtain it experimentally, because it shows that a part considered indivisible with respect to one observation is not indivisible with respect to other observations with more exact instruments or under other conditions. If we assume that an indivisible part exists, we see that we can also experimentally consider it indeterminate, and therefore smaller than any given part of the rectilinear object.[32]

Veronese thus considers that any given linear continuum *L* contains what may be termed "relative" indivisibles, i.e. parts which are, *with respect to a given means of observation*, smaller than any other given part. Any such indivisible *I* will be infinitesimal by comparison with the whole continuum *L*. On the other hand *I*, as a part of *L*, is itself a continuum and so subject to the same analysis as was the latter. So a more acute observational technique will yield parts *J* of *I* which are indivisible with respect to that technique and infinitesimal relative to *J*. Clearly no relative indivisible can be a mathematical point, since points are *absolutely* indivisible.

In fact, for Veronese points are nothing more than signs indicating "positions of the uniting of two parts" of a (rectilinear) continuum. They are, as they were for Aristotle,

---

[31]Quoted *ibid.*, pp. 136–137. In a footnote Veronese observes:

In order to establish the mathematical concepts, we can very well fall back on empirically obtained knowledge without therefore having to make any use of it later in the definitions themselves and in the proof.

[32]Quoted *ibid.*, p. 138.

> *...a product of the function of abstracting in our mind... not parts of the rectilinear object.*[33]

To elucidate the nature of points Veronese offers two thought experiments. In the first of these it is supposed that

> *...the part a of the rectilinear object is painted red, the remaining part a′ white, and suppose further that there is no other colour between the white and the red. That which separates the white from the red can be coloured neither white nor red, and therefore cannot be a part of the object, since by assumption all its parts are white or red. And this sign of separation of uniting can be considered as belonging either to the white or to the red, if one considers them independently of one another. If we now abstract from the colours, we can assume that the sign of separation between the parts a and a′ belongs to the object itself.*[34]

Accordingly a point can belong to a continuum thorough "assignment", but cannot be a part of it.[35]

In the second thought experiment, Veronese invites the reader to

> *cut a very fine thread at the place indicated by X with the blade of an extremely sharp knife,* [so that] *the two parts a and a′ separate* [fig. 1] *and we assume that one can put the thread back together without seeing where the cut was* [fig. 2], *in other words, without a particle of the thread being lost.*

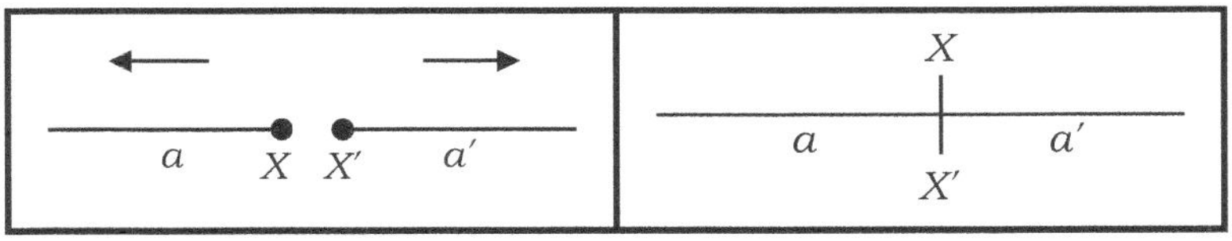

Figure 1 **Figure 2**

> *One produces this, apparently, if one looks at the thread from a certain distance. If one now considers the part a from right to left as the arrow above indicates, then what one sees of the cut is surely not part of the thread, just as what one sees from a body is not part of the body itself. It happens analogously if one looks at the part a′ from left to right. If the sign of separation X of the parts a and a′, which by assumption belongs to the thread itself, were part of the thread, then looking at a from right to left, one would not see all of this part, since that which separates the part a from a′ is only that which one sees in the way indicated above when one supposes the thread put back together.*[36]

From the "hypothesis" that the point is not part of the linear continuum, and is itself partless, Veronese then draws the inference:

---

[33] Quoted *ibid.*, p. 138.

[34] Quoted *ibid.*, p. 139

[35] *Ibid.*

[36] *Ibid.*, pp. 139–40.

that all the points we can imagine in it, however many that may be, do not constitute *the continuum when they are joined together and choosing a part (XX′)as small as one wants of the object (for time, an instant), however indeterminate, which is to say without X and X′ being fixed in our thoughts, intuition tells us that* this part is always continuous.[37]

As presented in the *Foundations,* the linear continuum is subject to what Veronese calls "the hypothesis on the existence of bounded infinitely large segments", namely that, if any segment is selected as unit, and one generates the *scale* based on it, consisting of the multiples of the unit segment by natural numbers, there is always an element of the continuum lying beyond the region covered by this scale.[38] In connection with this hypothesis Veronese writes:

In order to distinguish the segments bounded by ends which generate the region of a scale with arbitrary unit (AA_1) from those which don't generate the scale and are larger than them, we call the first finite *and the* second actually infinitely large or infinitely large *with respect to the unit [of the scale]. However, if the second is smaller than the first, we call it* actually infinitely small *or* infinitely small *with respect to the given unit. For example, the unit (AA_1) or an arbitrary bounded segment of a given scale is infinitely small with respect to an infinitely large segment (AA_∞).*[39]

Veronese's segments observe the expected order relations: for example, a segment is either finite or infinitely small or large with respect to a given segment, and the sum of two segments sharing just an endpoint which are finite or infinitely small or large with respect to a given segment bear the same relation to that segment.[40]

Veronese contrasts his own account of the continuum with that of Cantor and Dedekind in the following words:

Cantor and Dedekind...assert in their valuable works that ... the one-one relation between the points of [a] line and the points forming the real continuum is arbitrary. They certainly obtain this continuum by means of a sequence of abstract definitions of symbols which, although possible, are arbitrary... According to Dedekind, the numerical continuum is necessary in order to clarify the idea of the continuum of space. According to us, however, it is the intuitive rectilinear continuum which, by means of a point without parts, that serves to give us abstract definitions with respect to the continuum itself, of which the numerical continuum is only a special case. In this way, the definitions appear not as a force which keeps our mind in check but finds its complete justification in the perceptual representation of the continuum. One must take some account of this representation in the discussion of basic concepts, but without leaving the field of pure mathematics. Moreover, it would be truly marvellous if an abstract form as complicated as the numerical continuum obtained not only without being guided by the intuitive, but, as is done nowadays by some authors, from mere definitions of symbols, should then find itself in agreement with a representation as simple and primitive as that of the rectilinear continuum.[41]

Veronese's claim here is that Cantor and Dedekind's numerical continuum, which they regard as the "real" continuum, is itself no less arbitrary than the

---

[37]Quoted *ibid.*, p. 140.

[38]*Ibid.*, pp. 121–2.

[39]Quoted *ibid.*, p. 123. Here A_∞ is an element outside the scale generated by (AA_1).

[40]*Ibid.*, p. 123.

[41]Quoted *ibid.*, p. 142.

"arbitrary" correspondence they identify between the points of their continuum and those of a line. For Veronese the geometer it is the intuitive geometric continuum which must furnish the basis for any precise determination of the mathematical continuum, even though, as he says, such intuition "ought not [to] enter as a necessary component either in the statements of properties or of definitions, or in proofs."[42]

As for the points from which the Cantor-Dedekind discretized continuum is constructed, Veronese, again echoing Aristotle, has this to say:

> *The rectilinear continuum is independent of a system of points which we can imagine here. A system of points, if we think of a point as a sign of separation of two consecutive parts of a line or as the end of one of these parts, can never give the whole intuitive continuum, because a point has no parts. We find only that a system of points can represent* sufficiently *in geometrical investigations. The rectilinear continuum is never composed of its points but of segments, which the points join two by two, and which themselves are still continuous. In this way the* mystery *of continuity is pushed back from a given and constant part of the line to an indeterminate part as small as one likes, which is still always continuous, into which we are not permitted to enter with our representation. ... But it is well to mention that mathematically this mystery has no influence, because for us a determination of the continuum by means of a well-defined ordered system of points is sufficient. On the other hand, one should observe that a determination by points is incidental, because we have the intuition of the continuum just as well without it. If in fact one considers a point to be without parts, then... even if we make the points of a line correspond to starting from an origin, we don't get the whole continuum.*[43]

Finally Veronese remarks that, as far as he knows, "it has not yet been demonstrated that there are discontinuous systems of points which satisfy all the properties of space given by experience."[44] Given the likelihood of his knowing of Cantor's 1882 demonstration that continuous motion was possible in discontinuous spaces (see above), it would seem that Veronese did not regard the possibility of continuous motion alone as constituting a sufficient condition for a domain to possess all the properties of the space of experience. In any case, Veronese says, even if the properties of empirical space could be fully reproduced by some discontinuous system of points, "this would say nothing against the continuity of space."

5.3 Brentano

In his later years the Austrian philosopher **Franz Brentano** (1838–1917) became preoccupied with the nature of the continuous. Much of Brentano's philosophy has its starting-point in Aristotelian doctrine, and his conception of the continuum constitutes no exception. Aristotle's theory of the continuum, it will be recalled,

---

[42]Quoted *ibid.*, p. 144.

[43]Quoted *ibid.*, pp. 142–3.

[44]Quoted *ibid.*, pp. 144.

rests upon the assumption that all change is continuous and that continuous variation of quality, of quantity and of position are inherent features of perception and intuition. Aristotle considered it self-evident that a continuum cannot consist of points. Any pair of unextended points, he observes, are such that they either touch or are totally separated: in the first case, they yield just a single unextended point, in the second, there is a definite gap between the points. Aristotle held that any continuum—a continuous path, say, or a temporal duration, or a motion—may be divided *ad infinitum* into other continua but not into what might be called "discreta"—parts that cannot themselves be further subdivided. Accordingly, paths may be divided into shorter paths, but not into unextended points; durations into briefer durations but not into unextended instants; motions into smaller motions but not into unextended "stations". Nevertheless, this does not prevent a continuous line from being divided at a point constituting the common border of the line segments it divides. But such points are, according to Aristotle, just *boundaries,* and not to be regarded as actual *parts* of the continuum from which they spring. If two continua have a common boundary, that common border unites them into a single continuum. Such boundaries exist only *potentially*, since they come into being when they are, so to speak, marked out as connecting parts of a continuum; and the parts in their turn are similarly dependent as parts upon the existence of the continuum.

In its fundamentals Brentano's account of the continuous is akin to Aristotle's. Brentano regards continuity as something given in perception, primordial in nature, rather than a mathematical construction. He held that the idea of the continuous is a fundamental notion abstracted from sensible intuition:

> *Thus I affirm that... the concept of the continuous is acquired not through combinations of marks taken from different intuitions and experiences, but through abstraction from unitary intuitions...Every single one of our intuitions—both those of outer perception as also their accompaniments in inner perception, and therefore also those of memory—bring to appearance what is continuous.*[45]

Brentano suggests that the continuous is brought to appearance by sensible intuition in three stages. First, sensation presents us with objects having parts that coincide. From such objects the concept of *boundary* is abstracted in turn, and then one grasps that these objects actually *contain* coincident boundaries. Finally, it becomes clear that this is all that is required in order to understand the concept of a continuum.

Continuity is manifested in sensation in a variety of ways. In *visual* sensation, we are presented with *extension*, something possessing length and breadth, and hence with something such that between any two of its parts, provided these are separated, there is a third part. Every sensation possesses a certain *qualitative* continuity in that the object presented in the sensation could have a given manifested quality (colour, for example) in a greater or less degree, and between any two degrees of that quality lies still another degree of that quality. Finally, each sensation manifests *temporal* continuity: this is most evident when we perceive something as moving or at rest.

---

[45]Brentano (1988), p. 6.

Brentano recognizes that continua have *qualities* which cause them to possess *multiplicity*—a continuum may manifest continuity in several ways simultaneously. This led him to classify continua into *primary* and *secondary*: a secondary continuum being one whose manifestation is dependent upon another continuum. Here is Brentano himself on the matter:

> *Imagine, for example, a coloured surface. Its colour is something from which the geometer abstracts. For him there comes into consideration only the constantly changing manifold of spatial differences. But the colour, too, appears extended with the spatial surface, whether it manifests no specific colour-differences of its own—as in the case of a red colour which fills out a surface uniformly—or whether it varies in its colouring—perhaps in the manner of a rectangle which begins on one side with red and ends on the other side with blue, progressing uniformly through all colour-differences from violet to pure blue in between. In both cases we have to do with a multiple continuum, and it is the spatial continuum which appears thereby as primary, the colour-continuum as secondary. A similar double continuum can also be established in the case of a motion from place to place or of a rest, in which case it is a temporal continuum as such that is primary, the temporally constant or varying place that is the secondary continuum. Even when one considers a boundary of a mathematical body as such, for example a curved or straight line, a double continuity can be distinguished. The one presents itself in the totality of the differences of place that are given in the line, which always grows uniformly, whether in the case of straight, bent, or curved lines, and is that which determines the length of the line. The other resides in the direction of the line, and is either constant or alternating, and may vary continuously, or now more strongly, now less. It is constant in the case of the straight line, changing in the case of the broken line, and continuously varying in every line that is more or less curved. The direction-continuum here is to be compared with the colour-continuum discussed earlier and with the continuum of place in the case of rest or motion of a corporeal point in time. In the double continuum that presents itself to us in the line it is this continuum of directions that is to be referred to as the secondary, the manifold of differences of place as such as the primary continuum.*[46]

For Brentano the essential feature of a continuum is its inherent capacity to engender boundaries, and the fact that such boundaries can be grasped as coincident. Boundaries themselves possess a quality which Brentano calls *plerosis* ("fullness"). Plerosis is the measure of the number of directions in which the given boundary actually bounds. Thus, for example, within a temporal continuum the endpoint of a

[46]*Ibid.*, p. 21 f. Brentano's distinction of primary and secondary continua can be neatly represented within *category theory*: to put it succinctly, a *primary continuum* is a *domain*, a *secondary continuum* a *codomain*. We form a category $\mathscr{C}$—the *category of continua*—by taking continua as objects and *correlations* between continua as arrows. Then, given any arrow $f: A \rightarrow B$ in $\mathscr{C}$, the domain A of f may be taken as a "primary" continuum and its codomain B as a "secondary" continuum. In Brentano's example of a coloured surface, for instance, the primary continuum A is the given spatial surface, the secondary continuum B is the colour spectrum, and the correlation f assigns to each place in A its colour as a position in B. In the case of a corporeal point moving in space, the primary continuum A is an interval of time, the secondary continuum B a region of space, and the correlation f assigns to each instant in A the position in B occupied by the corporeal point. Finally, in the case of the varying direction of a curve the primary continuum A is the curve itself, the secondary continuum is the continuum of measures of angles, and the correlation f assigns to each point on the curve the slope of the tangent there: thus f is nothing other than the *first derivative* of the function associated with the curve.

past episode or the starting point of a future one bounds in a single direction, while the point marking the end of one episode and the beginning of another may be said to bound doubly. In the case of a spatial continuum there are numerous additional possibilities: here a boundary may bound in all the directions of which it is capable of bounding, or it may bound in only some of these directions. In the former case, the boundary is said to exist in *full plerosis*; in the latter, in *partial plerosis*. Brentano writes:

> *...the spatial nature of a point differs according to whether it serves as a limit in all or only in some directions. Thus a point located inside a physical thing serves as a limit in all directions, but a point on a surface or an edge or a vertex serves as a limit in only some direction. And the point in a vertex will differ in accordance with the directions of the edges that meet at the vertex... I call these specific distinctions differences of plerosis. Like any manifold variation, plerosis admits of a more and a less. The plerosis of the centre of a cone is more complete than that of a point on its surface; the plerosis of a point on its surface is more complete than that of a point on its edge, or that of its vertex. Even the plerosis of the vertex is the more complete the less the cone is pointed.*[47]

Brentano believed that the concept of plerosis enabled sense to be made of the idea that a boundary possesses "parts", even when the boundary lacks dimensions altogether, as in the case of a point. Thus, while the present or "now" is, according to Brentano, temporally unextended and exists only as a boundary between past and future, it still possesses two "parts" or aspects: it is both the end of the past and the beginning of the future. It is worth mentioning that for Brentano it was not just the "now" that existed only as a boundary; since, like Aristotle he held that "existence" in the strict sense means "existence *now*", it necessarily followed that existing things exist only as boundaries of what has existed or of what will exist, or both.

Brentano ascribes particular importance to the fact that points in a continuum can *coincide*. On this matter he writes:

> *Various other thorough studies could be* made [on the continuum concept] *such as a study of the impossibility of adjacent points and the* possibility of coincident points, *which have, despite their coincidence, distinctness and full relative independence.* [This] *has been and is misunderstood in many ways. It is commonly believed that if four different-coloured quadrants of a circular area touch each other at its centre, the centre belongs to only one of the coloured surfaces and must be that colour only. Galileo's judgment on the matter was more correct; he expressed his interpretation by saying paradoxically that the centre of the circle has as many parts as its periphery. Here we will only give some indication of these studies by commenting that everything which arises in this connection* follows from the point's relativity as involves a continuum and the fact that it is essential for it to belong to a continuum. *Just as the possibility of the coincidence of different points is connected with that fact, so is the existence of a point in diverse or more or less perfect* plerosis. *All of this is overlooked even today by those who understand the continuum to be an actual infinite multiplicity and who believe that we get the concept not by abstraction from spatial and temporal intuitions but from the combination of fractions between numbers, such as between 0 and 1.*[48]

---

[47]Quoted *ibid.*, p. xvii.

[48]Brentano (1974), p. 357.

Brentano's doctrines of plerosis and coincidence of points are well illustrated by applying them to the traditional philosophical problem of the initiation of motion: if a thing begins to move, is there a last moment of its being at rest or a first moment of its being in motion? The usual objection to the claim that both moments exist is that, if they did, there would be a time between the two moments, and at that time the thing could be said neither to be at rest nor to be in motion—in violation of the law of excluded middle. Brentano's response would be to say that both moments do exist, but that they *coincide*, so that there are no times between them; the violation of the law of excluded middle is thereby avoided. More exactly, Brentano would assert that the temporal boundary of the thing's being at rest—the end of its being at rest—is the same as the temporal boundary of the thing's being in motion—the beginning of its being in motion—, but the boundary is *twofold* in respect of its plerosis. The boundary is, in fact, in half plerosis at rest and in half plerosis in motion.

Brentano took a dim view of the efforts of mathematicians to construct the continuum from numbers. His attitude varied from rejecting such attempts as inadequate to according them the status of "fictions".[49] This is not surprising given his Aristotelian inclination to take mathematical and physical theories to be genuine descriptions of empirical phenomena rather than idealizations: in his view, if such theories were to be taken as literal descriptions of experience, they would amount to nothing better than "misrepresentations". Indeed, Brentano writes:

> *We must ask those who say that the continuum ultimately consists of points what they mean by a point. Many reply that a point is a cut which divides the continuum into two parts. The answer to this is that a cut cannot be called a thing and therefore cannot be a presentation in the strict and proper sense at all. We have, rather, only presentations of contiguous parts. ... The spatial point cannot exist or* be conceived of *in isolation. It is just as necessary for it to belong to a spatial continuum as for the moment of time to belong to a temporal continuum.*[50]

Concerning Poincaré's approach to the continuum[51] Brentano has this to say:

> *Poincaré ... follows extreme empiricists in the in the area of sensory psychology and therefore does not believe that there is granted to us an intuition of a continuous space. Poincaré's entire mode of procedure reveals that he also denies that we are in possession of an intuition of a continuous time. We saw how first of all he inserted between 0 and 1 fractions having a whole number as numerator and a whole power of 2 as denominator. In similar fashion, he then inserted all proper fractions whose denominator is a whole power of 3, and then also all those whose denominators are powers of every other whole number. He obtained thereby a series containing all rational fractions which, as he said, already has a certain continuity about it. He then inserted ... a series of irrational fractions. To these one now adds the series of fractions involving transcendental ratios... . Poincaré was prepared to admit that this process will never come to an end.... But he believed that he could be satisfied with the insertions already made. And nothing is more self-evident than*

[49]In a letter to Husserl drafted in 1905, Brentano asserts that "I regard it as absurd to interpret a continuum as a set of points."

[50]Brentano (1974), p. 354.

[51]See below.

that we have here a confession that the attempt to obtain a true continuum in this way has broken down.[52]

Dedekind's account of the continuum does not fare much better:

Dedekind differs from Poincaré already in the fact that he does not wish to deny that we have an intuition of a continuum—he simply does not want to make any use thereof. ... Dedekind's and Poincaré's constructions share in common that they fail to recognise the essential character of the continuum, namely that it allows the distinguishing of boundaries, which are nothing in themselves, but yet in conjunction make a contribution to the continuum. Dedekind believes that either the number ½ forms the beginning of the series ½ to 1, so that the series 0 to ½ would thereby be spared a final member, i.e. an end point which would belong to it, or conversely. But this is not how things are in the case of a true continuum. Rather it is the case that, when one divides a line, every part has a starting point, but in half plerosis.[53] *... If a red and a blue surface are in contact with each other then a red and a blue line coincide, each with different plerosis. And if a circular area is made up of three sectors, a red, a blue and a yellow, then the mid-point is a whole which consists to an equal extent of a red, a blue and a yellow part. According to Dedekind this point would belong to just one of the three colour-segments, and we should have to say that it could be separated from this while the segment in question remained otherwise unchanged. Indeed, the whole circular surface would then be conceivable as having been deprived of its mid-point, like Dedekind's number-series from which only the number ½ has fallen away. One sees immediately that this is absurd if one keeps in mind that the true concept of the continuum is obtained through abstraction from an intuition, and thus also that the entire conception has missed its target.*[54]

In conclusion,

One sees that in this entire putative construction of the concept of what is continuous the goal has been entirely missed; for that which is above all else characteristic of a continuum, namely the idea of a boundary in the strict sense (to which belongs the possibility of a coincidence of boundaries), will be sought after entirely in vain. Thus also the attempt to have the concept of what is continuous spring forth out of the combination of individual marks distilled from intuition is to be rejected as entirely mistaken, and this implies further that what is continuous must be given to us in individual intuition and must therefore have been extracted therefrom.[55]

Brentano's analysis of the continuum centred on its phenomenological and qualitative aspects, which are by their very nature incapable of reduction to the discrete. Brentano's rejection of the mathematicians' attempts to construct it in discrete terms is thus hardly surprising.

---

[52]Brentano (1988), p. 39.

[53]Here Brentano appears to be saying that when one divides a closed interval [*a*, b] at an intermediate point *c*, one necessarily obtains the closed intervals [*a*, *c*], [*c*, *b*], with the common point *c* (in half plerosis). In that case, Brentano have probably have regarded a continuous line as indecomposable, into disjoint intervals at least: see Appendix A.

[54]Brentano (1988), pp. 40–41. That Brentano considered "absurd" the idea of removing a single point from a continuum seems to indicate that his continuum has the same "syrupy" property as those of intuitionistic and smooth infinitesimal analysis. See Chaps. 9 and 10 below.

[55]Brentano (1988), pp. 4–5.

5.4 Peirce

The American philosopher-mathematician **Charles Sanders Peirce**'s (1839–1914) view of the continuum was, in a sense, intermediate between that of Brentano and the arithmetizers. Like Brentano, he held that the cohesiveness of a continuum rules out the possibility of it being a mere collection of discrete individuals, or points, in the usual sense:

> *The very word* continuity *implies that the instants of time or the points of a line are everywhere welded together.*
>
> *[The] continuum does not consist of indivisibles, or points, or instants, and does not contain any except insofar as its continuity is ruptured.*[56]

And even before Brouwer[57] Peirce seems to have been aware that a faithful account of the continuum will involve questioning the law of excluded middle:

> *Now if we are to accept the common idea of continuity ... we must either say that a continuous line contains no points or ... that the principle of excluded middle does not hold of these points. The principle of excluded middle applies only to an individual ... but places being mere possibilities without actual existence are not individuals.*[58]

But Peirce also held that any continuum harbours an *unboundedly large* collection of points—in his colourful terminology, a *supermultitudinous* collection—what we would today call a *proper class*. Peirce maintained that if "enough" points were to be crowded together by carrying insertion of new points between old to its ultimate limit they would—through a *logical* "transformation of quantity into quality"—lose their individual identity and become fused into a true continuum.[59] Here are his observations on the matter:

---

[56]Peirce (1976), p. 925.

[57]See below.

[58]Peirce (1976), p. xvi: the quotation is from a note written in 1903.

[59]In their Introduction to Peirce (1992), Ketner and Putnam characterize Peirce's conception of the continuum as "a possibility of repeated division which can never be exhausted in any possible world, not even in a possible world in which one can complete [nondenumerably] infinite processes." There is some resemblance between this conception and John Conway's system of *surreal numbers* (see Ehrlich 1994a). Conway's system may be characterized as being an η_α-field for every ordinal α, that is, a real-closed ordered field S which satisfies the condition that, for any pair of subsets X, Y for which every member of X is less than every member of Y, there is an element of S strictly between X and Y. (In their Introduction to Peirce [1992], Ketner and Putnam characterize Peirce's conception of the continuum as "a possibility of repeated division which can never be exhausted in any possible world, not even in a possible world in which one can complete [nondenumerably] infinite processes. This description would seem to apply equally well to Conway's conception.) It is not hard to show that, between any pair of members of S there is a *proper class* of members of S—in Peirce's terminology, a supermultitudinous collection. Nevertheless, S is still discrete: its elements, while supermultitudinous, remain distinct and unfused (were it not for this fact, Conway would scarcely be justified in calling the members of S "numbers"). On the face of it the discreteness of S would seem to imply that the presence of superabundant quantity in Peirce's sense is not enough to ensure continuity. Of course, Brentano would have dismissed this idea altogether, in view of his critical attitude towards any construction of the continuum by repeated insertion of points.

It is substantially proved by Euclid that there is but one assignable quantity which is the limit of a convergent series. That is, if there is an increasing convergent series, A *say, and a decreasing convergent series,* B *say, of which every approximation exceeds every approximation of* A*, and if there is no rational quantity which is at once greater than every approximation of* A *and less than every approximation of* B*, then there is but one surd quantity so intermediate...There is one surd quantity and only one for each convergent series, calling two series the same if their approximations all agree after a sufficient number of terms, or if their difference approximates toward zero. But this is only to say that the multitude of surds equals the multitude of denumerable sets of rational numbers which is... the* primipostnumeral[60] *multitude.*

...We remark that there is plenty of room to insert a secundipostnumeral *multitude of quantities between [a] convergent series and its limit. Any one of those quantities may likewise be separated from its neighbours, and we thus see that between it and its nearest neighbours there is ample room for a* tertiopostnumeral *multitude of other quantities, and so on through the whole denumerable series of postnumeral quantities.*

But if we suppose that all *such orders of systems of quantities have been inserted, there is no longer any room for inserting any more. For to do so we must select some quantity to be thus isolated in our representation. Now whatever one we take, there will always be quantities of higher order filling up the spaces on the two sides.*

We therefore see that such a supermultitudinous collection sticks together by logical necessity. Its constituent individuals are no longer distinct and independent subjects. They have no existence—no hypothetical existence—except in their relations to one another. They are not subjects, but phrases expressive of the properties of the continuum.

...Supposing a line to be a supermultitudinous collection of points, ... to sever a line in the middle is to disrupt the logical identity of the point there and make it two points. It is impossible to sever a continuum by separating the connections of the points, for the points only exist by virtue of those connections. The only way to sever a continuum is to burst it, that is, to convert what was one into two.[61]

Peirce's conception of the number continuum is also notable for the presence in it of an abundance of *infinitesimals*, a feature it shares with du Bois-Reymond's and Veronese's nonarchimedean number systems.[62] In defending infinitesimals, Peirce remarks that

It is singular that nobody objects to $\sqrt{-1}$ *as involving any contradiction, nor, since Cantor, are infinitely great quantities much objected to, but still the antique prejudice against infinitely small quantities remains.*[63]

[60]Peirce assumed what amounts to the generalized continuum hypothesis in supposing that each possible infinite set has one of the cardinalities $\aleph_0, 2^{\aleph_0}, 2^{2^{\aleph_0}}, \ldots$. These he termed *denumerable, primipostnumeral, secundipostnumeral,* etc.

[61]Peirce (1976), p. 95.

[62]I do not know whether Peirce was acquainted with their work.

[63]Peirce (1976), p. 123. In this connection it is worth quoting from a letter addressed by Peirce in 1900 to the editor of *Science* in which he defends his views on infinitesimals against the strictures of Josiah Royce:

Professor Royce remarks that my opinion that differentials may quite logically be considered as true infinitesimals, if we like, is shared by no mathematician "outside of Italy". *As a logician, I am more comforted by corroboration in the clear mental atmosphere of Italy than I could be by any seconding from a tobacco-clouded and bemused land (if any such there be) where no philosophical eccentricity misses its champion, but where sane logic has not found favor.*

Peirce actually held the view that the conception of infinitesimal is suggested by introspection—that the *specious present* is in fact an infinitesimal:

> *It is difficult to explain the fact of memory and our apparently perceiving the flow of time, unless we suppose immediate consciousness to extend beyond a single instant. Yet if we make such a supposition we fall into grave difficulties, unless we suppose the time of which we are immediately conscious to be strictly infinitesimal.*[64]

> *We are conscious of the present time, which is an instant, if there be any such thing as an instant. But in the present we are conscious of the flow of time. There is no flow in an instant. Hence, the present is not an instant.*[65]

Peirce championed the retention of the infinitesimal concept in the foundations of the calculus, both because of what he saw as the efficiency of infinitesimal methods, and because he regarded infinitesimals as constituting the "glue" causing points on a continuous line to lose their individual identity.

5.5 Poincaré

The idea of continuity played a central role in the thought of the great French mathematician **Henri Poincaré** (1854–1912). But sorting out his views on the continuum, concerning which he made numerous scattered remarks, is by no means an easy task. Indeed there seems to be an inconsistency in his attitude towards the set-theoretical, or arithmetized, continuum. On the one hand, he rejected actual infinity and impredicative[66] definition—both cornerstones of the Cantorian theory of sets which underpins the construction of the arithmetized continuum. And yet in his mathematical work he employs variables ranging over all the points of an interval of the set-theoretical continuum, and he "accepts the standard account of the least upper bound, which is impredicative."[67] But beneath this apparent inconsistency lies his belief that what ultimately underpins mathematics, creating its linkage with objective reality, is *intuition*—that "intuition is what bridges the gap between symbol and reality."[68] His view of the continuum, in particular, is informed by this credo. For Poincaré the continuum and the range of points on it is grasped in intuition in

[64]*Ibid.*, p. 124.

[65]*Ibid.*, p. 925.

[66]Impredicativity is a form of circularity: a definition of a term is impredicative if it contains a reference to a totality to which the term under definition belongs. See, e.g., Fraenkel et al. (1973), pp. 193–200.

[67]Folina (1992), p. xv.

[68]*Ibid.*, p. 113. And yet Poincaré also remarks, in connection with the continuous nowhere differentiable functions of analysis:

> *Instead of seeking to reconcile intuition with analysis, we have been content to sacrifice one of the two, and as analysis must remain impeccable, we have decided against intuition* (1946, p. 52).

something like the Kantian sense, and yet the continuum cannot be treated as a completed mathematical object, as a "mere set."[69]

Of the arithmetical continuum Poincaré remarks:

The continuum so conceived is only a collection of individuals ranged in a certain order, infinite to one another, it is true, but exterior *to one another. This is not the ordinary conception, wherein is supposed between the elements of the continuum a sort of intimate bond which makes of them a whole, where the point does not exist before the line, but the line before the point. Of the celebrated formula "the continuum is unity in multiplicity", only the multiplicity remains, the unity has disappeared. The analysts are none the less right in defining the continuum as they do, for they always reason on just this as soon as they pique themselves on their rigor. But this is enough to apprise us that the veritable mathematical continuum is a very different thing from that of the physicists and the metaphysicians.*[70]

But despite Poincaré's apparent acceptance of the arithmetic definition of the continuum, he questions the fact that (as with Dedekind and Cantor's formulations) the (irrational) numbers so produced are mere symbols, detached from their origins in intuition:

But to be content with this [fact] would be to forget too far the origin of these symbols; it remains to explain how we have been led to attribute to them a sort of concrete existence, and, besides, does not the difficulty begin even for the fractional numbers themselves? Should we have the notion of these numbers if we had not known a matter that we conceive as infinitely divisible, that is to say, a continuum?[71]

That being the case, Poincaré asks whether the notion of the mathematical continuum is "simply drawn from experience." To this he responds in the negative, for the reason that our sensations, the "raw data of experience", cannot be brought under an acceptable scheme of measurement:

It has been observed, for example, that a weight A of 10 grams and a weight B of 11 grams produce identical sensations, that the weight B is just as indistinguishable from a weight C of 12 grams, but that the weight A is easily distinguished from the weight C. Thus the raw results of experience may be expressed by the following relations:

$$A = B, B = C, A < C,$$

which may be regarded as the formula of the physical continuum.

According to Poincaré it is the "intolerable discord with the principle of contradiction" of this formula[72] which has forced the invention of the mathematical continuum. This latter is obtained in two stages. First, formerly indistinguishable terms are distinguished and a new term, indistinguishable from both, inserted between them. Repeating this procedure indefinitely gives rise to what Poincaré calls a *first-order*

[69]Folina (1992), p. xvi.

[70]Poincaré (1946), pp. 43–44.

[71]*Ibid.*, pp. 45–6.

[72]This formula ceases to be contradictory if the identity relation $=$ is replaced by a symmetric, reflexive, but *nontransitive* relation $\approx$: here $x \approx y$ is taken to assert that the sensations or perceptions x and y are indistinguishable. See Appendix C below.

continuum, in essence the rational number line. A second stage now becomes necessary because two first-order continua, for example the diagonal of a square and its inscribed circle, need not intersect. This second stage, in which are added all possible "boundary" points between first-order continua leads to the second-order or mathematical continuum. Here is how Poincaré describes the process:

> *But conceive of a straight line divided into two rays. Each of these rays will appear to our imagination as a band of a certain breadth; these bands moreover will encroach one on the other, since there must be no interval between them. The common part will appear to us as a point which will always remain when we try to imagine our bands narrower and narrower, so that we admit as an intuitive truth that if a straight line is cut into two rays their common boundary is a point; we recognize here the conception of Dedekind, in which an incommensurable number was regarded as the common boundary of two classes of rational numbers.*
>
> *Such is the origin of the continuum of second order, which is the mathematical continuum so called.*[73]

Poincaré goes on to discuss continua of higher dimensions. To obtain these he considers *aggregates* of sensations. As with single sensations, any given pair of these aggregates may or may not be distinguishable. He remarks that, while these aggregates, which he terms *elements*, are analogous to mathematical points, they are not in fact quite the same thing, for

> *we cannot say that our element is without extension, since we cannot distinguish it from neighbouring elements and it is thus surrounded by a sort of haze. If the astronomical comparison may be allowed, our 'elements' would be like nebulae, whereas the mathematical points would be like stars.*[74]

This leads to a definition of a physical continuum:

> *a system of elements will form a* continuum *if we can pass from any one of them to any other, by a series of consecutive elements such that each is indistinguishable from the preceding. This* linear series *is to the* line *of the mathematician what an isolated* element *was to the point.*[75]

Poincaré defines a *cut* in a physical continuum C to be a set of elements removed from it "which for an instant we shall regard as no longer belonging to this continuum." Such a cut may happen to subdivide C into several distinct continua, in which case C will contain two distinct elements A and B that must be regarded as belonging to two distinct continua. This becomes necessary

[73] *Ibid.*, p. 49.

[74] *Ibid.*, p. 52.

[75] *Ibid.*

> *because it will be impossible to find a linear series of consecutive elements of C, each of these elements indistinguishable from the preceding, the first being A and the last B,* without one of the elements of this series being indistinguishable from one of the elements of the cut.[76]

On the other hand, it may happen that the cut fails to subdivide the continuum *C*, in which case it becomes necessary to determine precisely which cuts will subdivide it. Poincaré calls a continuum *one-dimensional* if it can be subdivided by a cut which reduces to a finite number of elements all distinguishable from one another (and so forming neither a continuum nor several continua). When *C* can be subdivided only by cuts which are themselves continua, *C* is said to possess several dimensions:

> *If cuts which are continua of one dimension suffice, we shall say that C has two dimensions; if cuts of two dimensions suffice, we shall say that C has three dimensions, and so on.*[77]

Thus is defined the concept of a multidimensional physical continuum, based on "the very simple fact that two aggregates of sensations are distinguishable or indistinguishable."

Unlike Cantor, Poincaré accepted the infinitesimal, even if he did not regard all of the concept's manifestations as useful. This emerges from his answer to the question: "Is the creative power of the mind exhausted by the creation of the mathematical continuum?". He responds:

> *No; the works of Du Bois-Reymond demonstrate it in a striking way. We know the mathematicians distinguish between infinitesimals and that those of second order are infinitesimal not only in an absolute way, but also in relation to those of first order. It is not difficult to imagine infinitesimals of fractional and even irrational order, and thus we find again that scale of the mathematical continuum which has been dealt with in the preceding pages.*
>
> *Further, there are infinitesimals which are infinitely small in relation to those of the first order, and, on the contrary, infinitely great in relation to those of order $1 + \varepsilon$, and that however small ε may be. Here, then, are new terms intercalated in our series ... I shall say that thus has been created a sort of continuum of the third order.*
>
> *It would be easy to go further, but that would be idle; one would only be imagining symbols without possible application, and no one would think of doing that. The continuum of the third order, to which the consideration of the different orders of infinitesimals leads, is itself not useful enough to have won citizenship, and geometers regard it as a mere curiosity. The mind uses its creative faculty only when experience requires it.*[78]

Poincaré's attitude towards the continuum resembles in certain respects that of the intuitionists (see below): while the continuum exists, and is knowable intuitively, it is not a "completed" set-theoretical object. It is geometric intuition, not set theory, upon which the totality of real numbers is ultimately grounded.

---

[76] *Ibid.*, p. 53.

[77] *Ibid.*

[78] *Ibid.*, pp. 50–1.

5.6 Brouwer

The Dutch mathematician **L. E. J. Brouwer** (1881–1966) is best known as the founder of the philosophy of (neo)*intuitionism*. Brouwer's highly idealist views on mathematics bore some resemblance to Kant's. For Brouwer, mathematical concepts are admissible only if they are adequately grounded in intuition, mathematical theories are significant only if they concern entities which are constructed out of something given immediately in intuition, and mathematical demonstration is a form of construction in intuition. Brouwer's insistence that mathematical proof be constructive in this sense required the jettisoning of certain received principles of classical logic, notably the *law of excluded middle*: the assertion that, for any proposition *p*, either *p* or not *p*. Brouwer maintained, in fact, that the applicability of the law of excluded middle to mathematics

> *was caused historically by the fact that, first, classical logic was abstracted from the mathematics of the subsets of a definite finite set, that, secondly, an a priori existence independent of mathematics was ascribed to the logic, and that, finally, on the basis of this supposed apriority it was unjustifiably applied to the mathematics of infinite sets.*[79]

Brouwer held that much of modern mathematics is based on an illicit extension of procedures valid only in the restricted domain of the finite. He therefore embarked on the radical course of jettisoning virtually all of the mathematics of his day—in particular the set-theoretical construction of the continuum—and starting anew, using only concepts and modes of inference that could be given clear intuitive justification. In the process it would become clear precisely what are the logical laws that intuitive, or constructive, mathematical reasoning actually obeys, making possible a comparison of the resulting *intuitionistic*, or *constructive logic* with classical logic.[80]

While admitting that the emergence of noneuclidean geometry had discredited Kant's view of space, Brouwer maintained, in opposition to the logicists (whom he called "formalists") that arithmetic, and so all mathematics, must derive from *temporal intuition*. In his own words:

> *Neointuitionism considers the falling apart of moments of life into qualitatively different parts, to be reunited only while remaining separated by time, as the fundamental phenomenon of the human intellect, passing by abstracting from its emotional content into the fundamental phenomenon of mathematical thinking, the intuition of the bare two-oneness. This intuition of two-oneness, the basal intuition of mathematics, creates not only the numbers one and two, but also all finite ordinal numbers, inasmuch as one of the elements of the two-oneness may be thought of as a new two-oneness, which process may be repeated indefinitely; this gives rise still further to the smallest infinite ordinal* ω *. Finally this basal intuition of mathematics, in which the connected and the separate, the continuous and the discrete are united, gives rise immediately to the intuition of the linear continuum, i.e., of the "between", which is not exhaustible by the interposition of new units and which can*

---

[79]Quoted in Kneebone (1963),p. 246.

[80]This is not to say that Brouwer was primarily interested in *logic*, far from it: indeed, his distaste for formalization caused him to be quite dismissive of subsequent codifications of intuitionistic logic.

> *therefore never be thought of as a mere collection of units. In this way the apriority of time does not only qualify the properties of arithmetic as synthetic a priori judgments, but it does the same for those of geometry, and not only for elementary two- and three-dimensional geometry, but for non-euclidean and n-dimensional geometries as well. For since Descartes we have learned to reduce all these geometries to arithmetic by means of coordinates.*[81]

Brouwer maintained that it is the awakening of awareness of the temporal continuum in the subject, an event termed by him "The Primordial Happening" or "The Primordial Intuition of Time", that engenders the fundamental concepts and methods of mathematics. In "Mathematics, Science and Language" (1929), he describes how the notion of number—the discrete—emerges from the awareness of the continuous:

> Mathematical Attention *as an act of the will serves the instinct for self-preservation of individual man; it comes into being in two phases;* time awareness *and* causal attention. *The first phase is nothing but the fundamental intellectual phenomenon of the falling apart of a moment of life into two qualitatively different things of which one is experienced as giving away to the other and yet is retained by an act of memory. At the same time this split moment of life is separated from the Ego and moved into a world of its own, the world of perception. Temporal twoity, born from this time awareness, or the two-membered sequence of time phenomena, can itself again be taken as one of the elements of a new twoity, so creating temporal threeity, and so on. In this way, by means of the self-unfolding of the fundamental phenomenon of the intellect, a* time sequence of phenomena is created of arbitrary multiplicity.[82]

But in his doctoral dissertation of 1907 Brouwer regards continuity and discreteness as complementary notions, neither of which is reducible to the other:

> *...We shall go further into the basic intuition of mathematics (and of every intellectual activity) as the substratum, divested of all quality, of any perception of change, a unity of continuity and discreteness, a possibility of thinking together several entities, connected by a "between", which is never exhausted by the insertion of new entities. Since continuity and discreteness occur as inseparable complements, both having equal rights and being equally clear, it is impossible to avoid [regarding] each one of them as a primitive entity... Having recognized that the intuition of continuity, of "fluidity" is as primitive as that of several things conceived as forming together a unit, the latter being at the basis of every mathematical construction, we are able to state properties of the continuum as "a matrix of points to be thought of as a whole"*[83]

In that work Brouwer states unequivocally that the continuum is not constructible from discrete points:

> *...The* continuum as a whole [*is*] *given to us by intuition; a construction for it, an action which would create from the mathematical intuition 'all' its points as individuals, is inconceivable and impossible.*[84]

[81] Brouwer, *Intuitionism and Formalism,* in Benacerraf and Putnam (1977), p. 80.

[82] Brouwer, *Mathematics Science and Language.* In Mancosu (1998), p.45

[83] Brouwer (1975), p. 17.

[84] *Ibid.* p. 45.

Later Brouwer was to modify this doctrine. In his mature thought, he radically transformed the concept of point, endowing points with sufficient fluidity to enable them to serve as generators of a "true" continuum. This fluidity was achieved by admitting as "points", not only fully defined discrete numbers such as $\sqrt{2}$, π, e, and the like—which have, so to speak, already achieved "being"—but also "numbers" which are in a perpetual state of becoming in that their the entries in their decimal (or dyadic) expansions are the result of free acts of choice by a subject operating throughout an indefinitely extended time. The resulting *choice sequences* cannot be conceived as finished, completed objects: at any moment only an initial segment of each is known.[85] In this way Brouwer obtained the mathematical continuum in a way compatible with his belief in the primordial intuition of time—that is, as an unfinished, indeed unfinishable entity in a perpetual state of growth, a "medium of free development". In this conception, the mathematical continuum is indeed "constructed", not, however, by initially shattering, as did Cantor and Dedekind, an intuitive continuum into isolated points, but rather by assembling it from a complex of continually changing overlapping parts.

The mathematical continuum as conceived by Brouwer displays a number of features that seem bizarre to the classical eye. For example, in the Brouwerian continuum the usual law of comparability, namely that for any real numbers a, b either $a < b$ or $a = b$ or $a > b$, fails. Even more fundamental is the failure of the law of excluded middle in the form that for any real numbers a, b, either $a = b$ or $a \neq b$. The failure of these seemingly unquestionable principles in turn vitiates the proofs of a number of basic results of classical analysis, for example the Bolzano-Weierstrass theorem, as well as the theorems of monotone convergence, intermediate value, least upper bound, and maximum value for continuous functions.[86]

While the Brouwerian continuum may possess a number of negative features from the standpoint of the classical mathematician, it has the merit of corresponding more closely to the continuum of intuition than does its classical counterpart. Hermann Weyl pointed out a number of respects in which this is so:

> *In accordance with intuition, Brouwer sees the essential character of the continuum, not in the relation between element and set, but in that between part and whole. The continuum falls under the notion of the 'extensive whole', which Husserl characterizes as that "which permits a dismemberment of such a kind that the pieces are by their very nature of the same lowest species as is determined by the undivided whole.*[87]

Far from being bizarre, the failure of the law of excluded middle for points in the intuitionistic continuum is seen by Weyl as "fitting in well with the character of the intuitive continuum":

---

[85]For an illuminating informal account of choice sequences, see Fraenkel et al. (1973), pp. 255–261.

[86]The failure of these important results of classical analysis caused most mathematicians of the day to shun intuitionistic, and even constructive mathematics. It was not until the 1960s that adequate constructive versions were worked out. See Chap. 9 below.

[87]Weyl (1949), p. 52.

> *For there the separateness of two places, upon moving them toward each other, slowly and in vague gradations passes over into indiscernibility. In a continuum, according to Brouwer, there can be only continuous functions.* The continuum is not composed of parts.[88]

For Brouwer had indeed shown, in 1924, that every function defined on a closed interval of the continuum as he conceived it is uniformly continuous.[89] As a consequence, the intuitionistic continuum is *indecomposable* or *cohesive,* it is what we have termed an *Aristotelian* continuum.[90] In contrast with a discrete entity, the cohesive Brouwerian continuum cannot be composed of its parts. Brouwer's vision of the continuum has in recent years become the subject of intensive investigation by logicians and category-theorists.

5.7 Weyl

Hermann Weyl (1885–1955), one of most versatile mathematicians of the twentieth century, was unusual among scientists in being attracted to idealist philosophy. In his youth he inclined towards the idealism of Kant and Fichte, and later came to be influenced by Husserl's phenomenology. His idealist leanings can be seen particularly in his work on the foundations of mathematics.

Towards the end of his *Address on the Unity of Knowledge*, delivered at the 1954 Columbia University bicentennial celebrations, Weyl enumerates what he considers to be the essential constituents of knowledge. At the top of his list[91] comes

> . . .*intuition, mind's ordinary act of seeing what is given to it.*[92]

Throughout his life Weyl held to the view that intuition, or *insight*, not *proof,* furnishes the ultimate foundation of *mathematical* knowledge. Thus in *Das Kontinuum* of 1918 he writes:

> *In the Preface to Dedekind (1888) we read that "In science, whatever is provable must not be believed without proof." This remark is certainly characteristic of the way most mathematicians think. Nevertheless, it is a preposterous principle. As if such an indirect concatenation of grounds, call it a proof though we may, can awaken any "belief" apart from assuring ourselves through immediate insight that each individual step is correct. In all*

---

[88] *Ibid.*, p. 54.

[89] One might be inclined to regard this claim as impossible: is not a counterexample provided by, for example, the function f given by $f(0) = 0, f(x) = |x|/x$ otherwise? No, because from the intuitionistic standpoint this function is not everywhere defined on the interval [1, 1], being undefined at those arguments x for which it is unknown whether $x = 0$ or $x \neq 0$.

[90] See Chap. 9 and Appendix A below.

[91] The others, in order, are: *understanding and expression; thinking the possible;* and finally, in science, *the construction of symbols or measuring devices.*

[92] Weyl (1954), p. 629.

cases, this process of confirmation—and not the proof—remains the ultimate source from which knowledge derives its authority; it is the "experience of truth".[93]

While Weyl held that the roots of mathematics lay in the intuitively given, he recognized at the same time that it would be unreasonable to require all mathematical knowledge to possess intuitive immediacy. In *Das Kontinuum*, for example, he says:

The states of affairs with which mathematics deals are, apart from the very simplest ones, so complicated that it is practically impossible to bring them into full givenness in consciousness and in this way to grasp them completely.[94]

But Weyl did not think that this fact furnished justification for extending the bounds of mathematics to embrace notions which cannot be given fully in intuition *even in principle* (e.g., the actual infinite). He held, rather, that this extension of mathematics into the transcendent—the realm of being not fully accessible to intuition—is necessitated by the fact that mathematics plays an indispensable role in the physical sciences, where intuitive evidence is *necessarily* transcended. As he says in *The Open World*:

... if mathematics is taken by itself, one should restrict oneself with Brouwer to the intuitively cognizable truths ... nothing compels us to go farther. But in the natural sciences we are in contact with a sphere which is impervious to intuitive evidence; here cognition necessarily becomes symbolical construction. Hence we need no longer demand that when mathematics is taken into the process of theoretical construction in physics it should be possible to set apart the mathematical element as a special domain in which all judgments are intuitively certain; from this higher standpoint which makes the whole of science appear as one unit I consider Hilbert to be right.[95]

In *Consistency in Mathematics* (1929), Weyl characterized the mathematical method as

the a priori *construction of the possible in opposition to the* a posteriori *description of what is actually given.*[96]

The problem of mapping the limits on constructing "the possible" in this sense occupied Weyl a great deal. He was greatly exercised by the concept of the mathematical *infinite*, which he believed to elude "construction" in the idealized sense of set theory. Again to quote a passage from *Das Kontinuum:*

No one can describe an infinite set other than by indicating properties characteristic of the elements of the set. ... The notion that a set is a "gathering" brought together by infinitely many individual arbitrary acts of selection, assembled and then surveyed as a whole by consciousness, is nonsensical; "inexhaustibility" is essential to the infinite.[97]

---

[93]Weyl (1987), p. 119.
[94]*Ibid.*, p. 17.
[95]Weyl (1932), p. 82.
[96]Weyl (1929), p. 249.
[97]Weyl (1987), p. 23.

It is the necessity of bridging the gap between mathematics and external reality that compels the former to embody a conception of the actual infinite, as Weyl attests towards the end of *The Open World:*

> *The infinite is accessible to the mind intuitively in the form of a field of possibilities open to infinity, analogous to the sequence of numbers which can be continued indefinitely, but the completed, the actual infinite as a closed realm of actual existence is forever beyond its reach. Yet the demand for totality and the metaphysical belief in reality inevitably compel the mind to represent the infinite as closed being by symbolical construction.*[98]

As a fundamental, but at the same time perplexing "possible" in mathematics, the *continuum* became the subject of what was arguably Weyl's most searching mathematico-philosophical analysis. In his *Philosophy of Mathematics and Natural Science* he reflects on what he calls the "inwardly infinite" nature of a continuum:

> *The essential character of the continuum is clearly described in this fragment of Anaxagoras: "Among the small there is no smallest, but always something smaller. For what is cannot cease to be no matter how small it is being subdivided." The continuum is not composed of discrete elements which are "separated from one another as though chopped off by a hatchet." Space is infinite not only in the sense that it never comes to an end; but at every place it is, so to speak, inwardly infinite, inasmuch as a point can only be fixed as step-by-step by a process of subdivision which progresses* ad infinitum. *This is in contrast with the resting and complete existence that intuition ascribes to space. The "open" character is communicated by the continuous space and the continuously graded qualities to the things of the external world. A real thing can never be given adequately, its "inner horizon" is unfolded by an infinitely continued process of ever new and more exact experiences; it is, as emphasized by Husserl, a limiting idea in the Kantian sense. For this reason it is impossible to posit the real thing as existing, closed and complete in itself. The continuum problem thus drives one to epistemological idealism. Leibniz, among others, testifies that it was the search for a way out of the "labyrinth of the continuum" which first suggested to him the conception of space and time as orders of phenomena.*[99]

Weyl identifies three attempts in the history of thought "to conceive of the continuum as Being in itself.".[100] These are, respectively, atomism, the infinitely small, and set theory. In Weyl's view, despite atomism's brilliant success in unravelling the structure of matter, it had failed in that regard as to space, time, and mathematical extension because it "never achieved sufficient contact with reality." As for the infinitely small, it was not so much supplanted as rendered superfluous by the limit concept. Weyl saw the limit concept as providing the necessary link between the microcosm of the infinitely small and the realm of macroscopic objects. Without that link, the fact that the microcosm is governed by "elementary laws" making for ease of calculation, would remain entirely useless in drawing conclusions about the macrocosm.[101]

---

[98]Weyl (1932), p. 83.

[99]Weyl (1949), p. 41.

[100]*Ibid.*, p. 42

[101]But in this connection see the remarks on the Constancy Principle in Chap. 10.

Weyl believed that the ground of mathematics lies in what he calls *constructive cognition*, which unfolds in the three stages:

1. *We ascribe to that which is given certain characters which are not manifest in the phenomena but are arrived at as the result of certain mental operations. It is essential that the performance of these operations be universally possible and that their result is held to be uniquely determined by the given. But it is not essential that the operations which define the character be actually carried out.*
2. *By the introduction of symbols the assertions are split so that one part of the operations is shifted to the symbols and thereby made independent of the given and its continued existence. Thereby the free manipulation of concepts is contrasted with their application, ideas become detached from reality and acquire a relative independence.*
3. *Characters are not individually exhibited as they actually occur, but their symbols are projected onto the background of an ordered manifold of possibilities which can be generated by a fixed process and is open into infinity.*[102]

This threefold process is above all manifested in the generation of the infinite sequence of natural numbers. But then, says Weyl, cognition makes "a leap into the beyond" by turning the number sequence "that is never complete but remains open into the infinite into "a closed aggregate of objects existing in themselves."[103] This potentially dangerous move is compounded in the third attempt at hypostatizing the continuum, set theory, for it ascribes an analogous closure to "the places in the continuum, i.e. to the possible sequences or sets of natural numbers."[104] In Weyl's view this was a double error, for neither the aggregate of sets of natural numbers, nor (in general) individual such sets can be considered finished entities. Rather the continuum should be considered as an essentially incompletable "field of constructive possibilities"[105] To suppose otherwise is to risk running up against set-theoretic paradoxes such as Russell's.

During the period 1918–1921 Weyl wrestled with the problem of providing the continuum with an exact mathematical formulation free of objectionable set-theoretic assumptions. As he saw it in 1918, there is an unbridgeable gap between intuitively given continua (e.g. those of space, time and motion) on the one hand, and the discrete exact concepts of mathematics (e.g. that of real number) on the other. For Weyl the presence of this split meant that the construction of the mathematical continuum could not simply be "read off" from intuition. Rather, he believed at this time that the mathematical continuum must be treated as if it were an element of the transcendent realm, and so, in the end, justified in the same way as a physical theory. In Weyl's view, it was not enough that the mathematical theory be *consistent*; it must also be *reasonable*.

Das Kontinuum (1918) embodies Weyl's attempt at formulating a theory of the continuum which satisfies the first, and, as far as possible, the second, of these

---

[102] *Ibid.*, pp. 37–8.
[103] *Ibid.*, p. 38.
[104] *Ibid.*, p. 46.
[105] *Ibid.*, p. 50.

requirements. In the following passages from this work he acknowledges the difficulty of the task:

> ... *the conceptual world of mathematics is so foreign to what the intuitive continuum presents to us that the demand for coincidence between the two must be dismissed as absurd.*[106]

> ... *the continuity given to us immediately by intuition (in the flow of time and of motion) has yet to be grasped mathematically as a totality of discrete "stages" in accordance with that part of its content which can be conceptualized in an exact way.*[107]

> *Exact time- or space-points are not the ultimate, underlying atomic elements of the duration or extension given to us in experience. On the contrary, only reason, which thoroughly penetrates what is experientially given, is able to grasp these exact ideas. And only in the arithmetico-analytic concept of the real number belonging to the purely formal sphere do these ideas crystallize into full definiteness.*[108]

> *When our experience has turned into a real process in a real world and our phenomenal time has spread itself out over this world and assumed a cosmic dimension, we are not satisfied with replacing the continuum by the exact concept of the real number, in spite of the essential and undeniable inexactness arising from what is given.*[109]

However much he may have wished it, in *Das Kontinuum* Weyl did not aim to provide a mathematical formulation of the continuum as it is presented to intuition, which, as the quotations above show, he regarded as an impossibility (at that time at least). Rather, his goal was first to achieve *consistency* by putting the *arithmetical* notion of real number on a firm logical basis, and then to show that the resulting theory is *reasonable* by employing it as the foundation for a plausible account of continuous process in the objective physical world.[110]

Weyl had come to believe that mathematical analysis at the beginning of the twentieth century would not bear logical scrutiny, for its essential concepts and procedures involved vicious circles to such an extent that, as he says, "every cell (so to speak) of this mighty organism is permeated by contradiction." In *Das Kontinuum* he tries to overcome this by providing analysis with a *predicative* formulation—not, as Russell and Whitehead had attempted in their *Principia Mathematica*, by introducing a hierarchy of logically ramified types, which Weyl seems to have regarded as too complicated—but rather by confining the basic principle of set formation to formulas whose bound variables range over just the initial given entities (numbers). Thus he restricts analysis to what can be done in terms of natural numbers with the aid of three basic logical operations, together with the operation of substitution and the process of "iteration", i.e., primitive recursion. Weyl recognized that the effect of this restriction would be to render unprovable

---

106 Weyl (1987), p. 108.

107 *Ibid.*, p. 24.

108 *Ibid.*, p. 94.

109 *Ibid.*, p. 93.

110 The connection between mathematics and physics was of paramount importance for Weyl. His seminal work on relativity theory, *Space-Time-Matter*, was published in the same year (1918) as *Das Kontinuum*; the two works show subtle affinities.

many of the central results of classical analysis—e.g., Dirichlet's principle that any bounded set of real numbers has a least upper bound[111]—but he was prepared to accept this as part of the price that must be paid for the security of mathematics.

In section 6 of *Das Kontinuum* Weyl presents his conclusions as to the relationship between the intuitive and mathematical continua. He poses the question: Does the mathematical framework he has erected provide an adequate representation of physical or temporal continuity as it is *actually experienced*? He begins his investigation by noting that, according to his theory, if one asks whether a given function is continuous, the answer is not fixed once and for all, but is, rather, dependent on the extent of the domain of real numbers which have been defined up to the point at which the question is posed. Thus the continuity of a function must always remain *provisional*; the possibility always exists that a function deemed continuous *now* may, with the emergence of "new" real numbers, turn out to be discontinuous *in the future.*[112]

To reveal the discrepancy between this formal account of continuity based on real numbers and the properties of an intuitively given continuum, Weyl next considers the experience of seeing a pencil lying on a table before him throughout a certain time interval. The position of the pencil during this interval may be taken as a function of the time, and Weyl takes it as a fact of observation that during the time interval in question this function is continuous and that its values fall within a definite range. And so, he says,

> *This observation entitles me to assert that during a certain period this pencil was on the table; and even if my right to do so is not absolute, it is nevertheless reasonable and well-grounded. It is obviously absurd to suppose that this right can be undermined by "an expansion of our principles of definition"—as if new moments of time, overlooked by my intuition could be added to this interval, moments in which the pencil was, perhaps, in the vicinity of Sirius or who knows where. If the temporal continuum can be represented by a variable which "ranges over" the real numbers, then it appears to be determined thereby how narrowly or widely we must understand the concept "real number" and the decision about this must not be entrusted to* logical *deliberations over principles of definition and the like.*[113]

To drive the point home, Weyl focuses attention on the fundamental continuum of *immediately given phenomenal time*, that is, as he characterizes it,

---

[111]In this connection it is of interest to note that on 9 February 1918 Weyl and George Pólya made a bet in Zürich in the presence of twelve witnesses (all of whom were mathematicians) that "within 20 years, Pólya, or a majority of leading mathematicians, will come to recognize the falsity of the least upper bound property." When the bet was eventually called, everyone—with the single exception of Gödel—agreed that Pólya had won.

[112]This fact would seem to indicate that in Weyl's theory the domain of definition of a function is not unambiguously determined by the function, so that the continuity of such a "function" may vary with its domain of definition. (This would be a natural consequence of Weyl's definition of a function as a certain kind of relation.) A simple but striking example of this phenomenon is provided in classical analysis by the function *f* which takes value 1 at each rational number, and 0 at each irrational number. Considered as a function defined on the rational numbers, *f* is constant and so continuous; as a function defined on the real numbers, *f* fails to be continuous anywhere.

[113]Weyl (1987), p. 88.

> *... to that constant form of my experiences of consciousness by virtue of which they appear to me to flow by successively. (By "experiences" I mean what I experience, exactly as I experience it. I do not mean real psychical or even physical processes which occur in a definite psychic-somatic individual, belong to a real world, and, perhaps, correspond to the direct experiences.)*[114]

In order to correlate mathematical concepts with phenomenal time in this sense Weyl grants the possibility of introducing a rigidly punctate "now" and of identifying and exhibiting the resulting temporal points. On the collection of these temporal points is defined the relation of *earlier than* as well as a congruence relation of *equality of temporal intervals*, the basic constituents of a simple mathematical theory of time. Now Weyl observes that the discrepancy between phenomenal time and the concept of real number would vanish if the following pair of conditions could be shown to be satisfied:

1. *The immediate expression of the intuitive finding that during a certain period I saw the pencil lying there were construed in such a way that the phrase "during a certain period" was replaced by "in every temporal point which falls within a certain time span OE.* [Weyl goes on to say parenthetically here that he admits "that this no longer reproduces what is intuitively present, but one will have to let it pass, *if it is really legitimate to dissolve a period into temporal points.*"]
2. If P is a temporal point, then the domain of rational numbers to which l belongs if and only if there is a time point L earlier than P such that OL = l.OE *can be constructed arithmetically in pure number theory on the basis of our principles of definition, and is therefore a real number in our sense.*[115]

Condition 2 means that, if we take the time span *OE* as a unit, then each temporal point *P* is correlated with a definite real number. In an addendum Weyl also stipulates the converse.

But can temporal intuition itself provide evidence for the truth or falsity of these two conditions? Weyl thinks not. In fact, he states unequivocally that

> *... everything we are demanding here is obvious* nonsense: *to* these *questions, the intuition of time provides no answer—just as a man makes no reply to questions which clearly are addressed to him by mistake and, therefore, are unintelligible when addressed to him.*[116]

The grounds for this assertion are by no means immediately evident, but one gathers from the passages following it that Weyl regards the experienced *continuous flow* of phenomenal time as constituting an insuperable barrier to the whole enterprise of representing this continuum in terms of individual points, and even to the characterization of "individual temporal point" itself. As he says,

> *The view of a flow consisting of points and, therefore, also dissolving into points turns out to be mistaken: precisely what eludes us is the nature of the continuity, the flowing from point to point; in other words, the secret of how the continually enduring present can continually slip away into the receding past.*
>
> *Each one of us, at every moment, directly experiences the true character of this temporal continuity. But, because of the genuine primitiveness of phenomenal time, we cannot put our experiences into words. So we shall content ourselves with the following description. What I*

[114] *Ibid.*
[115] *Ibid.*, p. 89.
[116] *Ibid.*, p. 90.

> *am conscious of is for me both a being-now and, in its essence, something which, with its temporal position, slips away. In this way there arises the persisting factual extent, something ever new which endures and changes in consciousness.*[117]

Weyl sums up what he thinks can be affirmed about "objectively presented time"—by which I take it is meant "phenomenal time described in an objective manner"—in the following two assertions, which he claims apply equally, *mutatis mutandis,* to every intuitively given continuum, in particular, to the continuum of spatial extension:

1. *An individual point in it is non-independent, i.e., is pure nothingness when taken by itself, and exists only as a "point of transition" (which, of course, can in no way be understood mathematically);*
2. *it is due to the essence of time (and not to contingent imperfections in our medium) that a fixed temporal point cannot be exhibited in any way, that always only an approximate, never an exact determination is possible.*[118]

The fact that single points in a true continuum "cannot be exhibited" arises, Weyl continues, from the fact that they are not genuine individuals and so cannot be characterized by their properties. In the physical world they are never defined absolutely, but only in terms of a coordinate system, which, in an arresting metaphor, Weyl describes as "the unavoidable residue of the eradication of the ego." This metaphor, which Weyl was to employ more than once,[119] reflects the continuing influence of Husserlian phenomenological doctrine: in this case, the thesis that the existent is given in the first instance as the contents of a consciousness.[120] By 1919

[117] *Ibid.*, p. 91–92.

[118] *Ibid.*, p. 92.

[119] E.g. in Weyl (1950), 8 and (1949), p. 123.

[120] Many years later, in *Insight and Reflection*, Weyl expanded the metaphor into a full-fledged analogy: In Weyl (1969), objects, subjects (or egos), and the appearance of an object to a subject are correlated respectively with *points on a plane, (barycentric) coordinate systems in the plane,* and *coordinates of a point with respect to a such a coordinate system.* In Weyl's analogy, a coordinate system S consists of the vertices of a fixed nondegenerate triangle T; each point p in the plane determined by T is assigned a triple of numbers summing to 1—its *barycentric coordinates* relative to S—representing the magnitudes of masses of total weight 1 which, placed at the vertices of T, have centre of gravity at p. Thus objects, i.e. points, and subjects i.e., coordinate systems or triples of points belong to the same "sphere of reality." On the other hand, the *appearances* of an object to a subject, i.e., triples of numbers, lie, Weyl asserts, in a different sphere, that of *numbers.* These *number-appearances,* as Weyl calls them, correspond to the experiences of a subject, or of pure consciousness.

From the standpoint of naïv realism the points (objects) simply exist as such, but Weyl indicates the possibility of constructing geometry (which under the analogy corresponds to external reality) solely in terms of number-appearances, so representing the world in terms of the experiences of pure consciousness, that is, from the standpoint of idealism. Thus suppose that we are given a coordinate system S. Regarded as a subject or "consciousness", from its point of view a point or object now corresponds to what was originally an appearance of an object, that is, a triple of numbers summing to 1; and, analogously, any coordinate system S' (that is, another subject or "consciousness") corresponds to three such triples determined by the vertices of a nondegenerate triangle. Each point or object p may now be *identified* with its coordinates relative to S. The coordinates of p relative to any other coordinate system S' can be determined by a straightforward algebraic transformation: these coordinates represent the *appearance* of the object corresponding to p to the

Weyl had come to embrace Brouwer's views[121] on the intuitive continuum. The latter's influence looms large in Weyl's next paper on the subject, *On the New Foundational Crisis of Mathematics,* written in 1920. Here Weyl identifies two distinct views of the continuum: "atomistic" or "discrete"; and "continuous". In the first of these the continuum is composed of individual real numbers which are well-

subject represented by S'. Now these coordinates will, in general, *differ* from those assigned to p by our given coordinate system S, and will in fact coincide for all p if and only if S' is what is termed by Weyl the *absolute* coordinate system consisting of the three triples (1,0,0), (0,1,0), (0,0,1), that is, the coordinate system which corresponds to S itself. Thus, for this coordinate system, "object" and "appearance" coincide, which leads Weyl to term it the *Absolute I*. (This term Weyl borrows from Fichte, whom he quotes as follows: "The I demands that it comprise all reality and fill up infinity. This demand is based, as a matter of necessity, on the idea of the infinite I; this is the absolute I—which is not the I given in real awareness.")

Weyl points out that this argument takes place entirely within the realm of numbers, that is, for the purposes of the analogy, the *immanent consciousness.* In order to do justice to the claim of objectivity that all "I"s are equivalent, he suggests that only such numerical relations are to be declared of interest as remain unchanged under passage from an "absolute" to an arbitrary coordinate system, that is, those which are invariant under arbitrary linear coordinate transformations. When this scheme is given a purely axiomatic formulation, Weyl sees a third viewpoint emerging in addition to that of realism and idealism, namely, a *transcendentalism* which "postulates a transcendental reality but is satisfied with modelling it in symbols."

Interestingly, by the time this was written, Weyl seems to have moved away somewhat from the phenomenology that originally suggested the geometric analogy. For he asserts that a number of Husserl's theses become "demonstratively false" when translated into the context of the analogy, "something which," he opines, "gives serious cause for suspecting them." Unfortunately, he does not specify which of Husserl's theses he has in mind.

Weyl goes on to emphasize:

> *Beyond this, it is expected of me that I recognize the other I—the* you*—not only by observing in my thought the abstract norm of invariance or objectivity, but absolutely: you are* for you, *once again, what* I *am for* myself: *not just an existing but a conscious carrier of the world of appearances.*

This recognition of the *Thou,* according to Weyl, can be presented within his geometric analogy only if it is furnished with a purely *axiomatic* formulation. In taking this step Weyl sees a third viewpoint emerging in addition to that of realism and idealism, namely, a *transcendentalism* which "postulates a transcendental reality but is satisfied with modelling it in symbols."

But Weyl, ever-sensitive to the claims of subjectivity, hastens to point out that this scheme by no means resolves the enigma of selfhood. In this connection he refers to Leibniz's attempt to resolve the conflict between human freedom and divine predestination by having God select for existence, on the grounds of sufficient reason, certain beings, such as Judas and St. Peter, whose nature thereafter determines their entire history. Concerning this solution Weyl remarks characteristically:

> *it] may be objectively adequate, but it is shattered by the desperate cry of Judas: Why did* I *have to be Judas! The impossibility of an objective formulation to this question strikes home, and no answer in the form of an objective insight can be given. Knowledge cannot bring the light that is I into coincidence with the murky, erring human being that is cast out into an individual fate.*

For further discussion of Weyl's philosophical views see Bell (2003).

[121]Weyl described Brouwer's system of mathematics as refraining from making the "leap into the beyond" required by classical set theory (Weyl 1949, p. 50).

defined and can be sharply distinguished. Weyl describes his earlier attempt at reconstructing analysis in Das Kontinuum as atomistic in this sense:

> *Existential questions concerning real numbers only become meaningful if we analyze the concept of real number in this extensionally determining and delimiting manner. Through this conceptual restriction, an ensemble of individual points is, so to speak, picked out from the fluid paste of the continuum. The continuum is broken up into isolated elements, and the flowing-into-each other of its parts is replaced by certain conceptual relations between these elements, based on the "larger-smaller" relationship. This is why I speak of the* atomistic *conception of the continuum.*[122]

Weyl now repudiated atomistic theories of the continuum, including that of *Das Kontinuum*. He writes:

> *In traditional analysis, the continuum appeared as the set of its points; it was considered merely as a special case of the basic logical relationship of* element and set. *Who would not have already noticed that, up to now, there was no place in mathematics for the equally fundamental relationship of* part and whole? *The fact, however,* that it has parts, *is a fundamental property of the continuum; and so (in harmony with intuition, so drastically offended against by today's "atomism") this relationship is taken as the mathematical basis for the continuum by Brouwer's theory. This is the real reason why the method used in delimiting subcontinua and in forming continuous functions starts out from* intervals *and not* points *as the primary elements of construction. Admittedly a set also has parts. Yet what distinguishes the parts of sets in the realm of the "divisible" is the existence of "elements" in the set-theoretical sense, that is, the existence of* parts that themselves do not contain any further parts. *And indeed, every part contains at least one "element". In contrast, it is inherent in the nature of the continuum that every part of it can be further divided without limitation. The concept of a point must be seen as an idea of a limit, "point" is the idea of a* limit *of a division extending* in infinitum. *To represent the continuous connection of the points, traditional analysis, given its shattering of the continuum into isolated points, had to have recourse to the concept of a* neighbourhood. *Yet, because the concept of continuous function remained mathematically sterile in the resulting generality, it became necessary to introduce the possibility of "triangulation" as a restrictive condition.*[123]

Like Brentano, Weyl knew that to "shatter a continuum into isolated points" would be to eradicate the very feature which characterizes a continuum—the fact that its cohesiveness is inherited by every one of its (connected) parts.

While intuitive considerations, together with Brouwer's influence, must certainly have fuelled Weyl's rejection of atomistic conceptions of the continuum, it also had a *logical* basis. For Weyl had come to regard as meaningless the formal procedure—employed in *Das Kontinuum*—of negating universal and existential statements concerning real numbers conceived as developing sequences or as sets of rationals. This had the effect of undermining the whole basis on which his theory had been erected, and at the same time rendered impossible the very formulation of a "law of excluded middle" for such statements. Thus Weyl found himself espousing a position considerably more radical than that of Brouwer, for whom negations of

[122] Weyl (1998a, b), p. 91.

[123] Weyl (1998a, b), p. 115.

quantified statements had a perfectly clear constructive meaning, under which the law of excluded middle is simply not generally affirmable.

Of existential statements Weyl says:

> *An existential statement—e.g., "there is an even number"—is not a judgement in the proper sense at all, which asserts a state of affairs; existential states of affairs are the empty invention of logicians.*[124]

Weyl termed such pseudostatements "judgement abstracts", likening them to "a piece of paper which announces the presence of a treasure, without divulging its location." Universal statements, although possessing greater substance than existential ones, are still mere intimations of judgements, "judgement instructions", for which Weyl provides the following metaphorical description:

> *If knowledge be compared to a fruit and the realization of that knowledge to the consumption of the fruit, then a universal statement is to be compared to a hard shell filled with fruit. It is, obviously, of some value, however, not as a shell by itself, but only for its content of fruit. It is of no use to me as long as I do not open it and actually take out a fruit and eat it.*[125]

Above and beyond the claims of logic, Weyl welcomed Brouwer's construction of the continuum by means of sequences generated by free acts of choice, thus identifying it as a "medium of free Becoming" which "does not dissolve into a set of real numbers as finished entities".[126] Weyl felt that Brouwer's intuitionistic approach had brought him closer than anyone else to bridging that "unbridgeable chasm" between the intuitive and mathematical continua. In particular, he found compelling the fact that the Brouwerian continuum is not the union of two disjoint nonempty parts—that it is *indecomposable* or *cohesive*.[127] "A genuine continuum," Weyl says, "cannot be divided into separate fragments." In 1921 Weyl observed:

> *...if we pick out a specific point, say, $x = 0$, on the number line C (i.e., on the variable range of a real variable x), then one cannot, under any circumstance, claim either coincides with it or is disjoint from it. The point $x = 0$ thus does not at all split the continuum C into two parts $C^-: x < 0$ and $C^+: x > 0$, in the sense that C would consist of the union of C^-, C^+ and the one point 0 ... If this appears offensive to present-day mathematicians with their atomistic thought habits, it was in earlier times a self-evident view held by everyone: Within a continuum, one can very well generate subcontinua by introducing boundaries; yet it is irrational to claim that the total continuum is made up of the boundaries and the subcontinua. The point is, a genuine continuum is something connected in itself, and it cannot be divided into separate fragments; this conflicts with its nature.*[128]

In later publications he expresses this more colourfully by quoting Anaxagoras to the effect that a continuum "defies the chopping off of its parts with a hatchet."[129]

---

[124] *Ibid.*, p. 97.

[125] *Ibid.*, p. 98.

[126] See Chap. 9 below.

[127] See Appendix A.

[128] Weyl (1921), p. 111.

[129] See Chap. 1 above.

Weyl also agrees with Brouwer that all functions everywhere defined on a continuum are continuous, but here certain subtle differences of viewpoint emerge. Weyl contends that what mathematicians had taken to be discontinuous functions actually consist of several continuous functions defined on separated continua. For example, the "discontinuous" function defined by $f(x) = 0$ for $x < 0$ and $f(x) = 1$ for $x \geq 0$ in fact consists of the two functions with constant values 0 and 1 respectively defined on the separated continua $\{x: x < 0\}$ and $\{x: x \geq 0\}$. The union of these two continua fails to be the whole of the real continuum because of the failure of the law of excluded middle: it is not the case that, for be any real number x, either $x < 0$ or $x \geq 0$.

Brouwer, on the other hand, had not dismissed the possibility that discontinuous functions could be defined on proper parts of a continuum, and still seems to have been searching for an appropriate way of formulating this idea.[130] In particular, at that time Brouwer would probably have been inclined to regard the above function f as a genuinely discontinuous function defined on a *proper part* of the real continuum. For Weyl, it seems to have been a self-evident fact that all functions defined on a continuum are continuous, but this is because Weyl confines attention to functions which turn out to be continuous by definition. Brouwer's concept of function is less restrictive than Weyl's and it is by no means immediately evident that such functions must always be continuous.

Weyl defined real functions as mappings correlating each interval in the choice sequence determining the argument with an interval in the choice sequence determining the value "interval by interval" as it were, the idea being that approximations to the input of the function should lead effectively to corresponding approximations to the input. Such functions are continuous by definition. Brouwer, on the other hand, considers real functions as correlating choice sequences with choice sequences. and the continuity of these is by no means obvious. The fact that Weyl refused to grant (free) choice sequences—whose identity is in no way predetermined—sufficient individuality to admit them as arguments of functions perhaps betokens a commitment to the conception of the continuum as a "medium of free Becoming" even deeper than that of Brouwer.

There thus being only minor differences between Weyl's and Brouwer's accounts of the continuum, Weyl abandoned his earlier attempt at the reconstruction of analysis and "joined Brouwer." At the same time, however, Weyl recognized that the resulting gain in intuitive clarity had been bought at a considerable price:

> *Mathematics with Brouwer gains its highest intuitive clarity. He succeeds in developing the beginnings of analysis in a natural manner, all the time preserving the contact with intuition much more closely than had been done before. It cannot be denied, however, that in advancing to higher and more general theories the inapplicability of the simple laws of classical logic eventually results in an almost unbearable awkwardness. And the*

[130]Brouwer established the continuity of functions fully defined on a continuum in 1904 but did not publish a definitive account until 1927. In that account he also considers the possibility of partially defined functions.

mathematician watches with pain the greater part of his towering edifice which he believed to be built of concrete blocks dissolve into mist before his eyes.[131]

Although he later practiced intuitionistic mathematics very rarely, Weyl remained an admirer of intuitionism. And the "riddle of the continuum" retained its fascination for him, as is attested to by the observation he made one of his last papers, *Axiomatic and Constructive Procedures in Mathematics,* written in 1954, and with which we conclude Part I:

... the constructive transition to the continuum of real numbers is a serious affair... and I am bold enough to say that not even to this day are the logical issues involved in that constructive concept completely clarified and settled.[132]

[131]Weyl (1949), p. 54.
[132]Weyl (1985), p. 17.

Part II
Continuity and Infinitesimals in Today's Mathematics

Chapter 6
Topology

6.1 Topological Spaces

In the late nineteenth and early twentieth century the investigation of continuity led to the creation of *topology*,[1] a major new branch of mathematics conferring on the idea of the continuous a vast generality. The origins of topology lie both in Cantor's theory of sets of points as well as the idea, which had first emerged in the calculus of variations, of treating functions as points of a space.[2] Central to topology is the concept of *topological space*. A topological space is a domain equipped with sufficient structure to enable functions between such spaces to be identified as continuous. Now intuitively, a continuous function is one with no "jumps", that is, it always sends "neighbouring" points of its domain to "neighbouring" points of its range. In order to support the idea of a continuous function in this intuitive sense, the concept of topological space must accordingly embody some notion of *neighbourhood*. It was in terms of the neighbourhood concept that **Felix Hausdorff** (1868–1942) first introduced, in 1914, the concept of *topological space*.

Suppose we are given a set X of elements which may be such entities as points in n-dimensional space, plane or space curves, real or complex numbers, although we make no assumptions as to their exact nature. The members of X will be called *points*. Now suppose in addition that corresponding to each point x of X we are given a collection $\mathcal{N}_x$ of subsets of X, the members of which, called *basic neighbourhoods of* x are subject to the following conditions:

(i) each member of $\mathcal{N}_x$ contains x;
(ii) the intersection of any pair of members of $\mathcal{N}_x$ includes a member of $\mathcal{N}_x$;

[1] From Greek *topos*, "place" . Here we shall be concerned with what has become known as *general topology*, as opposed to *algebraic* or *combinatorial* topology.

[2] The study of abstract spaces, which led to the definition of topological space was initiated by **Maurice Fréchet** (1878–1903) in 1906.

J. L. Bell, *The Continuous, the Discrete and the Infinitesimal in Philosophy and Mathematics*, The Western Ontario Series in Philosophy of Science 82,
https://doi.org/10.1007/978-3-030-18707-1_6

(iii) for any $U \in \mathcal{N}_x$ and any $y \in U$, there is $V \in \mathcal{N}_x$ such that $V \subseteq U$.

We may think of each $U \in N_x$ as determining a notion of proximity to x: the points of U are accordingly said to be *U-close* to x. Using this terminology, (i) may be construed as saying that, for any basic neighbourhood U of x, x is U-close to itself; (ii) that, for any basic neighbourhoods U and V of x, there is a basic neighbourhood W of x such that any point W-close to x is both U-close and V-close to x; and (iii) that, for any basic neighbourhood V of x, any point which is U-close to x itself has a basic neighbourhood all of whose points are U-close to x. A subset of S which includes a basic neighbourhood of x is called a *neighbourhood* of x.

A *topological space*, in brief, a *space*, may now be defined to be a set X together with an assignment, to each point x of X, of a collection $\mathcal{N}_x$ of subsets of X satisfying conditions (i)–(iii). As examples of topological spaces we have: the real line $\mathbb{R}$ with $\mathcal{N}_x$ consisting of all open intervals with rational radii centred on x; the Euclidean plane with $\mathcal{N}_x$ consisting of all open discs with rational radii centred on x; and, in general, n-dimensional Euclidean space with $\mathcal{N}_x$ consisting of all open n-spheres with rational radii centred on x. Any subset A of any of these spaces becomes a topological space by taking as neighbourhoods the intersections with A of the neighbourhoods in the containing space.

Hausdorff's original definition of topological space included the condition:

(iv) if $x \neq y$, then there are members U of $\mathcal{N}_x$ and V of $\mathcal{N}_y$ such that $U \cap V = \emptyset$.

A topological space satisfying this condition is called a *Hausdorff* space[3]; all the above mentioned spaces are Hausdorff spaces, but there are many examples of non-Hausdorff spaces.[4]

Most topological notions can be defined entirely in terms of neighbourhoods. Thus a *limit point* of a set of points in a topological space is a point each of whose neighbourhoods contains points of the set: a limit point of a set is then a point which, while not necessarily in the set, is nevertheless "arbitrarily close" to it (lying on its boundary, for instance). The *boundary* of a subset A of a topological space X is the collection of points $x \in X$ which are limit points of both A and its complement $X - A$. A set is *open* if it includes a neighbourhood of each of its points and *closed* if it contains all its limit points: it is easily shown that the closed sets are precisely the complements of open sets.

Topological spaces can also be defined in terms of open sets. Thus we define a *topology* on a set X to be a family $\mathcal{T}$ of subsets of X satisfying the following conditions:

(a) the union of any subfamily of $\mathcal{T}$ belongs to $\mathcal{T}$;
(b) the intersection of any pair of members of $\mathcal{T}$ belongs to $\mathcal{T}$;
(c) $X \in \mathcal{T}$ and $\emptyset \in \mathcal{T}$.

---

[3]Thus points in a Hausdorff space are, as it were, "housed off".

[4]For example, the real line $\mathbb{R}$ with basic neighbourhoods at each point x reduced to the single set $\{y: x \leq y\}$.

Equipped with a topology $\mathscr{T}$, X is called a *topological space* (in this second sense) or simply a *space*; the members of T are called T *-open*, *open in X*, or simply *open*, *sets*. Given such a space X, we define a basic neighbourhood of a point x to be an open subset U containing x. It is then readily shown that the families $\mathscr{N}_x$ of basic neighbourhoods, in this sense, of points x of X satisfy the conditions (i), (ii), (iii) originally laid down for basic neighbourhoods.

A *base* for a topological space X is a family $\mathscr{U}$ of open sets in X such that every open set in X is a union of members of $\mathscr{U}$. For example, the family of all open intervals with rational endpoints is a base for $\mathbb{R}$, which accordingly has a *countable* base.

A concept of discreteness can be introduced for topological spaces: thus a space is *discrete* if each point in it (more exactly each singleton) is an open set. In a discrete spaces points possess the maximum degree of separation, and the space itself possesses the minimum degree of cohesion.

In a topological space, the set-theoretical complement of an open set is closed but not usually open, so within the topology the "negation" of an open set is the *interior* of—the largest open set included in—the complement. This implies that the "double negation" of an open set U is not in general equal to U. It follows that, as observed first by Stone and Tarski in the 1930s, the algebra of open sets is not Boolean or classical, but instead obeys the rules of intuitionistic logic. In acknowledgment that these rules were first formulated by A. Heyting, such an algebra is called a *Heyting algebra.*

The notion of continuous function, or *map*, between topological spaces, can be defined both in terms of neighbourhoods and open sets. A function $f: X \to Y$ between two topological spaces X and Y is said to be *continuous*, if, for any point x in X, and any neighbourhood V of the image $f(x)$ of x, a neighbourhood U of x can be found whose image under f is included in V, in other words, such that the images of any point U-close to x is V-close to $f(x)$. In the case when X and Y are both the real line $\mathbb{R}$, this condition translates into Weierstrass's criterion for continuity: for any x and for any $\varepsilon > 0$, there is $\delta > 0$ such that $|f(x) - f(y)| < \varepsilon$ whenever $|x - y| < \delta$. In terms of open sets, a function $f: X \to Y$ is continuous if and only if, for any open set U in Y, the inverse image $f^{-1}[U]$ is open in X.

A *topological equivalence* or *homeomorphism* between topological spaces is a biunique function which is continuous in both directions, that is, both it and its inverse are continuous. Two spaces are *homeomorphic* if there exists a homeomorphism between them: intuitively, this means that each space can be continuously and reversibly deformed into the other. A property of a space is *topological* if, when possessed by a given space, it is also possessed by all spaces homeomorphic to the given one. *Topology* may now be broadly defined as the study of topological properties.[5] If we consider geometric figures such as triangles and circles as spaces,

[5]In view of the fact that a doughnut and a coffee cup (with a handle) are topologically equivalent, John Kelley (in Kelley 1955) famously defined a *topologist* to be someone who cannot tell the two apart.

we see immediately that, in general, geometric properties such as, e.g., being a triangle or a straight line are not topological, since a triangle is evidently homeomorphic to any simple closed curve and a straight line to any open curve. (To see this, imagine both triangle and line made from cooked pasta or modelling clay.) Topological properties are grosser than geometric ones, since they must stand up under arbitrary continuous deformations.[6] So, for example, a topological property of the triangle is not its triangularity but the property of dividing the plane into two—"inside" or "outside"— regions, as well as the property that, if two points are removed, it falls into two pieces, while if only a single point is removed, one piece remains. As another example we may consider the properties of *one-sidedness* or *two-sidedness* of a surface. The standard one-sided surface—the so-called *Möbius strip* (or band), discovered independently in 1858 by **A. F. Möbius** (1790–1868) and **J. B. Listing**[7] (1806–1882)—may be constructed by gluing together the two ends of a strip of paper after giving one of the ends a half twist. Both one- and two-sidedness are topological properties.

Metric spaces constitute an important class of topological spaces. A *metric* on a set X is a function d which assigns to each pair (x, y) of elements of X a nonnegative real number $d(x, y)$ in such a way that:

$$d(x,y) = d(y,x), \quad d(x,y) + d(y,z) \geq\geq d(x,z), \quad d(x,y) = 0 \text{ if and only if } x = y.$$

We think of $d(x, y)$ as the *distance* between x and y. The first of these conditions then expresses the symmetry of distance, and the second—the *triangle inequality*—corresponds to the familiar fact that the sum of the lengths of two sides of a triangle is never exceeded by that of the remaining side. A set equipped with a metric is called a *metric space*. Each metric space may be considered a topological space in which a basic neighbourhood of a point x is the subset of X consisting of all points y at distance $< r$, for positive r. All of the examples of topological spaces given above are metric spaces, but not every topological space is a metric space. Metric spaces are the topological generalizations of Euclidean spaces.

One of the most important properties a subset of a topological space may possess is that of *compactness*. To define this concept, we introduce the notion of an *open covering* of a subset A of a space X: this is defined to be a family of open subsets of X whose union contains A. Now A is compact if every open covering of A has a *finite* subfamily which is also an open covering of A. Clearly any finite subset of a topological space is compact; compactness may be seen as a topological version of finiteness, or, more generally, boundedness. It can in fact be shown that the compact subsets of any Euclidean space are precisely the closed bounded subsets.

There are a number of different ways of describing *connectedness*, or *cohesion* in topological terms. The standard characterization is to call a subset of a space

---

[6]That is, topological properties are chosen so as to satisfy Leibniz's Principle of Continuity.

[7]It was also Listing who, in his work *Vorstudien zur Topologie* of 1848, first uses the term "topology"; the subject being known prior to this as *analysis situs*, "positional analysis".

Fig. 6.1

connected if no matter how it is split into two disjoint sets, at least one of these contains limit points of the other. For example, any interval on the real line, and any Euclidean space, is connected in this sense. It is easy to prove that a space is connected if and only if it is not the union of two disjoint nonempty open (or. equivalently, closed) sets.

Another characterization of cohesion derives from Cantor's original definition of connectedness for point sets in Euclidean spaces. Given two points a and b of a space X, a *simple chain from a to b* is a collection $\{A_1, \ldots, A_n\}$ of subsets of B such that A_1 (and only A_1) contains a, A_n (and only A_n) contains b, and $A_i \cap A_j$ is nonempty if and only if i and j differ by at most 1. Thus each link of the chain intersects just its immediate predecessor and successor (as well as itself), as in Fig. 6.1. A space X is then said to be *consolidated* if for any pair of points a, b and any open covering $\mathcal{O}$ of X, $\mathcal{O}$ contains a simple chain from a to b. It can then be shown[8] that any connected space is consolidated.

Arcwise connectedness is another, stronger version of cohesion. We say that a space is *arcwise connected* if every pair of its points can be joined by an arc—that is, a homeomorphic image of a closed interval—in the space. Every arcwise connected space is connected, but not conversely. For metric spaces imposing certain additional conditions along with connectedness ensure that the space is arcwise connected. A space is said to be *locally connected* if every neighbourhood of any point contains a connected open neighbourhood of that point. Local connectedness means connectedness "in the small". Now in any space that is both connected and locally connected, each pair of points can be joined by a simple chain of connected sets; such a simple chain can be regarded as an approximation to an arc. When such a space is also a metric space, and is in addition compact, then these simple chains can be refined into arcs, yielding the conclusion[9] that any compact, connected, and locally connected metric space[10] is also arcwise connected.

A (topological) *continuum* is defined to be a compact connected subset of a topological space. Recalling that Cantor defined a continuum as a perfect connected (in his sense) set of points, it is significant that within any *bounded* region of a Euclidean space Cantor's continua coincide with continua in the topological sense.

---

[8]Hocking and Young (1961), p. 108.

[9]*Ibid.*, p. 116.

[10]A metric space with these properties is called a *Peano space* or *Peano continuum.*

The study of topological continua has led to a number of intuitively satisfying results. Let us call two subsets of a topological space *separated* if neither contains limit points of the other; it is then the case that a set is disconnected if and only if it is the union of two nonempty separated subsets. If X is a connected space, we define a *cut point* of X to be a point x of X such that $X - \{x\}$ is disconnected; otherwise x is said to be a *non-cut point* of X. Thus a cut point of a space is one whose removal disconnects the space. For example, every point of $\mathbb{R}$ is a cut point, while the end points of a closed interval are its only non-cut points. On the other hand, neither a circle nor any Euclidean space of dimension ≥ 2 has cut points. It can be shown[11] that every continuum with at least two points has at least two non-cut points. Moreover, if a metric continuum M has exactly two non-cut points, it is homeomorphic to a closed interval[12]; and if, for any two point x and y, $M - \{x, y\}$ is disconnected, then M is homeomorphic to a circle.

In Euclidean spaces there is a natural definition of *dimension*, namely, the number of coordinates required to identify each point of the space. In a general topological space there is no mention of coordinates and so this definition is not applicable. In 1912 Poincaré formulated the first definition of dimension for a topological space; this was later refined by Brouwer, **Paul Urysohn** (1898–1924) and **Karl Menger** (1902–1985). The definition in general use today is formulated in terms of the boundary[13] of a subset of a topological space. Now the topological dimension of a subset M of a topological space is defined inductively as follows: the empty set is assigned dimension -1; and M is said to be n-dimensional at a point p if n is the least number for which there are arbitrarily small neighbourhoods of p whose boundaries in M all have dimension $< n$. The set M has topological dimension n if its dimension at all of its points is $\leq n$ but is equal to n at one point at least.

Dimension thus defined is a topological property. Menger and Urysohn defined a curve as a one-dimensional closed connected set of points, so rendering the property of being a curve a topological property. In 1911 Brouwer proved the important result that n-dimensional Euclidean space has topological dimension n, so establishing once and for all the invariance of dimension of Euclidean space under continuous mappings.

6.2 Manifolds

Although the origins of the concept of a manifold may be traced to Gauss's investigations of the intrinsic properties of surfaces, it was Riemann who, in the mid -nineteenth century, first explicitly introduced the idea in a general form. Riemann had conceived of an n - dimensional manifold as an abstract space locally

---

[11]Hocking and Young (1961), p. 49.

[12]*Ibid.*, p. 54.

[13]Defined above.

resembling n-dimensional Euclidean space in the sense that in the vicinity of each point an n-dimensional coordinate system can be introduced and a distance function defined between points whose coordinates are infinitesimally close.[14] This evolved into the more general conception of a *differentiable* manifold, that is, a manifold possessing a differentiable structure at each point. It is our purpose here to describe a special type of differentiable manifold—the so-called *smooth* manifolds .[15]

The modern concept of manifold is based on the idea of a *chart.* Given a topological space T, let U be a nonempty open subspace[16] of T which is homeomorphic to an open subspace X of the n-dimensional Euclidean space $\mathbb{R}^n$. A homeomorphism $\sigma : p \mapsto p^{\sigma}$[17] of U with X is called a *chart* on U (or in T). In a given chart σ on U, each point p of U corresponds to a point $x = p^{\sigma}$ of X, so that p may be identified with $(x_1,\ldots, x_n)$, the coordinates of x. The real numbers x_i are called the *coordinates of p* (in the chart σ), and n is the *dimension* of the chart.

Now suppose that there is a homeomorphism Φ of X with a another open subspace Y of $\mathbb{R}^n$ and let Ψ be its inverse. If $y = x^{\Phi}$ is a general point of Y with coordinates y_i $(i = 1,\ldots, n)$, then Φ and Ψ may be described by means of continuous functions φ_i, ψ_i: $R^n \to R$:

$$\left.\begin{array}{l} y_i = \varphi_i(x_1,\ldots,x_n) \\ x_i = \psi_i(y_1,\ldots,y_n) \end{array}\right\} \quad (i = 1, \ldots, n),$$

or simply, writing x for $(x_1,\ldots, x_n)$ and y for $(y_1,\ldots, y_n)$,

$$y_i = \varphi_i(x) \quad (x \in X) \quad x_i = \psi_i(y) \quad (y \in Y). \tag{6.1}$$

---

[14]The distance ds between points $(x_1,\ldots, x_n)$ and points $(x_1 + \mathrm{d}x_1,\ldots, x_n + \mathrm{d}x_n)$ being given by

$$ds^2 = \sum_{i=1}^{n}\sum_{j=1}^{n} g_{ij}\, dx_i\, dx_j$$

where the g_{ij} are functions of the coordinates $(x_1,\ldots, x_n)$, $g_{ij} = g_{ji}$ and the right-hand side is always positive. This expression for ds^2 is a generalization of the usual Euclidean distance formula

$$ds^2 = dx_1^2 + \ldots + dx_n^2.$$

[15]My account has been adapted from Ch. 1 of Cohn (1957).

[16]If T is a topological space, and A a subset of T, the topology on T induces a natural topology on A—the *relative topology*—defined by identifying the open subsets of A as precisely the intersections with A of the open subsets of T. The subset A, with the relative topology so defined, is called a *subspace* of T.

[17]For convenience we write p^{σ} for the value $\sigma(p)$ of σ at p, and similarly below.

Notice that the composite $\Phi \circ \sigma$ is a homeomorphism of U with Y, and therefore also a chart on U.

Conversely, if σ and τ are any two charts on the same subspace U of T, mapping U into X and Y respectively, then $\Phi = \tau \circ \sigma^{-1}$ is a homeomorphism of X with Y having inverse $\Psi = \sigma \circ \tau^{-1}$ so that the coordinates x and y of corresponding points in X and Y are related by equations of the form (6.1). This shows that, if we regard the passage from x to y as a *change of coordinates*, then the Eq. (6.1)—with continuous φ_i and ψ_i —are the most general equations defining a change of coordinates.

A function from a subspace of $\mathbb{R}^n$ to $\mathbb{R}$ is said to be *smooth* if it has partial derivatives of arbitrarily high orders. Two charts in T whose coordinates are related by the Eq. (6.1) are said to be *smoothly related at a point p* of T, if they are defined on a neighbourhood of p and if the functions φ_i and ψ_i occurring in (6.1) are smooth. If two charts are smoothly related at every point of T at which they are defined, they are said to be *smoothly related.*

A topological space is said to be *locally Euclidean* at a point p if there exists a chart σ on a neighbourhood of p. A *manifold* is a Hausdorff space which is locally Euclidean at each of its points. It follows that in a manifold M each point has a chart defined on some neighbourhood, so that the family of charts in M may be said to *cover M*.

We now define a *smooth structure* on a Hausdorff space M to be a family of charts $\mathcal{F}$ in M satisfying the three following conditions:

(i) *At each point of M there is a chart which belongs to* $\mathcal{F}$;
(ii) *Any two charts of* $\mathcal{F}$ *are smoothly related;*
(iii) *Any chart in M which is smoothly related to every chart of* $\mathcal{F}$ *itself belongs to* $\mathcal{F}$.

Conditions (i) and (ii) are naturally expressed by saying that $\mathcal{F}$ is *smooth* and *maximal*, respectively. A smooth structure on M is accordingly a maximal smooth family of charts covering M. It is plain that a Hausdorff space with a smooth structure is necessarily a manifold—with this structure the space is called a *smooth manifold.* Members of the smooth structure are called *admissible charts.* It can be shown[18] that any smooth covering family of charts on a Hausdorff space is contained in a unique maximal smooth family of charts.

Let M be a manifold and f a real-valued function defined on a subspace (possibly all) of M: we shall express this by saying that f is defined *in M.* If σ is a chart on some subspace U on which f is defined, the latter determines a function $f*$ from a subspace of $\mathbb{R}^n$ to $\mathbb{R}$ by defining $f*(x) = f(p)$, where $x = p^\sigma$. If M is a smooth manifold, a real-valued function f in M is said to be *smooth* at a point p if it is defined on some neighbourhood of p and its expression $f*$ in terms of an admissible chart σ is a smooth function. (This definition does not depend on the choice of the chart σ.) A function which is smooth at every point at which it is defined is said to be *smooth.*

Finally, suppose we are given two smooth manifolds M and N and a map Φ: $M \to N$. For each function f defined in N we define the function f_Φ in M by:

---

[18]Cohn (1957)., Theorem 1.2.1.

$$f_{\Phi}(p) = f\left(p^{\Phi}\right) \ (p \in M).$$

Then the map Φ: $M \to N$ is said to be *smooth* if f_{Φ} is a smooth function in M whenever f is a smooth function in N.

Manifolds and topological spaces give rise to *categories*, the topic of our next chapter.

Chapter 7
Category/Topos Theory

7.1 Categories and Functors

Category theory is a framework for the investigation of mathematical form and structure in their most general manifestations. Central to it is the concept of *structure-preserving map*, or *transformation*. While the importance of this notion was long recognized in geometry (witness, for example, Klein's *Erlanger Programm* of 1872),[1] its pervasiveness in mathematics did not really begin to be appreciated until the rise of abstract algebra in the 1920s and 30s,[2] where, in the form of homomorphism, the idea had been central from the beginning. So emerged the view that the essence of a mathematical structure lies not in its internal constitution as a set-theoretical construct, but rather in the nature of its relationship with other structures of the same kind through the network of transformations between them.

This idea achieved its fullest expression in the *theory of categories* introduced in 1945 by **S. Eilenberg** (1913–1998) and **S. Mac Lane** (1909–2005).[3] They created an axiomatic framework within which the notion of map and preservation of structure are primitive, that is, are not defined in terms of anything else. As in biology, the viewpoint of category theory is that mathematical structures fall naturally into species or *categories*. But a category is specified not just by identifying the species of mathematical structure which constitute its *objects*; one must also specify the transformations linking these objects.

[1]The kernel of Klein's *Erlanger Programm* was the characterization of geometry as the study of those properties of figures that remain invariant under a particular group of transformations, See, e.g. E.T. Bell (1945), pp. 443–9.

[2]See, e.g., Kline (1972), Ch. 49

[3]Eilenberg, S. and Mac Lane, S., *General Theory of Natural Equivalences,* Transactions of the American Mathematical Society **58**, 1945, pp. 231–294.

J. L. Bell, *The Continuous, the Discrete and the Infinitesimal in Philosophy and Mathematics*, The Western Ontario Series in Philosophy of Science 82,
https://doi.org/10.1007/978-3-030-18707-1_7

Formally, then, a category consists of two collections of entities called *objects* and *maps*. Each map f is assigned an object A called its *domain* and an object B called its *codomain*: this fact is indicated by writing $f: A \to B$. Each object A is assigned a map $1_A: A \to A$ called the *identity map* on A. For any pair of maps $f: A \to B$, $g: B \to C$ (i.e., such that the domain of g coincides with the codomain of f; such pairs of maps are called *composable*), there is defined a *composite* map $g \circ f: A \to C$. These prescriptions are subject to the *associative* and *identity* laws, viz., given three maps $f: A \to B$, $g: B \to C$, $h: C \to D$, the composites $h \circ (g \circ f)$ and $(h \circ g) \circ f$ coincide; and for any map $f: A \to B$ the composites $f \circ 1_A$ and $1_B \circ f$ both coincide with f.

There are a number of ways of envisioning categories. They may, for instance, be considered as frameworks for the *analysis of variation*. Thus we suppose given.

- *domains of variation* or *types* $A, B, C, \ldots$
- *transformations* or *correlations* $f: A \to B$ or $A \xrightarrow{f} B$ between such domains: A and B are said to be *correlated* by f; A is the *domain*, B the *codomain* of f.

As concrete examples we may consider:

Space → *Time*	Analogue clocks
Natural numbers → *Time*	Digital clocks
Time → *Space*	Motions
Space → *Rational/Real numbers*	Thermometers, barometers, speedometers

A correlation $A \to B$ may be thought of as a *B-valued quantity varying over A*. As such, correlations may be *composed*:

$$\frac{A \xrightarrow{f} B \xrightarrow{g} C}{A \xrightarrow{g \circ f} C}$$

E.g. the use of a digital stopwatch amounts to the composite correlation.

Natural numbers → *Time* → *Space*

Composition of correlations is *associative* in the evident sense.

Associated with each domain A is an *identity* correlation $A \xrightarrow{1_A} A$ satisfying $f \circ 1_A = f$, $g \circ 1_A = g$ for any $A \xrightarrow{f} B$, $C \xrightarrow{g} A$.

These specifications are just the basic data of a category.

A category may also be thought of as the objective presentation of a mathematical form (or idea), with objects as structures manifesting the associated form, and maps as form-preserving transformations between structures. Now topological spaces and continuous functions constitute the objects and maps, respectively, of a category, the *category* **Top** of *topological spaces*. **Top** may be thought of as the objective presentation of the *form of continuity*. Other important examples of categories include the category **Set** of sets, with (arbitrary) sets as objects and (arbitrary) functions as maps; the category **Grp** of groups, with groups as objects and group homomorphisms as maps; and the category **Man** of manifolds, with differentiable manifolds as objects and smooth functions between them as maps. For **Set**, the associated form is *pure discreteness*; for **Grp**, it is *composition-inversion*, and for **Man**, it is *smoothness*.

A concept central to category theory is that of functor. A *functor* F: $\mathbf{C} \to \mathbf{D}$ from a category **C** to a category **D** is an assignment, to each object A of **C,** of an object FA of **D** and to each map f: $A \to B$ of **C** a map Ff: $FA \to FB$ of **D** such a way that composites and identities are preserved, i.e. $F(g \circ f) = Fg \circ Ff$ for composable f, g, and $F(1_A) = 1_{FA}$ for any object A. As examples of functors we have the so-called "forgetful" functors **Top** $\to$ **Set, Grp** $\to$ **Set,** and **Man** $\to$ **Set** that assign to each space, group, or manifold its underlying set of points (i.e., "forgets" the structure); the functor **Top** $\to$ **Grp** that assigns to each topological space its homology group[4] of a prescribed dimension; and the functor **Man** $\to$ **Man** that assigns to each manifold its tangent bundle.

If categories are associated with mathematical forms, then functors may be conceived of as devices for effecting a *change of form*.[5] Thus, for example, the three "forgetful" functors just mentioned take the forms of continuity, composition-inversion and smoothness, respectively, to the form of pure discreteness. Observe that functors, or "changes of form" can be *composed* in the evident way. This gives rise to the "meta-category **Cat** with categories as objects and functors as maps. **Cat** may be regarded as the category-theoretic embodiment of the idea of mathematical form, and its maps, the functors between categories, as vehicles for inducing morphological variation.

7.2 Pointless Topology

Another way of presenting topology in category-theoretic terms is through the development of what has become known as *pointless topology*.[6] Here the idea is to deal directly with the topology of a space, bypassing the underlying set which supports that topology: the approach may be seen as a way of investigating the continuous (topologies) without reduction to the discrete (sets of points). Pointless topology originates with the observation that, under the partial order of inclusion, a topology is a certain kind of *complete lattice*, that is, a partially ordered structure $(L, \leq)$ possessing a largest element **1,** and in which each subset X has a least upper bound, or *join,* $\vee X$, and a greatest lower bound, or *meet*, $\wedge X$. (In the case of a

[4]It was the attempt to create a tractable theory of homology groups that first led Eilenberg and Mac Lane to formulate the idea of a functor, and then of a category.

[5]At the same time, functors embody a version of the Principle of Continuity in that composites and identities are preserved.

[6]See Johnstone (1983).

topology on a set X, **1** is X, joins are set-theoretic unions and the meet of a family $\mathcal{U}$ of open sets is the interior of the intersection of $\mathcal{U}$, that is, the union of all the open sets included in the intersection of $\mathcal{U}$.) Additionally, in a topology meets distribute over joins in the sense that[7]

$$a \wedge \bigvee X = \bigvee \{a \wedge x : x \in X\}.$$

A complete lattice satisfying this condition is called a *locale*. A locale may be regarded as a topology not built on an underlying set of points.[8] Pointless topology is obtained by enlarging attention from topologies to locales.

We note that an *implication* operation $\Rightarrow$ may be introduced in a locale L by defining

$$x \Rightarrow y = \vee \{z : x \wedge z \leq y\}.$$

It is then not hard to show that $x \Rightarrow y$ is the *largest* element z for which $x \wedge z \leq y$. A complete lattice in which an implication operation is so definable is called a *complete Heyting algebra;* thus any locale is one.

What are the counterparts, for locales, of continuous functions between topological spaces? Recall that a continuous function f: $X \to Y$ between topological spaces is characterized by the property that the inverse image $f^{-1}[U]$ of any open set U in Y is open in X. The inverse image operation can accordingly be seen as a function from the topology on Y to that on X. Now the inverse image operation also preserves **1,** arbitrary unions and pairwise intersections of subsets. So it is natural to take a continuous map f between locales to be one satisfying the analogous conditions:

$$f(\mathbf{1}) = \mathbf{1}; \quad f(x \wedge y) = f(x) \wedge f(y); \quad f(\vee X) = f(\vee X).$$

Because inverse images and functions "run in opposite directions" we agree that, given two locales L and L', a continuous map $L \to L'$ will be a function f: $L' \to L$— in the opposite direction— satisfying the above conditions.

Locales and continuous functions in the sense just defined can now be regarded as constituting the objects and maps of **Loc**, the *category of locales*. Associating each topological space with its topology and each continuous map between spaces with its inverse image operation now gives rise to a functor **Top** $\to$ **Loc** which embeds the former in the latter. If **Top** can be seen as an embodiment of the idea of continuity but still dependent on the point concept, **Loc** can be seen as a more general embodiment of the idea of continuity, now freed of all dependence on the point concept.

[7]Here we write $x \wedge y$ for $\{x, y\}$.

[8]It should be mentioned, however, that the concept of a point can be defined for locales and locales arising as topologies characterized in terms of that notion. See Johnstone (1982), Ch. 2.

7.3 Sheaves and Toposes

The most far-reaching generalization of the concept of topological space, and hence the most general embodiment of the idea of continuity, is the category-theoretic concept of *topos*. The idea of a topos developed from the concept of a *sheaf* on a topological space. Mac Lane and Moerdijk present the idea underlying the latter in the following terms:

> *A sheaf [on a space X] is a way of describing a class of functions on X—especially classes of "good" functions, such as the functions on (parts of) X which are continuous or differentiable. The description tells the way in which a function f defined on an open subset U of X can be restricted to functions $f|_V$ on open subsets $V \subseteq U$ and then can be recovered by piecing together (collating) the restrictions to the open subsets V_i of a covering of U. The restriction-collation description applies not just to functions, but to other mathematical structures defined "locally" on a space X.*[9]

Here is the formal definition. We first define a *presheaf*[10] on a topological space X to be an assignment to each open set U of a set $F(U)$ and to each pair of open sets U, V such that $V \subseteq U$ of a "restriction" map F_{UV}: $F(U) \rightarrow F(V)$ such that, whenever $W \subseteq U \subseteq V$, $F_{UW} = F_{VW} \circ F_{UV}$ and F_{UU} is the identity map on $F(U)$. If $s \in F(U)$, write $s|_V$ for $F_{UV}(s)$—the *restriction* of s to V. A presheaf F is a *sheaf* if it satisfies the following "covering" condition: whenever $U = \bigcup_{i \in I} U_i$ and we are given a set $\{s_i: i \in I\}$ such that $s_i \in F(U_i)$ for all $i \in I$ and $s_i|_{U_i \cap U_j} = s_j|_{U_i \cap U_j}$ for all $i, j \in I$, then there is a *unique* $s \in F(U)$ such that $s|_{U_i} = s_i$ for all $i \in I$. For example, $C(U)$ = set of continuous real-valued functions on U, and $s|_V$ = restriction of s to V defines the sheaf of continuous real-valued functions on X.

Given two presheaves F, G on X, a *natural transformation* η: $F \rightarrow G$ is an assignment, to each open set U, of a map η_U: $F(U) \rightarrow G(U)$ which is compatible with restrictions, i.e. satisfies for $V \subseteq U$, $\eta_V \circ F_{UV} = G_{UV} \circ \eta_U$. Presheaves, or sheaves, on X, together with natural transformations, form categories **Pshv(X), Shv (X)**—the *category of presheaves*, or *sheaves*,[11] respectively. on X. **Shv**(X) may be regarded as a category-theoretic embodiment of X: it encodes all the information about locally defined structures on X.

A sheaf on a topological space arises by the imposition of a covering condition on a presheaf. In the early 1960s **Alexandre Grothendieck** (1928–2014) extended the

[9] Mac Lane and Moerdijk (1992), p. 64. Grothendieck describes a sheaf on a space as a "ruler that can be used for taking measurements on it." (1986, Promenade 13).

[10] A presheaf on X may be seen as a functor of a certain kind. Any partially ordered set (P, $\leq$) can be regarded as a category in which the members of P constitute the objects and, for each p, $q \in P$, there is given exactly one map $p \rightarrow q$ just when $p \leq q$. In particular the family **O**(X) of open sets of a topological space X, partially ordered by inclusion, may be regarded as a category. The same family, partially ordered by *reverse* inclusion, yields the "opposite" category $\mathbf{O}(X)^{\mathbf{op}}$. Functors from the latter to **Set** are just the presheaves on X.

[11] Grothendieck sees the category of sheaves on a space as a "superstructure of measurement", which may be "taken to incorporate all that is most essential about that space." (1986, Promenade 13).

notion of covering, and so also the concept of sheaf to *arbitrary categories*, thus conferring on both concepts a vast new generality.

To see how this was effected, we require the notion of a *set varying over* a category **C** (or **C**-*variable* set), which is defined simply to be a functor from **C** to the category **Set,** and the notion of a *natural transformation* between variable sets. A natural transformation between two **C**- variable sets F and G is a function assigning to each object A of **C** a map η_A: $FA \rightarrow GA$ in such a way that, for each map f: $A \rightarrow B$, we have $Gf \circ \eta_A = \eta_B \circ Ff$. The category $\mathbf{Set}^{\mathbf{C}}$ of sets varying over **C** then has as objects **C**-variable sets and as arrows natural transformations between them.

Now we can define a *presheaf* on **C** to be a set varying over the category $\mathbf{C}^{\text{op}}$—the opposite category to **C** in which all maps are "reversed".[12] The category $\mathbf{Set}^{\mathbf{C}^{\text{op}}}$ is called the *category of preheaves* on **C.** There is a natural embedding[13]—the *Yoneda embedding*—of **C** into $\mathbf{Set}^{\mathbf{C}^{\text{op}}}$ whose action on objects is defined as follows. For any object C of **C**, YC is the presheaf on **C** which assigns, to each object X of **C**, the set Hom(X, C) of maps $X \rightarrow C$ in **C**. YC—the *Yoneda presheaf* of C—is the natural presheaf representative of C; the two are usually identified.

To see how the notion of covering may be extended to arbitrary categories, we re-examine the concept in the case of a topological space X. Given an open set U, we define a *sieve* on U to be a family S of open subsets of U such that any open subset of a member of S is itself a member of S; if $\bigcup S = U$, the sieve S is called a *covering sieve* on U. We write $J(U)$ for the set of all covering sieves on U. It is easily seen that $J(U)$ satisfies the following conditions:

(1) *the maximal sieve consisting of all open subsets of U is a member of* $J(U)$;
(2) *if* $S \in J(U)$ *and V is open in X, then the sieve* $V{*}S = \{W: W \in S \text{ and } W \subseteq V\}$ *is a member of* $J(V)$; *that is, the restriction of a covering sieve to a smaller open set is a covering sieve;*
(3) *if* $S \in J(U)$ *and R is a sieve on U such that, for each* $V \in S$ *we have* $V{*}S \in J(V)$, *then* $R \in J(U)$; *that is, if the restriction of a sieve R to each member of a covering sieve S is a covering sieve, then R is a covering sieve.*

These three conditions were taken by Grothendieck as being characteristic of the notion of covering; they can be generalized to an arbitrary category **C** in the following way, giving rise to what has become known as a *Grothendieck topology.*

First, by analogy with the above definition of sieve, we define a *sieve* on an object C of **C** to be a family S of maps in **C,** each with codomain C,[14] closed under composition on the right, that is, satisfying $f \in S \Rightarrow f \circ g \in S$ for any map g composable with f on the right. Equivalently, a sieve on C may be defined as a

[12] To be precise, $\mathbf{C}^{\text{op}}$ has the same objects as **C;** and in $\mathbf{C}^{\text{op}}$ the maps from an object A to an object B correspond precisely to the maps $B \rightarrow A$ in **C**, with composites and identity maps defined analogously.

[13] A functor F: $\mathbf{C} \rightarrow \mathbf{D}$ is an *embedding* if, for any objects A, B of **C**, any map $FA \rightarrow FB$ in **D** is of the form Ff for a unique f: $A \rightarrow B$ in **C.**

[14] Here we view a map f: $D \rightarrow C$ as an "inclusion" of D in C.

subfunctor[15] of *C's* Yoneda presheaf *YC*. Note that, if *S* is a sieve on *C* and $h: D \to C$ is any map with codomain *C*, then the set

$$h^*(S) = \{g : \text{codomain}(g) = D \ \& \ h \circ g \in S\}$$

is a sieve on *D*. A *Grothendieck topology,* or *covering system,* on **C** is a function *J* which assigns to each object *C* a collection *J* (*C*) of sieves—called *J-covering sieves*—on *C* in such a way that the following conditions (i), (ii), (iii) are satisfied:

(i) *for any C, the maximal sieve* $\{h: \text{codomain } (h) = C\}$ *is a member of J(C);*
(ii) *if* $S \in J(C)$ *and* $h: D \to C$*, then the sieve* $h{*}S$ *is a member of J(D);*
(iii) *if* $S \in J(C)$ *and R is a sieve on U such that, for each* member $h: D \to C$ *of S we have* $h{*}R \in J(D)$*, then* $R \in J(C)$*.*

Clearly these conditions are immediate generalizations of (1), (2), (3) to the category **C.** The Grothendieck topology *Trv* on **C** in which only the maximal sieves on an object are covering sieves is called the *trivial* topology.

A category equipped with a covering system was termed by Grothendieck a *site.*[16] If *X* is a topological space, and *J* is defined on the category **O**(*X*) of open sets by taking *J*(*U*) to be the family of all sieves *S* on *U* in the above sense such that $\bigcup S = U$, then **O**(*X*), equipped with the Grothendieck topology *J,* is a site. It is called the site *canonically associated* with *X.* In view of this a site may be regarded as a "generalized topological space", or indeed as a new embodiment of the idea of a pointless space.

Now we can sketch how Grothendieck went on to extend the concept of sheaf to arbitrary sites.[17] Recall that any sieve *S* on an object *C* of **C** may be regarded as a subfunctor of *C*'s Yoneda functor *YC.* Now suppose we are given an object *F* of $\mathbf{Set}^{\mathbf{C}^{op}}$—a presheaf on **C**—and a map $f: S \to F$ in $\mathbf{Set}^{\mathbf{C}^{op}}$—a natural transformation from *S* to *F.* A map $g: YC \to F$ is called an *extension* of *f* to *YC* if its restriction (in the evident sense) to the subfunctor *S* coincides with *f.* The presheaf *F* is then a *(J-)sheaf* if, for any object *C* and any *J*-covering sieve *S* on *C,* each map $f: S \to F$ in $\mathbf{Set}^{\mathbf{C}^{op}}$ has a *unique* extension to *YC.* Speaking figuratively, a *J*-sheaf is a presheaf which "believes" that (the Yoneda presheaf of) any object *C* is "really covered" by any of its *J*-covering sieves, in the sense that, in $\mathbf{Set}^{\mathbf{C}^{op}}$ is, any map from a *J*-covering sieve to *F* fully determines a map from *YC* to *F.* Note that every presheaf is a *Trv*-sheaf.

Now the category $\mathbf{Shv}_J(\mathbf{C})$ of sheaves on a site (**C**, *J*) has objects all sheaves and as maps all natural transformations between them. So in particular $\mathbf{Shv}_{Trv}(\mathbf{C})$ is just the category of presheaves on **C.** There is a natural functor $L: \mathbf{Set}^{\mathbf{C}^{op}} \to \mathbf{Shv}_J(\mathbf{C})$

[15]If *F* and *G* are two functors $\mathbf{C}^{op} \to \mathbf{Set}$, *F* is called a *subfunctor* of *G* if $FX \subseteq GX$ for any object *X* of **C.**

[16]Strictly speaking, in defining a site one should specify that the underling category **C** be *small*, that is, its collections of objects and maps should both be sets rather than proper classes in the sense of Gödel-Bernays set theory.

[17]For a detailed account, see Mac Lane and Moerdijk (1992), Ch. III.

called the *associated sheaf functor* which sends each presheaf F to the to the sheaf LF which "best approximates" it in an appropriate sense.[18]

Grothendieck assigned the term "topos"[19] to any category of sheaves associated with a site. So in particular all categories of presheaves, or variable sets, are toposes. Grothendieck saw the topos concept as uniting the continuous and the discrete:

> *This idea encapsulates, in a single topological intuition, both the traditional topological spaces, incarnation of the world of the continuous quantity ... and a huge number of other sorts of structures which ... had appeared to belong irrevocably to the "arithmetic world" of "discontinuous" or "discrete" aggregates.*[20]

F. William Lawvere and **Myles Tierney** later formulated a simple and decisive list of axioms satisfied by Grothendieck's toposes. These turned out to be the same as certain key axioms satisfied by the category of sets: aremarkable instance of the unity of the Continuous and the Discrete. Lawvere and Tierney introduced the term *elementary topos* (subsequently just "topos") for a category **E** possessing the following properties, each of which, suitably expressed in category-theoretic language, is clearly possessed by **Set**:

(a) **E** has a "terminal" object 1 such that, for any object A, there is a unique map $A \to 1$. (In **Set,** 1 may be taken to be any singleton, in particular $\{0\}$. Note that elements of a set A correspond to maps $1 \to A$.)[21]
(b) Any pair of objects A, B of **E** has a (Cartesian) product $A \times B$ in **E.**
(c) For any pair of objects A, B in **E** one can form in **E** the 'exponential" object B^A of all maps $A \to B$.
(d) **E** has a "truth-value" object Ω containing a distinguished element true such that for each object A there is a natural correspondence between subobjects of A and maps $A \to \Omega$. (In **Set**, Ω may be taken to be the set $2 = \{0, 1\}$, and true the element 1; maps $A \to \Omega$ are then *characteristic functions* on A and the exponential Ω^A corresponds to the *power set* of X.)

---

[18]See, e.g., Mac Lane and Moerdijk (1992), p. 87.

[19]The term is a back-formation from the word "topology" to its original Greek source "topos", "place".

[20]Grothendieck (1986), Promenade 13. He goes on:

> *As is often the case in mathematics, we've succeeded ... in expressing a certain idea—that of a "space" in this instance—in terms of another one—that of a "category". Each time the discovery of such a* translation *from one notion (representing one kind of situation) to another (which corresponds to a different situation) enriches our understanding of both notions, owing to the unanticipated confluence of specific intuitions which relate first to one and then to the other. Thus we see that a situation said to have a "topological" character (embodied in some given space) has been translated into a situation whose character is "algebraic" (embodied in the category); or, if you wish, "continuity" (as present in the space) finds itself "translated" or "expressed by a categorical structure of an algebraic character, which until then had been understood in terms of something "discrete" or "discontinuous".*

[21]More generally, *elements* or *points* of an object A of a category are maps from the terminal object to A.

(e) **E** has a "natural numbers" system, that is, a triple (N, s, o) consisting of an object N and maps $s\colon N \to N$, $o\colon 1 \to N$. These are subject to the *Peano-Lawvere axiom*

For any diagram $1 \xrightarrow{a} X \xrightarrow{g} X$ in **E**, *there is a unique map* $N \xrightarrow{f} X$ *for which the diagram*

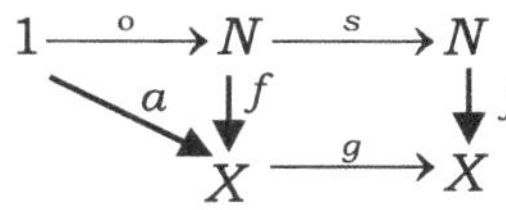

commutes.

Thus **Set** and all categories of "sets undergoing variation" such as sheaves and presheaves are toposes.

Toposes can also arise as categories of sets with a *generalized equality relation*||, with arrows preserving that relation in an appropriate sense. Some of the most important examples in this regard are the *categories of Heyting algebra-valued sets.* Given a complete Heyting algebra H, an *H-valued set* is a pair (I, δ) consisting of a set I and a map $\delta\colon I \times I \to H$ (the generalized equality relation on I) satisfying the following conditions, in which we write $\delta_{ii'}$ for $\delta(i, I')$ (and similarly below):

$$\delta_{ii'} = \delta_{i'i} \quad \text{(symmetry)}$$

$$\delta_{ii'} \wedge \delta_{i'i''} \leq \delta_{ii''} \quad \text{(substitutivity)}$$

The category $\mathbf{Set}_H$ of *H-valued sets* has as objects all H-valued sets. A $\mathbf{Set}_H$ -arrow f: $(I, \delta) \to (J, \varepsilon)$ is a map $f\colon I \times J \to H$ such that

$$\delta_{ii'} \wedge fij \leq f_{i'j} f_{ij} \wedge \varepsilon_{jj'} \leq f_{ij'} \quad \text{(preservation of identity)}$$

$$f_{ij} \wedge f_{ij'} \leq \varepsilon_{jj'} \quad \text{(single-valuedness)}$$

$$\vee_{i \in I} f_{ij} = \delta_{ii} \ \text{(denined on I)}$$

The composite $g \circ f$ of two arrows $f\colon (I, \delta) \to (J, \varepsilon)$ and $g\colon (J, \varepsilon) \to (K, \eta)$ is given by $(g \circ f)_{ik} = \vee_{j \in J}(f_{ij} \wedge g_{jk})$.Then $\mathbf{Set}_H$ is a topos in which the subobject classifier is the H-valued set $(H, \Leftrightarrow)$.

There are three principal types of topos:

Type I: categories of "sets undergoing variation These include sheaves and presheaves.
Type II: categories of sets with a generalized equality relation.
Type III: categories arising as models of intuitionistic many- sorted theories (see below)

Lawvere and Tierney established a striking connection between topos theory and *logic*. This originates with the observation that in the category of sets we have the usual logical operations $\wedge$ (conjunction), $\vee$ (disjunction), $\neg$ (negation), $\to$ (implication), $\forall$ (universal quantification), $\exists$ (existential quantification) defined on the object $2 = \{0, 1\}$ of truth values. The richness of a topos **E**'s internal structure enables analogous "logical operations" to be defined on its object Ω of truth values.

The remarkable thing is that the body of laws satisfied by the logical operations in a topos—its "internal logic"—does not, in general, correspond to classical logic, but rather to the *intuitionistic* or *constructive* logic of Brouwer and Heyting in which the law of excluded middle is not affirmed.[22]

To be precise, to each topos **E** can be associated a certain formal language $\mathcal{L}(\mathbf{E})$—its *internal language*[23]—which resembles the usual language of set theory in that included among its primitive signs are equality (=), membership (∈) and the set formation operator ({:}). $\mathcal{L}(\mathbf{E})$ is a *many-sorted* language—let us call it a *topos* language— with each of its sorts corresponding to an object of **E.** Thus, in particular, for each object A of **E**, $\mathcal{L}(\mathbf{E})$ contains a list of variables of sort A. Each term t of $\mathcal{L}(\mathbf{E})$ with free variables $x_1, \ldots, x_n$, of sorts $A_1, \ldots, A_n$ is then assigned an object B as a sort in such a way as to enable a map $[\![t]\!] : A_1 \times \ldots \times A_n \to B$—the *interpretation* of t in **E**—to be defined. Terms of sort Ω are called *formulas,* or *propositions.* A formula φ is said to be *true*, or to *hold*, in **E** if its interpretation $[\![\varphi]\!] : A_1 \times \ldots \times A_n \to \Omega$ has constant value true. It can be shown then that all the axioms and rules of inference of (free) intuitionistic logic[24] formulated in $\mathcal{L}(\mathbf{E})$ are true in **E.** It is in this sense that the "internal logic" of a topos is intuitionistic or constructive.[25]

The set of all sentences of $\mathcal{L}(\mathbf{E})$ which are true in **E** is called the (*internal*) *theory* *Th*(**E**) of **E.** *Th*(**E**) is an intuitioistic many-sorted theory. A topos **E** is then, in a natural sense, a model of *Th*(**E**)**.** Within the internal language and the associated theory of a topos mathematical concepts can be formulated, arguments carried out and constructions performed much as one does in "ordinary" set theory, only observing the rules of intuitionistic logic. In fact, any topos may be regarded as a model of *intuitionistic set theory* **IST**.[26] This means that, within a topos, mathematical constructions can be carried out as if within **IST.** In particular, the rational numbers can be defined as usual and the real numbers oconstructed by employing either Dedekind's procedure of making cuts in the rationals or Cantor's procedure employing equivalence classes of Cauchy sequences of rationals. While classically these two constructions lead to isomorphic results, this is *not true* in constructive logic: indeed, a number of toposes have been constructed in which (in the topos's internal logic), the ordered rings of Dedekind and Cantor reals fail to be isomorphic.[27]

---

[22]See Chap. 9.

[23]See Bell (1988).

[24]For these see, e.g., Bell (1998), Ch. 8.

[25]Of course, many important toposes such as the topos of sets have an internal logic that is classical. These are exceptions, however.

[26]By intuitionistic set theory **IST** we mean the theory in intuitionistic first-order logic whose axioms are the "usual" axioms of Zermelo set theory (without the axiom of choice), namely: Extensionality, Pairing, Union, Power set, Infinity and Separation. For an exposition of **IST**, see Bell (2014).

[27]In fact, this is the case for the topos Shv ($\mathbb{R}$) of sheaves on the usual (classical) topological space $\mathbb{R}$ of real numbers. For an account see Johnstone (2002), §D4.7.

Accordingly, in a topos-theoretic or constructive universe there is more than one candidate[28] for the role of the mathematical continuum. The classical view that the linear continuum is a uniquely determined entity is replaced by a pluralistic conception under which the continuum has a number of realizations with essentially different properties. And even if one decides, for example, to choose the Dedekind reals as one's continuum, its properties may fall far short of those possessed by its classical counterpart. For example, in constructive logic it cannot be proved that the Dedekind reals satisfy the least upper bound property. It can be shown, in fact,[29] that, in a topos's internal theory, the Dedekind reals possess this property exactly when the logical law[30] $\neg(\alpha \wedge \beta) \rightarrow (\neg\alpha \vee \neg\beta)$ holds there. This is an arresting instance of the connection between logic and the properties of the mathematical continuum made visible by the shift from classical to constructive logic.

We have mentioned that in 1924 Brouwer proved from his intuitionistic principles that every real-valued function on a closed interval of the real numbers is uniformly continuous. A number of toposes have been constructed in which the corresponding statement for Dedekind reals holds. One such[31] is the topos of sheaves on the space of irrational numbers. In any topos in which Brouwer's theorem holds all closed intervals are cohesive.[32]

Topos theory thus allows a remarkable flexibility in the handling of the continuum. In Chap. 10 we shall describe a still more remarkable representation of the continuum made possible by topos theory, one in which the infinitesimal plays a key role.

---

[28]Others are mentioned in Johnstone, *op. cit.*

[29]*Op. cit.*, Thm. 4.7.11.

[30]This is the only one of De Morgan's laws which does not generally hold in constructive logic. It fails, in particular, in Shv($\mathbb{R}$).

[31]As shown by Dana Scott in his paper *Extending the topological interpretation to intuitionistic analysis II*, pp. 235–255 of Buffalo Conference in Proof Theory and Intuitionism, North-Holland, 1970. Scott's result was extended by Martin Hyland in *Continuity and spatial toposes*, in Fourman, Mulvey and Scott (1979), pp. 440–465. In Mac Lane and Moerdijk (1992), Ch. VI, §9 a topos is constructed in which all real-valued functions defined on the whole real line are continuous. See Appendix C for further discussion.

[32]See Appendix A.

Chapter 8
Nonstandard Analysis

Once the continuum had been provided with a set-theoretic foundation, the use of the infinitesimal in mathematical analysis was largely abandoned.[1] And so the situation remained for a number of years. The first signs of a revival of the infinitesimal approach to analysis surfaced in 1958 with a paper by A. H. Laugwitz and C. Schmieden.[2] But the major breakthrough came in 1960 when it occurred to the mathematical logician **Abraham Robinson** (1918–1974) that "the concepts and methods of contemporary Mathematical Logic are capable of providing a suitable framework for the development of the Differential and Integral Calculus by means of infinitely small and infinitely large numbers."[3] This insight led to the creation of *nonstandard analysis* (**NSA**),[4] which Robinson regarded as realizing Leibniz's conception of infinitesimals and infinities as ideal numbers possessing the same properties as ordinary real numbers. In the introduction to his book on the subject he writes:

> *It is shown in this book that Leibniz's ideas can be fully vindicated and that they lead to a novel and fruitful approach to classical Analysis and to many other branches of mathematics. The key to our method is provided by the detailed analysis of the relation between mathematical languages and mathematical structures which lies at the bottom of contemporary model theory.*[5]

---

[1]However, nonarchimedean ordered structures, containing infinite and infinitesimal elements, continued to be studied, chiefly by algebraists, notably Hölder, Hahn, Baer, and Birkhoff. See Fuchs (1963) and Ehrlich (2006).

[2]Laugwitz, A.H. and Schmieden, C., *Eine Erweiterung der Infinitesimalrechnung*, Mathematische Zeitschrift **69**, pp. 1–39.

[3]Robinson [1996], p. xiii.

[4]So-called, Robinson says, because his theory "involves and was, in part, inspired by the so-called Non-standard models of Arithmetic whose existence was first pointed out by T. Skolem." (*Ibid.*)

[5]Robinson, *op. cit.*, p. 2.

J. L. Bell, *The Continuous, the Discrete and the Infinitesimal in Philosophy and Mathematics*, The Western Ontario Series in Philosophy of Science 82,
https://doi.org/10.1007/978-3-030-18707-1_8

After Robinson's initial insight, a number of ways of presenting nonstandard analysis were developed. Here is a sketch of one of them.[6]

Starting with the classical real line $\mathbb{R}$, a set-theoretic universe—the *standard universe*—is first constructed over it: here by such a universe is meant a set U containing $\mathbb{R}$ which is closed under the usual set-theoretic operations of union, power set, Cartesian products and subsets. Now write $\mathbf{U}$ for the structure $(U, \in)$, where $\in$ is the usual membership relation on U: associated with this is the extension L (U) of the first-order language of set theory to include a name u for each element u of U. Now, using the well-known compactness theorem for first-order logic, $\mathbf{U}$ is extended to a new structure $^*\mathbf{U} = (^*U, {}^*\in)$, called a *nonstandard universe*, satisfying the following key principle:

> ***Saturation Principle.*** *Let* $\mathbf{\Phi}$ *be a collection of* L(U)*-formulas with exactly one free variable. If* $\mathbf{\Phi}$ *is* finitely satisfiable *in* u, *that is, if for any finite subset* $\mathbf{\Phi}'$ *of* $\mathbf{\Phi}$ *there is an element of* U *which satisfies all the formulas of* $\mathbf{\Phi}'$ in $\mathbf{U}$, *then there is an element of* $*U$ *which satisfies all the formulas of* $\mathbf{\Phi}$ *in* $^*\mathbf{U}$.

The saturation property expresses the intuitive idea that the nonstandard universe is very rich in comparison to the standard one. Indeed, while there may exist, for each finite subcollection $\boldsymbol{F}$ of a given collection of properties $\boldsymbol{P}$, an element of U satisfying the members of $\boldsymbol{F}$ in u, there may not necessarily be an element of U satisfying *all* the members of $\boldsymbol{P}$. The saturation of $^*\boldsymbol{U}$ guarantees the existence of an element of $*U$ which satisfies, in $^*\boldsymbol{U}$, all the members of $\boldsymbol{P}$. For example, suppose the set $\mathbb{N}$ of natural numbers is a member of U; for each $n \in \mathbb{N}$ let $P_n(x)$ be the property $x \in \mathbb{N}$ & $n < x$. Then clearly, while each finite subcollection of the collection $\boldsymbol{P} = \{P_n\colon n \in \mathbb{N}\}$ is satisfiable in u, the whole collection is not. An element of $*U$ satisfying all the members of $\boldsymbol{P}$ in $*$u will then be an "natural number" greater than every member of $\mathbb{N}$, that is, an *infinite* number.

From the saturation property it follows that $^*\boldsymbol{U}$ satisfies the important

> ***Transfer Principle.*** *If* σ *is any sentence of* L(U), *then* σ *holds in* $\boldsymbol{U}$ *if and only if it holds in* $^*\boldsymbol{U}$.

The transfer principle may be seen as a version of Leibniz's continuity principle: it asserts that all first-order properties are preserved in the passage to or "transfer" from the standard to the nonstandard universe.

The members of U are called *standard sets,* or *standard objects*; those in $*U - U$ *nonstandard sets* or *nonstandard objects*: $*U$ thus consists of both standard and nonstandard objects. The members of $*U$ will also be referred to as $*$*-sets* or $*$*-objects* Since $U \subseteq *U$, under this convention every set (object) is also a $*$-set (object). The $*$*-members* of a $*$-set A are the $*$-objects x for which $x * \in A$

If A is a standard set, we may consider the collection $A^\bigstar$ —the *inflate* of A— consisting of all the $*$-members of A: this is not necessarily a standard set nor even a $*$-set. The inflate of a standard set may be regarded as the same set viewed from a nonstandard vantage point. While clearly $A \subseteq A^\bigstar$, $A^\bigstar$ may contain "nonstandard" elements not in A. It can in fact be shown that *infinite* standard sets always get

[6]Based on Ch. 11 of Bell and Machover (1977), and Keisler (1994).

"inflated" in this way. Using the transfer principle, any function f between standard sets automatically extends to a function—also written f—between their inflates.

Each mathematical structure $\mathbf{A} = (A, R)$ has an inflate $\mathbf{A}^\bigstar = (A^\bigstar, R^\bigstar)$. From the transfer principle it follows that $\mathbf{A}$ and $\mathbf{A}^\bigstar$ have precisely the same first-order properties.

Now suppose that the set $\mathbb{N}$ of natural numbers is a member of U. Then so is the set $\mathbb{R}$ of real numbers, since each real number may be identified with a set of natural numbers. $\mathbb{R}$ may be regarded as an ordered field, and the same is therefore true of its inflate $\mathbb{R}^\bigstar$.This, the *hyperreal line*, has precisely the same first-order properties as $\mathbb{R}$. The members of $\mathbb{R}^\bigstar$ are called *hyperreals*. A standard hyperreal is then just a real, to which we shall refer for emphasis as a *standard real*. Since $\mathbb{R}$ is infinite, nonstandard hyperreals must exist. The saturation principle implies that there must be an *infinite* (nonstandard) hyperreal,[7] that is, a hyperreal a such that $a > n$ for every $n \in \mathbb{N}$. In that case its reciprocal $\frac{1}{a}$ is infinitesimal in the sense of exceeding 0 and yet being smaller than $\frac{1}{n+1}$ for every $n \in \mathbb{N}$. In general. we call a hyperreal *a infinitesimal* if its absolute value $|a|$ is $< \frac{1}{n+1}$ for every $n \in \mathbb{N}$. In that case the set I of infinitesimals contains not just 0 but a substantial number (in fact, infinitely many) of other elements. Clearly I is an additive subgroup of $\mathbb{R}$, that is, if $a, b \in I$, then $a - b \in I$.

The **members** of the inflate $\mathbb{N}^\bigstar$ of $\mathbb{N}$ are called *hypernatural numbers*. As for the hyperreals, it can be shown that $\mathbb{N}^\bigstar$ also contains nonstandard elements which must exceed every member of $\mathbb{N}$: these are called *infinite* hypernatural numbers.

For hyperreals a, b we define $a \approx b$ and say that a and b are *infinitesimally close* if $a - b \in I$. This is an equivalence relation on the hyperreal line: for each hyperrreal a we write $\mu(a)$ for the equivalence class of a under this relation and call it the *monad* of a. The monad of a hyperreal a thus consists of all the hyperreals that are infinitesimally close to a: it may thought of as a small cloud centred at a. Note also that $\mu(0) = I$.

A hyperreal a is *finite* if it is not infinite; this means that $|a| < n$ for some $n \in \mathbb{N}$. It is not difficult to show that finiteness is equivalent to the condition of *near-standardness*: here a hyperreal a is *near-standard* if $a \approx r$ for some standard real r.

Much of the usefulness of nonstandard analysis stems from the fact that statements of classical analysis involving limits or the (ε, δ) criterion admit succinct, intuitive translations into statements involving infinitesimals or infinite numbers, in turn enabling comparatively straightforward proofs to be given of classical theorems. Here are some examples of such translations[8]:

---

[7]It follows that $\mathbb{R}^\bigstar$ is a nonarchimedean ordered field. One might question whether this is compatible with the facts that $\mathbb{R}^\bigstar$ and $\mathbb{R}$ share the same first-order properties, but the latter is archimedean. These data are consistent because the archimedean property is not first-order. However, while $\mathbb{R}^\bigstar$ is nonarchimedean, it is**-archimedean* in the sense that, for any $a \in \mathbb{R}^\bigstar$ there is $n \in \mathbb{R}^\bigstar$ for which $a < n$.

[8]Robinson (1996), Ch. 3. A number of "nonstandard" proofs of classical theorems may also be found there.

- *Let* $< s_n >$ be *a standard infinite sequence of real numbers and let s be a standard real number. Then s is the limit of* $< s_n >$ *within* $\mathbb{R}$, $\lim_{n \to \infty} s_n = s$ *in the classical sense, if and only if* $s_n \approx s$ *for all* infinite *subscripts n.*
- *A standard sequence* $< s_n >$ *converges if and only if* $s_n \approx s_m$ *for all infinite n and m.* (Cauchy's criterion for convergence.)

Now suppose that *f* is a real-valued function defined on some open interval (*a, b*). We have remarked above that *f* automatically extends to a function—also written *f*—on $(a, b)^*$.

- *In order that the standard real number c be the limit of f(x) as x approaches* x_0, $\lim_{x \to x_0} f(x) = c$, *with* x_0 *a standard real number in (a, b), it is necessary and sufficient that* $f(x) \approx f(x_0)$ *for all* $x \approx x_0$.
- *The function f is continuous at a standard real number* x_0 *in (a, b) if and only if* $f(x) \approx f(x_0)$ *for all* $x \approx x_0$. (This is *equivalent to saying that f maps the monad of* x_0 *into the monad of* $f(x_0)$.
- *In order that the standard number c be the derivative of f ay* x_0 *it is necessary and sufficient that*

$$\frac{f(x) - f(x_0)}{x - x_0} \approx c$$

for all $x \neq x_0$ *in the monad of* x_0.

Many other branches of mathematics admit neat and fruitful nonstandard formulations. We end this chapter with a sketch of how *topology* looks from a nonstandard point of view.

We fix a nonempty $X \in U$, and a topology $\boldsymbol{T}$ on *X*: necessarily $\boldsymbol{T} \in U$. For each point $p \in X$ let $\mathcal{N}_p$ be the collection of neighbourhoods of *p* with respect to the topology $\boldsymbol{T}$. We define the *monad* of *p* to be the collection

$$\mu(p) = \left\{ q \in X^\star : q \in V^\star \text{ for all } V \in \mathcal{N}_p \right\}.$$

The mapping μ thus defined, which maps *X* into the collection of all subcollections of $X^\star$, is called the *monadology* of the space *X* (that is, of the space $(X, \boldsymbol{T})$). If $q \in \mu(p)$ we write $q \approx p$ and say that *q* is *near p*. If $q \approx p$ for some $p \in X$, we say that *q* is *near-standard.* Clearly every point of *X* is near-standard, but there may be near-standard points of $X^\star$ which are not standard.

One can now proceed to formulate nonstandard characterizations of various standard topological notions. For example:

- *A subset A of X is open if and only if* $p \in A \Rightarrow \mu(p) \subseteq A^\star$; *A is closed if and only if* $\mu(p) \cap A^\star \neq \varnothing \Rightarrow p \in A$. (It follows from this that the topology of a space can be recovered from its monadology.)
- *X is a Hausdorff space if and only if the monads of distinct points are disjoint.*

- *X is compact if and only if every point of $X^\star$ is near-standard.*
- *Let f: $X \to Y$ be a map of X into a topological space Y. Then f is continuous if, for all $p \in X$, $q \in X^\star$, $q \approx p \Rightarrow f(q) \approx f(p)$.*

Thus $f:X \to Y$ is continuous if, for any point $p \in X$, points of $X^\star$ near p are mapped by f to points of $Y^\star$ near $f(p)$. Indeed, "this is what mathematicians always wanted to say about continuity, but didn't quite know how. In nonstandard analysis, this highly intuitive characterization is at the same time completely rigorous."[9]

Finally, it should be pointed out that while the usual models of nonstandard analysis are obtained using highly nonconstructive tools, other methods have been developed for producing such models which are constructively acceptable.[10]

[9]Bell and Machover (1977), p. 560.

[10]See Martin-Löf (1990), Moerdijk (1995) and Palmgren (1998, 2001).

Chapter 9
The Continuum in Constructive and Intuitionistic Mathematics

In constructive mathematics, a problem is counted as solved only if an explicit solution can, in principle at least, be produced. Thus, for example, "There is an x such that $P(x)$" *means* that, in principle at least, we can explicitly produce an x such that $P(x)$. If the solution to the problem involves parameters, we must be able to present the solution explicitly by means of some *algorithm* or *rule* when given values of the parameters. That is, "for every x there is a y such that $P(x, y)$ *means* that, we possess an explicit method of determining, for any given x, a y for which $P(x, y)$. This leads us to examine what it means for a mathematical object to be explicitly given.

9.1 The Constructive Real Number Line[1]

To begin with, everybody knows what it means to give an *integer* explicitly. For example, $7 \cdot 10^4$ is given explicitly, while the number n defined to be 0 if an odd perfect number exists, and 1 if an odd perfect number does not exist, is not given explicitly. The number of primes less than, say, $10^{1000000}$ is given explicitly, in the sense intended here, since we could, *in principle at least,* calculate this number. Constructive mathematics as we shall understand it is not concerned with questions of feasibility, nor in particular with what can actually be computed in real time by actual computers. *Rational numbers* may be defined as pairs of integers (a, b) without a common divisor (where $b > 0$ and a may be positive or negative, or a is 0 and b is 1). The usual arithmetic operations on the rationals, together with the operation of taking the absolute value, are then easily supplied with explicit definitions. Accordingly it is clear what it means to give a rational number explicitly. To

[1]My account here draws on Bishop and Bridges (1985).

J. L. Bell, *The Continuous, the Discrete and the Infinitesimal in Philosophy and Mathematics*, The Western Ontario Series in Philosophy of Science 82,
https://doi.org/10.1007/978-3-030-18707-1_9

specify exactly what is meant by giving a *real number* explicitly is not quite so simple. For a real number is by its nature an infinite object, but one normally regards only finite objects as capable of being given explicitly. This difficulty may be overcome by stipulating that, to be given a real number, we must be given a (finite) *rule* or *explicit procedure* for calculating it to any desired degree of accuracy. Intuitively speaking, to be given a real number r is to be given a method of computing, for each positive integer n, a rational number r_n such that

$$| r - r_n | \quad < \frac{1}{n}.$$

These r_n will then obey the law

$$| r_m - r_n | \quad \leq \frac{1}{m} + \frac{1}{n}.$$

So, given any numbers k, p, we have, setting $n = 2k$,

$$| r_{n+p} - r_n | \quad \leq \frac{1}{n+p} + \frac{1}{n} \leq \frac{2}{n} = \frac{1}{k}.$$

One is thus led to *define* a (constructive) real number to be a sequence of rationals $(r_n) = r_1, r_2, \ldots$ such that, for any k, a number n can be found such that[2]

$$| r_{n+p} - r_n | \quad \leq \frac{1}{k}$$

Here we understand that to be given a *sequence* we must be in possession of a *rule* or explicit method for generating its members. Each rational number α may be regarded as a real number by identifying it with the real number $(\alpha, \alpha, \ldots)$. The set of all real numbers will be denoted, as usual, by $\mathbb{R}$. Now of course, for any "given" real number there are a variety of ways of giving explicit approximating sequences for it. Thus it is necessary to define an equivalence relation, "equality on the reals". The correct definition here is: $r =_{\mathbb{R}} s$ if and only if for any k, a number n can be found so that

$$| r_{n+p} - s_{n+p} | \leq \frac{1}{k} \text{ for all } p.$$

When we say that two real numbers are equal we shall mean that they are equivalent in this sense, and so write simply "=" for "$=_{\mathbb{R}}$". To assert the *in*equality $r \neq s$ of two real numbers r, s is to assert that the equality $r = s$ leads to a contradiction. Inequality in this sense is constructively weak. In constructive mathematics a stronger notion of

[2]It will be observed that in defining a constructive real number in this way we are following Cantor's, rather than Dedekind's characterization.

inequality, that of *apartness*, is normally used instead. We say that r and s are *apart,* or *distinguishable*, written $r \# s$, if n and k can actually be found so that $|r_{n+p} - s_{n+p}| > \frac{1}{k}$ for all p. Clearly $r \# s$ implies $r \neq s$, but the converse cannot be affirmed constructively.[3]

Is it constructively the case that for any real numbers x and y, we have $x = y \vee x \neq y$? The answer is no. For if this assertion were constructively true, then, in particular, we would have a method of deciding whether, for any given rational number r, whether $r = \pi^{\sqrt{2}}$ or not. But at present no such method is known—it is not known, in fact, whether $\pi^{\sqrt{2}}$ is rational or irrational. We can, of course, calculate $\pi^{\sqrt{2}}$ to as many decimal places as we please, and if in actuality it is unequal to a given rational number r, we shall discover this fact after a sufficient amount of calculation. If, however, $\pi^{\sqrt{2}}$ is *equal* to r, even several centuries of computation cannot make this fact certain; we can be sure only that its value is very close to r. We have no method which will tell us, in finite time, whether $\pi^{\sqrt{2}}$ exactly coincides with r or not. This situation may be summarized by saying that equality on the reals is *not decidable.* (By contrast, equality on the integers or rational numbers is decidable.) Observe that this does *not* mean $\neg(x = y \vee x \neq y)$. We have not actually derived a *contradiction* from the assumption $x = y \vee x \neq y$; we have only given an example showing its implausibility. It is natural to ask whether it can actually be *refuted.* For this it would be necessary to make some assumption concerning the real numbers which contradicts classical mathematics. Certain schools of constructive mathematics are willing to make such assumptions; but the majority of constructivists confine themselves to methods which are also classically correct.[4]

Despite the fact that constructive equality of real numbers is not a decidable relation, it is *stable*[5] in the sense of satisfying the *law of double negation* $\neg(r \neq s) \Rightarrow r = s$. In fact, we can prove the stronger assertion that $\neg(r \# s) \Rightarrow r = s$. For, given k, we may choose n so that $|r_{n+p} - r_n| \leq \frac{1}{4k}$ and $|s_{n+p} - s_n| \leq \frac{1}{4k}$ for all p. If $|r_n - s_n| \geq \frac{1}{k}$, then we would have $|r_{n+p} - s_{n+p}| \geq \frac{1}{2k}$ for all p, which entails $r \neq s$. If $\neg(r \# s)$, it follows that $|r_n - s_n| < \frac{1}{k}$ and $|r_{n+p} - s_{n+p}| \leq \frac{2}{k}$ for every p. Since for every k we can find n so that this inequality holds for every p, it follows that $r = s$.[6]

One should not, however, conclude from the stability of equality that the law of double negation $\neg\neg A \rightarrow A$ is generally affirmable. That it is not can be seen from the following example. Write the decimal expansion of π and below the decimal expansion $\rho = 0.333\ldots$, terminating it as soon as a sequence of digits

---

[3]In fact the converse is equivalent to *Markov's Principle*, which asserts that, if, for each n, $x_n = 0$ or 1, and if it is contradictory that $x_n = 0$ for all n, then there exists n for which $x_n = 1$. This thesis is accepted by some, but not all schools of constructivism.

[4]In Chap. 10 we shall describe a model of the real line in which the decidability of equality can be refuted.

[5]The stability of equality on the reals in constructive and intuitionistic analysis blocks the possibility of defining infinitesimals as real numbers ε which are "indistinguishable from 0" in the sense that $\neg(\varepsilon \neq 0)$. In smooth infinitesimal analysis (see Chap. 10) infinitesimals can, and indeed are, defined in this way.

[6]From the facts that $\neg(r \# s) \Rightarrow r = s$ it follows easily that $r \neq s \Leftrightarrow \neg\neg(r \# s)$.

0123456789 has appeared in π. Then if the 9 of the first sequence 0123456789 in π is the kth digit after the decimal point, $\rho = \frac{10^k - 1}{3 \cdot 10^k}$. Now suppose that ρ were not rational; then $\rho = \frac{10^k - 1}{3 \cdot 10^k}$ would be impossible and no sequence 0123456789 could appear in π, so that $\rho = \frac{1}{3}$, which is also impossible. Thus the assumption that ρ is not rational leads to a contradiction; yet we are not warranted to assert that ρ is rational, for this would mean that we could calculate integers m and n for which $\rho = \frac{m}{n}$. But this evidently requires that we can produce a sequence 0123456789 in π or demonstrate that no such sequence can appear, and at present we can do neither.

9.2 Constructive Meaning of the Logical Operators

An examination of constructive mathematical reasoning leads naturally to the following interpretations of the logical operators:

- $p \wedge q$: we have a proof of p and a proof of q.
- $p \vee q$: we have either a proof of p or a proof of q.
- $\neg p$: we can derive a contradiction (such as $0 = 1$) from p.
- $p \rightarrow q$: we can convert any proof of p into a proof of q.
- $\exists x p(x)$: we have a procedure for producing both an object a and a proof that $P(a)$ holds
- $\forall x{\in}A\ P(x)$: we have a procedure which, applied to an object a and a proof that $a \in A$, demonstrates that $P(a)$ holds.

Combining the constructive meaning of negation with that of disjunction yields the constructive meaning of the *law of excluded middle:* $p \vee \neg p$ is now seen to express the nontrivial claim that we have a method of deciding which of p or $\neg p$ holds, that is, a method of either proving p or deducing a contradiction from p. If p is an unsolved problem, this claim is dubious at best.

If the symbol "$\exists$" is taken to mean "explicit existence" in the sense indicated above, it cannot be expected to obey the laws of classical logic. For example, $\neg\forall$ is classically equivalent to $\exists\neg$, but the mere knowledge that something cannot always occur does not enable us actually to *determine* a location where it fails to occur. This is generally the case with existence proofs by contradiction. For instance, consider the following standard proof of the *Fundamental Theorem of Algebra:* every polynomial p of degree >0 has a (complex) zero. If p lacks a zero, then $\frac{1}{p}$ is entire and bounded, and so by Liouville's theorem must be constant. This proof gives no hint of how actually to construct a zero.[7]

On the basis of these interpretations, in 1930 **A. Heyting** formulated the axioms of *intuitionistic* (or *constructive*) *logic.*[8] This may be presented as a formal system with the following axioms and rules of inference:

[7]But constructive proofs of this theorem are known.

[8]See Heyting (1956).

Axioms

$$p \to (q \to p)$$

$$[p \to (q \to r) \to [(p \to q) \to (p \to r)]$$

$$p \to (q \to p \wedge q)$$

$$p \wedge q \to p \qquad p \wedge q \to q$$

$$p \to p \vee q \qquad q \to p \vee q$$

$$(p \to r) \to [(q \to r) \to (p \vee q \to r)]$$

$$(p \to q) \to [(p \to \neg q) \to \neg p]$$

$$\neg p \to (p \to q)$$

$$p(t) \to \exists x p(x) \qquad \forall x p(x) \to p(y) \qquad (x \text{ free in } \alpha \text{ and } t \text{ free for } x \quad \text{in } p)$$

$$x = x \qquad p(x) \wedge x = y \to p(y)$$

Rules of Inference

$$\frac{p,\ p \to q}{q}$$

$$\frac{q \to p(x)}{q \to \forall x p(x)} \qquad \frac{p(x) \to q}{\exists x p(x) \to q} \qquad (x \text{ not free in } \beta)$$

Classical logic is obtained from intuitionistic logic either by adding as an axiom the *law of excluded middle* $p \vee \neg p$, or the *law of double negation* $\neg\neg p \to p$. The chief difference between classical and intuitionistic logic is that in the latter *neither of these two laws are generally affirmed.*

9.3 Order on the Constructive Reals

The *order relation* on the reals is given constructively by stipulating that $r < s$ is to mean that we have an explicit lower bound on the distance between r and s. That is,

$$r < s \Leftrightarrow n \textit{ and } k \textit{ can be found so that } s_{n+p} - r_{n+p} > 1/k \textit{ for all } p.$$

It can readily be shown that, for any real numbers x, y such that $x < y$, there is a rational number α such that $x < \alpha < y$.

We observe that $r \# s \Leftrightarrow r < s \vee s < r$. The implication from right to left is clear. Conversely, suppose that $r \# s$. Find n and k so that $| r_{n+p} - s_{n+p} | > \frac{1}{k}$ for every p, and determine $m > n$ so that $| r_m - r_{m+p} | < \frac{1}{4k}$ and $| s_m - s_{m+p} | < \frac{1}{4k}$ for every p. Either $r_m - s_m > \frac{1}{k}$ or $s_m - r_m > \frac{1}{k}$; in the first case $r_{m+p} - s_{m+p} > \frac{1}{2k}$ for every p, whence $s < r$; similarly, in the second case, we obtain $r < s$.

We define $r \leq s$ to mean that $s < r$ is false. Notice that $r \leq s$ is not the same as $r < s$ or $r = s$: in the case of the real number ρ defined above, for instance, clearly $\rho \leq \frac{1}{3}$; but we do not know whether $\rho < \frac{1}{3}$ or $\rho = \frac{1}{3}$. Still, it is true that $r \leq s \wedge s \leq r \Rightarrow r = s$. For the premise is the negation of $r < s \vee s < r$, which, by the above, is equivalent to $\neg r \# s$. But we have already seen that this last implies $r = s$.

There are several common properties of the order relation on real numbers which hold classically but which cannot be established constructively. Consider, for example, the trichotomy law $x < y \vee x = y \vee y < x$. Suppose we had a method enabling us to decide which of the three alternatives holds. Applying it to the case $y = 0$, $x = \pi^{\sqrt{2}} - r$ for rational r would yield an algorithm for determining whether $\pi^{\sqrt{2}} = r$ or not, which we have already observed is an open problem. One can also demonstrate the failure of the trichotomy law (as well as other classical laws) by the use of "fugitive sequences". Here one picks an unsolved problem of the form $\forall n P(n)$, where P is a decidable property of integers—for example, Goldbach's conjecture that every even number ≥ 4 is the sum of two odd primes. Now one defines a sequence—a "fugitive" sequence—of integers (n_k) by $n_k = 0$ if $2k$ is the sum of two primes and $n_k = 1$ otherwise. Let r be the real number defined by $r_k = 0$ if $n_k = 0$ for all $j \leq k$, and $r_k = 1/m$ otherwise, where m is the least positive integer such that $n_m = 1$. It is then easy to check that $r \geq 0$ and $r = 0$ if and only if Goldbach's conjecture holds. Accordingly the correctness of the trichotomy law would imply that we could resolve Goldbach's conjecture. Of course, Goldbach's conjecture might be resolved in the future, in which case we would merely choose another unsolved problem of a similar form to define our fugitive sequence.

A similar argument shows that the law $r \leq s \vee s \leq r$ also fails constructively: define the real number s by $s_k = 0$ if $n_k = 0$ for all $j \leq k$; $s_k = 1/m$ if m is the least positive integer such that $n_m = 1$, and m is even; $s_k = -1/m$ if m is the least positive integer such that $n_m = 1$, and m is odd. Then $s \leq 0$ (resp. $0 \leq s$) would mean that there is no number of the form $2 \cdot 2k$ (resp. $2 \cdot (2k + 1)$) which is not the sum of two primes. Since neither claim is at present known to be correct, we cannot assert the disjunction $s \leq 0 \vee 0 \leq s$.

In constructive mathematics there is a convenient substitute for trichotomy known as the *comparison principle*. This is the assertion

$$r < t \Rightarrow r < s \vee s < t.$$

Its validity can be established in a manner similar to the foregoing.

9.4 Algebraic Operations on the Constructive Reals

The fundamental operations +, –, ·, $^{-1}$ and $|\cdot|$ are defined for real numbers as one would expect, viz.

- $r + s$ is the sequence $(r_n + s_n)$
- $r - s$ is the sequence $(r_n - s_n)$
- $r \cdot s$ or rs is the sequence $(r_n s_n)$
- if $r \# 0$, r^{-1} is the sequence (t_n), where $t_n = r_n^{-1}$ if $t_n \neq 0$ and $t_n = 0$ if $r_n = 0$
- $|r|$ is the sequence $(|r_n|)$

It is then easily shown that $rs \# 0 \Leftrightarrow r \# 0 \wedge s \# 0$. For if $r \# 0 \wedge s \# 0$, we can find k and n such that $|r_{n+p}| > \frac{1}{k}$ and $|s_{n+p}| > \frac{1}{k}$ for every p, so that $|r_{n+p}s_{n+p}| > \frac{1}{k^2}$ for every p, and $rs \# 0$. Conversely, if $rs \# 0$, then we can find k and n so that

$$|r_{n+p}s_{n+p}| > \frac{1}{k}, |r_{n+p} - r_n| < 1, |s_{n+p} - s_n| < 1$$

for every p. It follows that

$$|r_{n+p}| > \frac{1}{k}(|s_n| + 1) \text{ and } |s_{n+p}| > \frac{1}{k}(|r_n| + 1)$$

for every p, whence $r \# 0 \wedge s \# 0$.

But it is not constructively true that, if $rs = 0$, then $r = 0$ or $s = 0$! To see this, use the following prescription to define two real numbers r and s. If in the first n decimals of π no sequence 0123456789 occurs, put $r_n = s_n = 2^{-n}$; if a sequence of this kind does occur in the first n decimals, suppose the 9 in the first such sequence is the kth digit. If k is odd, put $r_n = 2^{-k}$, $s_n = 2^{-n}$; if k is even, put $r_n = 2^{-n}$, $s_n = 2^{-k}$. Then we are unable to decide whether $r = 0$ or $s = 0$. But $rs = 0$. For in the first case above $r_n s_n = 2^{-2n}$; in the second $r_n s_n = 2^{-k-n}$. In either case $|r_n s_n| < \frac{1}{m}$ for $n > m$, so that $rs = 0$.

9.5 Convergence of Sequences and Completeness of the Constructive Reals

As usual, a sequence (a_n) of real numbers is said to *converge* to a real number b, or to have *limit* s if, given any natural number k, a natural number n can be found so that for every natural number p,

$$|b - a_{n+p}| < 2^{-k}.$$

As in classical analysis, a constructive necessary and condition that a sequence (a_n) of real numbers be convergent is that it be a *Cauchy* sequence, that is, if, given any given any natural number k, a natural number n can be found so that for every natural number p,

$$\left|a_{n+p} - a_n\right| < 2^{-k}.$$

But some classical theorems concerning convergent sequences are no longer valid constructively. For example, a bounded momotone sequence need no longer be convergent. A simple counterexample is provided by the sequence (a_n) defined as follows: $a_n = 1\text{–}2^{-n}$ if among the first n digits in the decimal expansion of π no sequence 0123456789 occurs, while $a_n = 2\text{–}2^{-n}$ if among these n digits such a sequence does occur. Since it is not known whether the limit of this sequence, if it exists, is 1 or 2, we cannot claim that that this limit exists as a well defined real number.

In classical analysis $\mathbb{R}$ is *complete* in the sense that every nonempty set of real numbers that is bounded above has a supremum. As it stands, this assertion is constructively incorrect. For consider the set A of members $\{x_1, x_2, \ldots\}$ of any fugitive sequence of 0 s and 1 s. Clearly A is bounded above, and its supremum would be either 0 or 1. If we knew which, we would also know whether $x_n = 0$ for all n, and the sequence would no longer be fugitive.

However, the completeness of $\mathbb{R}$ can be salvaged by defining suprema and infima somewhat more delicately than is customary in classical mathematics. A nonempty set A of real numbers is *bounded above* if there exists a real number b, called an *upper bound* for A, such that $x \leq b$ for all $x \in A$. A real number b is called a *supremum*, or *least upper bound*, of A if it is an upper bound for A and if for each $\varepsilon > 0$ there exists $x \in A$ with $x > b - \varepsilon$. We say that A is *bounded below* if there exists a real number b, called a *lower bound* for A, such that $b \leq x$ for all $x \in A$. A real number b is called an *infimum,* or *greatest lower bound*, of A if it is a lower bound for A and if for each $\varepsilon > 0$ there exists $x \in A$ with $x < b + \varepsilon$. The supremum (respectively, infimum) of A, is unique if it exists and is written sup A (respectively, inf A).

Let us prove the *constructive least upper bound principle.*

Theorem Let A be a nonempty set of real numbers that is bounded above. Then sup A exists if and only if for all $x, y \in \mathbb{R}$ with $x < y$, either y is an upper bound for A or there exists $a \in A$ with $x < a$.

Proof If sup A exists and $x < y$, then either sup $A < y$ or $x <$ sup A; in the latter case we can find $a \in A$ with sup A – (sup $A - x) < a$, and hence $x < a$. Thus the stated condition is necessary.

Conversely, suppose the stated condition holds. Let a_1 be an element of A and choose an upper bound b_1 for A with $b_1 > a_1$. We construct recursively a sequence (a_n) in A and (b_n) of upper bounds for A such that, for each $n \geq 0$,

$$a_n \leq a_{n+1} \leq b_{n+1} \leq b_n$$

and

$$b_{n+1} - a_{n+1} \leq \frac{3}{4}(b_n - a_n).$$

Having found $a_1, \ldots, a_n$ and $b_1, \ldots, b_n$, if $a_n + \frac{3}{4}(b_n - a_n)$ is an upper bound for A, put $b_{n+1} = a_n + \frac{3}{4}(b_n - a_n)$ and $a_{n+1} = a_n$; while if there exists $a \in A$ with $a > a_n + \frac{3}{4}(b_n - a_n)$, we set $a_{n+1} = a$ and $b_{n+1} = b_n$. This completes the recursive construction.

From (i) and (ii) we have

$$0 \leq b_n - a_n \leq \left(\frac{3}{4}\right)^{n-1}(b_1 - a_1).$$

It follows that the sequences (a_n) and (b_n) converge to a common limit c: with $a_n \leq c \leq b_n$ for $n \geq 1$. Since each b_n is an upper bound for A, so is c. On the other hand, given $\varepsilon > 0$, we can choose n so that $c \geq a_n > c - \varepsilon$, where $a_n \in A$. Hence $c = \sup A$.

An analogous result for infima can be stated and proved in a similar way.

9.6 Functions on the Constructive Reals

Considered constructively, a *function* from $\mathbb{R}$ to $\mathbb{R}$ is a rule F which enables us, when given a real number x, to compute another real number $F(x)$ in such a way that, if $x = y$, then $F(x) = F(y)$. It is easy to check that every polynomial is a function in this sense, and that various power series and integrals, for example those defining tan x and e^x, also determine functions.

Viewed constructively, some classically defined "functions" on $\mathbb{R}$ can no longer be considered to be defined on the whole of $\mathbb{R}$. Consider, for example, the "blip" function B defined by $B(x) = 0$ if $x \neq 0$ and $B(0) = 0$. Here the domain of the function is $\{x \in \mathbb{R}: x = 0 \vee x \neq 0\}$. But we have seen that we cannot assert dom$(B) = \mathbb{R}$. Accordingly, the blip function is not well defined as a function from $\mathbb{R}$ to $\mathbb{R}$. Of course, classically, B is the simplest *discontinuous* function defined on $\mathbb{R}$. The fact that the simplest possible discontinuous function fails to be defined on the whole of $\mathbb{R}$ gives grounds for the suspicion that *no* function defined on $\mathbb{R}$ can be discontinuous; in other words, that, constructively speaking, *all functions defined on* $\mathbb{R}$ *are continuous*.[9] The claim is plausible. For if a function F is well-defined on all reals x, it must be possible to compute the value for all rules x determining real numbers, that is, determining their sequences of rational approximations $x_1, x_2, \ldots$.

[9]As remarked in Chap. 5, this was a central tenet of intuitionism's founder, Brouwer.

Now $F(x)$ must be computed to accuracy ε in a finite number of steps—the number of steps depending on ε. This means that only finitely many approximations can be used, i.e., $F(x)$ can be computed to within ε only when x is known within δ for some δ. Thus F should indeed be continuous. In fact all known examples of constructive functions defined on $\mathbb{R}$ are continuous.

Constructively, a real-valued function f is *continuous* if for each $\varepsilon > 0$ we can find a specific $\delta > 0$ such that $|f(x) - f(y)| \leq \varepsilon$ whenever $|x - y| < \delta$. If all functions on $\mathbb{R}$ are continuous, then a subset A of $\mathbb{R}$ may fail to be genuinely *complemented*: that is, there may be no subset B of $\mathbb{R}$ disjoint from A such that $\mathbb{R} = A \cup B$.

To see this, we first suppose that A, B are disjoint subsets of $\mathbb{R}$ and that there is a point $a \in A$ which can be approached arbitrarily closely by points of B (or vice-versa). Then, assuming all functions on $\mathbb{R}$ are continuous, it cannot be the case that $\mathbb{R} = A \cup B$. For if so, we may define the function f on $\mathbb{R}$ by $f(x) = 0$ if $x \in A$, $f(x) = 1$ if $x \in B$. Then for all $\delta > 0$ there is $b \in B$ for which $|b - a| < \delta$, but $|f(b) - f(a)| = 1$. Thus f fails to be continuous at a, and we conclude that $\mathbb{R} \neq A \cup B$.

Now if we take A to be any fini te set of real numbers, any union of open or closed intervals, or the set $\mathbb{Q}$ of rational numbers, then in each case the set B of points "outside" A satisfies the above condition. Accordingly, for each such subset A, $\mathbb{R}$ is not "decomposable" into A and the set of points "outside" A, in the sense that these two sets of points together do not exhaust $\mathbb{R}$. This fact indicates that the constructive continuum is a great deal more "cohesive" than its classical counterpart. For classically, the continuum is merely *connected* in the sense that it is not (nontrivially) decomposable into two open (or closed) subsets. Constructively, however, $\mathbb{R}$ is indecomposable into subsets which are neither open nor closed. Indeed, in some formulations of constructive analysis,[10] $\mathbb{R}$ (and all of its intervals) is *cohesive* in the stronger sense that it cannot be decomposed in *any way whatsoever*.[11] In this sense the constructive real line can be brought close to the ideal of a true continuum.

Certain well-known theorems of classical analysis concerning continuous functions fail in constructive analysis. One such is the *theorem of the maximum*: a uniformly continuous function on a closed interval assumes its maximum at some point. For consider, as Fig. 9.1 below, a function f: $[0,1] \to \mathbb{R}$ with two relative maxima, one at $x = \frac{1}{3}$ and the other at $x = \frac{2}{3}$ and of approximately the same value. Now arrange things so that $f\left(\frac{1}{3}\right) = 1$ and $f\left(\frac{2}{3}\right) = 1 + t$, where t is some small parameter. If we could tell where f assumes its absolute maximum, clearly we could also determine whether $t \leq 0$ or $t \geq 0$, which, as we have seen, is not, in general, possible. Nevertheless, it can be shown that from f we can in fact calculate the maximum value itself, so that at least one can assert the existence of that maximum, even if one can't tell exactly where it is assumed.

Another classical result that fails to hold constructively in its usual form is the well-known *intermediate value theorem*. This is the assertion that, for any continuous function f from the unit interval $[0, 1]$ to $\mathbb{R}$, such that $f(0) = -1$ and $f(1) = 1$,

[10] E.g. in intuitionistic analysis (see below) and smooth infinitesimal analysis (see Chap. 10).

[11] See Appendix A.

Fig. 9.1

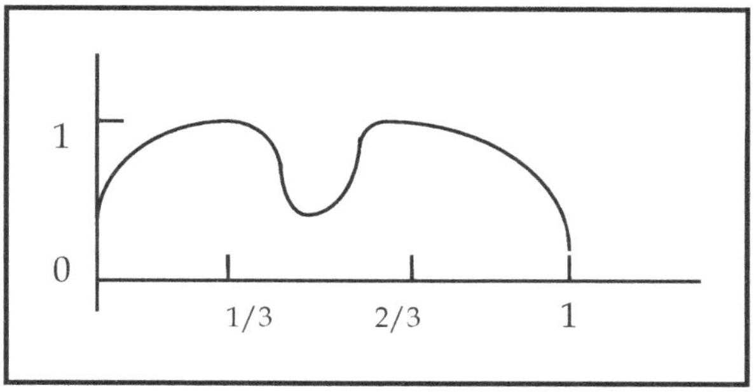

Fig. 9.2

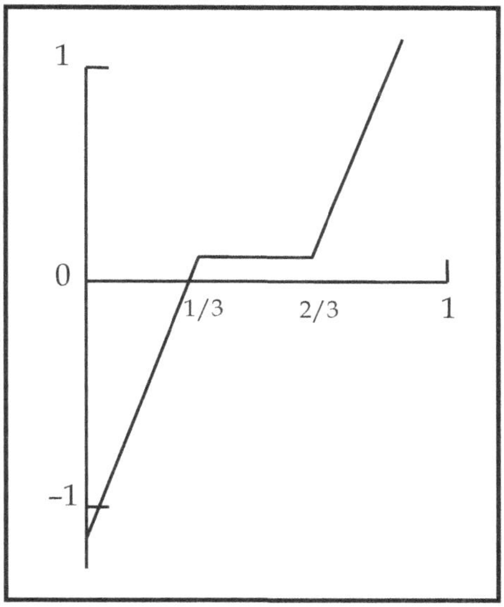

there exists a real number $a \in [0,1]$ for which $f(a) = 0$. To see that this fails constructively, consider the function f depicted in Fig. 9.2 above: here f is piecewise linear, taking the value t (a small parameter) between $x = \frac{1}{3}$ and $x = \frac{2}{3}$. If the intermediate value theorem held, we could determine a for which $f(a) = 0$. Then either $a < \frac{2}{3}$ or $a > \frac{1}{3}$; in the former case $t \geq 0$; in the latter $t \leq 0$. Thus we would be able to decide whether $t \geq 0$ or $t \leq 0$; but we have seen that this is not constructively possible in general.

However, it can be shown that, constructively, the intermediate value theorem is "almost" true in the sense that given any continuous function f satisfying the conditions of the Intermediate Value Theorem, then, for every $\varepsilon > 0$, there is a real number a in [0. 1] such that $|fr(a)| < \varepsilon$. And also in the sense that, if we write $P(f)$ for

$$\forall b\ \forall a < b \exists c\ (a < c < b \wedge f\ (c) \neq\neq 0),$$

then

$$\forall f[P(f) \rightarrow \exists x(f(x) = 0)].$$

This example illustrates how a single classical theorem "refracts" into several constructive theorems.

9.7 Axiomatizing the Constructive Reals

The constructive reals can be furnished with an axiomatic description.[12] We begin by assuming the existence of a set ***R*** with.

- a binary relation > (*greater than*)
- a corresponding *apartness relation* # defined by $x \# y \Leftrightarrow x > y$ or $y > x$
- a unary operation $x \mapsto -x$
- binary operations $(x, y) \mapsto x + y$ (*addition*) and $(x, y) \mapsto xy$ (*multiplication*)
- distinguished elements 0 (*zero*) and 1 (*one*) with $0 \neq 1$
- a unary operation $x \mapsto x^{-1}$ on the set of elements $\neq 0$.

The elements of ***R*** are called *real numbers.* A real number x is *positive* if $x > 0$ and *negative* if $-x > 0$. The relation $\geq$ (*greater than or equal to*) is defined by

$$x \geq y \quad \Leftrightarrow \forall z(y > z \Rightarrow x > z).$$

The relations < and $\leq$ are defined in the usual way; x is *nonnegative* if $0 \leq x$.

The sets ***N*** of natural numbers, $\boldsymbol{N}^+$ of positive integers, ***Z*** of integers and ***Q*** of rational numbers are identified with the usual subsets of ***R***; for instance $\boldsymbol{N}^+$ is identified with the set of elements of *R* of the form $1 + 1 + \cdots + 1$.

These relations and operations are subject to the following three groups of axioms, which, taken together, form the system **CA** of axioms for *constructive analysis,* or the *constructive real numbers.*

Field Axioms

$$x + y = y + x \quad (x + y) + z = x + y + z) \quad 0 + x = x \quad x + (-x) = 0 \quad xy = yx$$

$$(xy)z = x(yz) \quad 1x = x \quad x\#0 \rightarrow xx^{-1} = 1 \quad x(y + z) = xy + xz$$

[12]My account here is based on Bridges (1999).

Order Axioms

$$(x \geq y \wedge y \geq x) \leftrightarrow x = y.$$

$$\neg\ (x > y \wedge y > x) \quad x > y \to \forall z(x > z \vee z > y) \quad \neg(x\#y) \to x = y$$

$$x > y \to \forall z(x + z > y + z) \quad (x > 0 \wedge y > 0) \to xy > 0.$$

Notice that from the fourth of the order axioms it follows that $\neg(x \neq y) \to x = y$,[13] that is, the equality relation is *stable.*

The final two axioms introduce special properties of > and ≥. In the second of these the notions *bounded above, bounded below, and bounded* are defined as in classical mathematics, and the *least upper bound,* if it exists, of a nonempty[14] set S of real numbers is the unique real number b such that

- b is an upper bound for S, and
- for each $c < b$ there exists $s \in S$ with $s > c$.

Special Properties of >

Archimedean axiom. For each $x \in \boldsymbol{R}$ such that $x \geq 0$ there exists $n \in \boldsymbol{N}$ such that $x < n$.

The least upper bound principle. Let S be a nonempty subset of R that is bounded above, such that for all real numbers a, b with $a < b$, either b is an upper bound for S or else there exists $s \in S$ with $s > a$. Then S has a least upper bound.

The following basic properties of > and ≥ can then be established.[15]

$$\neg(x > x) \quad x \geq x \; (x > y \wedge y > z) \to x > z \; \neg(x > y \wedge y \geq x) \; (x > y \geq z) \to x > z$$

$$\neg(x > y) \leftrightarrow y \geq x \quad \neg(x \geq y) \leftrightarrow \neg\neg(y > x) \quad (x \geq y \geq z) \to x \geq z$$

$$\neg(x > y \wedge x = y) \quad x \geq 0 \to (x = 0 \leftrightarrow \forall \varepsilon > 0\ (x < \varepsilon)) \quad x + y > 0$$
$$\to (x > 0 \vee y > 0) \quad x > 0 \to -x < 0$$

$$(x > y \wedge z < 0) \to yz > xz \quad x\#0 \leftrightarrow x^2 \ > 0 \quad 1 > 0 \quad 0 < x < 1 \to x > \ x^2$$

$$n \in N^+ \ \to n^{-1} > 0$$

$$\textit{if} \ \ x > \ 0 \ \textit{and} \ y \geq 0, \ \textit{then there exists} \ \in \mathbf{Z} \ \textit{such that} \ nx \ > \ y$$

$$x > 0 \to x^{-1} > 0 \qquad xy > 0 \to (x \neq 0 \vee y \neq 0)$$

[13]Here and in the sequel $x \neq y$ is an abbreviation for $\neg(x = y)$.

[14]Here and in the remainder of this chapter "nonempty" has the stronger constructive meaning of being inhabited, to wit, that an element of the set in question can actually be constructed.

[15]Bridges (1999), pp. 103–5.

if $a < b$, *then there exists* $r \in \mathbf{Q}$ *such that* $a < r < b$

The constructive real line $\mathbb{R}$ as introduced above is a model of **CA.** Are there any other models, that is, models not isomorphic to $\mathbb{R}$? If classical logic is assumed, **CA** is a categorical theory and so the answer is no. But this is not the case within intuitionistic logic, for it can be shown that, in a topos, both the Dedekind and Cantor reals are models of **CA**, while, as has been pointed out above, these may fail to be isomorphic.

9.8 The Intuitionistic Continuum

In constructive analysis, a real number is an infinite (convergent) sequence of rational numbers generated by an effective rule, so that the constructive real line is essentially just a restriction of its classical counterpart. Brouwerian intuitionism takes a more imaginative view of the matter, resulting in a considerable enrichment of the arithmetical continuum over the version offered by strict constructivism. As conceived by intutionism, the arithmetical continuum admits as real numbers "not only infinite sequences determined in advance by an effective rule for computing their terms, but also ones in whose generation free selection plays a part."[16] The latter are called *(free) choice sequences.* Without loss of generality we may and shall assume that the entries in choice sequences are natural numbers.

Hermann Weyl describes Brouwer's conception of choice sequences in the following terms.

> *In Brouwer's analysis, the individual place in the* continuum, *the real number, is to be defined not by a set but by a* sequence *of natural numbers, namely, by a law which correlates with every natural number n a natural number* $\varphi(n)$*... How then do assertions arise which concern... all real numbers, i.e., all values of a real variable? Brouwer shows that frequently statements of this form in traditional analysis, when correctly interpreted, simply concern the totality of natural numbers. In cases where they do not, the notion of sequence changes its meaning: it no longer signifies a sequence determined by some law or other, but rather one that is created* step by step by free acts of choice, *and thus necessarily remains* in statu nascendi. *This "becoming"* selective sequence (werdende Wahlfolge) *represents the continuum, or the variable, while the sequence determined* ad infinitum *by a law represents the individual real number in the continuum. The continuum no longer appears, to use Leibniz's language, as an aggregate of fixed elements but as a medium of free "becoming". Of a selective sequence* in statu nascendi, *naturally only those properties can be meaningfully asserted which already admit of a yes-or-no decision (as to whether or not the property applies to the sequence) when the sequence has been carried to a certain point; while the continuation of the sequence beyond this point, no matter how it turns out, is incapable of overthrowing that decision.*[17]

While constructive analysis does not formally contradict classical analysis and may in fact be regarded as a subtheory of the latter, a number of intuitionistically plausible principles have been proposed for the theory of choice sequences which

[16]Dummett (1977), p. 62.

[17]Weyl (1949), p. 52.

render intuitionistic analysis divergent from its classical counterpart.[18] One such principle is *Brouwer's Continuity Principle*. This assderts that, given a relation $R(\alpha, n)$ between choice sequences α and numbers n, if for each α a number n may be determined for which $R(\alpha, n)$ holds, then n can already be determined on the basis of the knowledge of a finite number of terms of α.[19] From this one can prove a weak version of the Continuity Theorem, namely, that every function from $\mathbb{R}$ to $\mathbb{R}$ is continuous.[20] Another such principle is *Bar Induction*, a certain form of induction for well-founded sets of finite sequences.[21] Brouwer used Bar Induction and the Continuity Principle in proving his Continuity Theorem that every real-valued function defined on a closed interval is uniformly continuous, from which it follows that the intuitionistic continuum, and all of its closed intervals, are cohesive.[22]

Brouwer gave the intuitionistic conception of mathematics an explicitly subjective twist by introducing the *creative subject*. The creative subject was conceived as a kind of idealized mathematician for whom time is divided into discrete sequential stages, during each of which he may test various propositions, attempt to construct proofs, and so on. In particular, it can always be determined whether or not at stage n the creative subject has a proof of a particular mathematical proposition p. While the theory of the creative subject remains controversial, its purely mathematical consequences can be obtained by a simple postulate which is entirely free of subjective and temporal elements.

The creative subject allows us to define, for a given proposition p, a binary sequence $<a_n>$ by:

$a_n = 1$ if the creative subject has a proof of p at stage n;
$a_n = 0$ otherwise.

Now if the construction of these sequences is the only use made of the creative subject, then references to the latter may be avoided by postulating the principle known as *Kripke's Scheme:*

For each proposition p there exists an increasing binary sequence $<a_n>$ *such that p holds if and only if* $a_n = 1$ *for some n.*

Taken together, these principles have been shown[23] to have remarkable consequences for the cohesiveness[24] of subsets of the continuum. Not only is the

---

[18]Intuitionistic analysis, nevertheless, an extension of **CA.**

[19]This may be seen to be plausible if one considers that the according to Brouwer the construction of a choice sequence is incompletable; at any given moment we can know nothing about it outside the identities of a finite number of its entries. Brouwer's Continuity Principle amounts to the assertion that every function from $\mathbb{N}^{\mathbb{N}}$ to $\mathbb{N}$ is continuous.

[20]Bridges and Richman (1987), p. 109.

[21]For an explicit statement of the principle of Bar Induction, see Ch. 3 of Dummett (1977), or Ch. 5 of Bridges and Richman (1987).

[22]See Appendix A.

[23]Van Dalen (1997).

[24]See Appendix A.

intuitionistic continumm cohesive, but, assuming Brouwer's Continuity Principle and Kripke's Scheme, it remains cohesive even if one pricks it with a pin.[25] "The [intuitionistic] continuum has, as it were, a syrupy nature, one cannot simply take away one point."[26] If in addition Bar Induction is assumed, then, even more surprisingly, cohesiveness is maintained even when all the rational points are removed from the continuum.

9.9 An Intuitionistic Theory of Infinitesimals

In 1980 Richard Vesley showed that a natural notion of infinitesimal can be developed within intuitionistic mathematics.[27] His idea was that an infinitesimal should be a "very small" real number in the sense of not being known to be distinguishable—that is, strictly greater than or less than—zero.

Let α be a variable ranging over choice sequences, x, y variables ranging over real numbers in the intuitionistic continuum R and n a variable ranging over natural numbers. Vesley observes that from Kripke's scheme one can prove

$$\forall\alpha\exists x[(x > 0 \Leftrightarrow \forall n\, \alpha(n) = 0) \wedge (x > 0 \Leftrightarrow \neg\forall n\, \alpha(n) = 0)],$$

from which it follows that

$$\forall\alpha\exists x\big[\big(x\#0 \Leftrightarrow (\forall n\, \alpha(n) = 0 \vee \neg\forall n\, \alpha(n) = 0)\big].$$

Now for each choice sequence α define the subsets $L(\alpha)$ and $M(\alpha)$ of the by the condition

$$x \in L(\alpha) \Leftrightarrow \big[\big(x\#0 \Leftrightarrow (\forall n\, \alpha(n) = 0 \vee \neg\forall n\, \alpha(n) = 0)\big]$$
$$x \in M(\alpha) \Leftrightarrow \exists y \in L(\alpha)\ |x| \leq |y|$$

The members of $L(\alpha)$ may be considered very small in the sense that they cannot be distinguished from zero unless a decision is made as to whether α is identically zero or not, something which in general cannot be guaranteed. The members of M (α), the α-*infinitesimals*, are those real numbers which are bounded in absolute value by members of $L(\alpha)$. Notice that the property of being an α-infinitesimal is quite unstable, since in the event of such a decision being made as to whether α is identically zero or not, $L(\alpha)$ automatically becomes identical with the set of reals which can be distinguished from zero and so $M(\alpha)$ with the set of all reals.

---

[25]More exactly, for any real number a, the complement $\mathbb{R} - \{a\}$ of $\{a\}$ is cohesive.

[26]*Ibid.* There the classical continuum is described as the "frozen intuitionistic continuum".

[27]Vesley (1981).

Each $M(\alpha)$ can then be shown to be an ideal in R containing at least one nonzero[28] element. Moreover, the $M(\alpha)$ violate the archimedean property in the weak sense that the following can be proved:

$$\neg\forall\alpha\exists x \in M(\alpha)\exists n\; n.x \;>\; 1.$$

The derivative of a function can then be defined in the following way: given f: $\mathbb{R} \to \mathbb{R}$, the derivative of f at $x \in \mathbb{R}$ is $a \in \mathbb{R}$ if there is a function g: $\mathbb{R} \to \mathbb{R}$ carrying α-infinitesimals to α-infinitesimals for every α and which also satisfies

$$\forall z \mid f(x+z) - f(x) - a.z \mid = g(z). \mid z \mid .$$

The function g measures the discrepancy between the derivative and the difference quotient of f. Using this definition the familiar formulas for differentiation can be derived.[29]

[28] *i.e.* $\neq 0$, not $\# 0$.

[29] At the end of the paper the author asks whether the calculus can be treated fully along these lines, and whether such an approach has advantages. The question appears to be open.

Chapter 10
Smooth Infinitesimal Analysis/Synthetic Differential Geometry

10.1 Smooth Worlds

Mathematicians have developed two methods of deriving the theorems of geometry: the *analytic* and the *synthetic*. While the analytical method is based on the introduction of numerical coordinates, and so on the theory of real numbers, the (much older) idea behind the synthetic approach is to furnish the subject of geometry with an autonomous foundation within which the theorems become deducible by logical means from an initial body of postulates.

The most familiar examples of the synthetic geometry are classical Euclidean geometry and the synthetic projective geometry introduced by Desargues in the seventeenth century and revived and developed by Poncelet, Steiner and others in the nineteenth century.

The power of analytic geometry derives very largely from the fact that it permits the methods of the calculus, and, more generally, of mathematical analysis, to be introduced into geometry, leading in particular to *differential geometry*. That being the case, the notion of a "synthetic" differential geometry appears elusive: how can differential geometry be placed on a "purely geometric" or "axiomatic" foundation when the apparatus of the calculus seems to be inextricably involved?

There have been (at least) two attempts to develop a synthetic differential geometry. The first was initiated by Herbert Busemann[1] in the 1940s, building on earlier work of Paul Finsler. Here the idea was to build a differential geometry that, in its author's words, "requires no derivatives": the basic objects in Busemann's approach are not differentiable manifolds, but metric spaces of a certain type in which the notion of a geodesic can be defined in an intrinsic manner.

The second approach, the focus of our attention here, was first proposed in the 1960s by **F. W. Lawvere**, in his pursuit of a decisive axiomatic framework for

[1]Busemann (1955).

J. L. Bell, *The Continuous, the Discrete and the Infinitesimal in Philosophy and Mathematics*, The Western Ontario Series in Philosophy of Science 82,
https://doi.org/10.1007/978-3-030-18707-1_10

continuum mechanics. His ideas have led to the development of a theory now claiming, with good reason, exclusive title to the appellation *synthetic differential geometry* (**SDG**).

Since differential geometry "lives" in the category **Man** of manifolds, it might be supposed that in formulating a "synthetic differential geometry" the category-theorist's goal would be to find an axiomatic description of **Man** itself. But in fact the category **Man** has a couple of "deficiencies" which make it unsuitable as an object of axiomatic description:

1. It lacks exponentials: that is, the "space of all smooth maps" from one manifold to another in general fails to be a manifold. And even if it did—
2. It also lacks "infinitesimal objects"; in particular, there is no "infinitesimal" manifold[2] $\mathbf{\Delta}$ for which the tangent bundle **Tan***M* of an arbitrary manifold *M* can be identified as the exponential "manifold" $M^{\mathbf{\Delta}}$ of all "infinitesimal paths" in *M*.[3]

Lawvere's idea was to enlarge **Man** to a category **Space**—a *smooth category* or a *smooth world,* with objects called *smooth spaces*—through whose introduction these two deficiencies are surmounted, which admits a simple axiomatic description, and is at the same time sufficiently similar to **Set** to allow mathematical construction and calculation to proceed in the familiar way. Smooth categories are the *natural models* of **SDG.**

The essential features of **Space** are these:

- In enlarging **Man** to **Space**—in contradistinction to **Set**—no "new" maps between manifolds are introduced, that is, all maps in **Space** between objects of **Man** are smooth[4] differentiable arbitrarily many times, It is for this reason that analysis in **Space** is called *smooth infinitesimal analysis* (**SIA**).
- Nevertheless, **Space,** like **Set,** is a *topos.*[5]
- But unlike **Set, Space** *has infinitesimal objects.* Let **R** be the *smooth real line*, that is, the real line considered as an object of **Man,** and hence also of **Space**. Then there is a nondegenerate infinitesimal segment $\mathbf{\Delta}$ of **R** around 0 which is *rigid,* that is, remains *straight* and *unbroken* under *any* map in **Space.** In other words, $\mathbf{\Delta}$ is subject in **Space** to *Euclidean motions only.* If we identify maps on **R** as curves, then the images of $\mathbf{\Delta}$ can be considered as "infinitesimal straight segments" of curves.

SDG provides an image of the world in which the continuous is an autonomous notion, not explicable in terms of the discrete. In **SDG** *all* functions or correlations between spaces are smooth, and so in particular continuous. Accordingly **SDG** realizes in a very strong way Leibniz's Principle of Continuity: *natura non facit saltus.*

---

[2]An *Incredible Shrinking Man(ifold),* no less.

[3]It is this deficiency that makes the construction of the tangent bundle in **Man** something of a headache: see Spivak (1975).

[4]That is, differentiable arbitrarily many times

[5]See Chap. 7.

Fig. 10.1

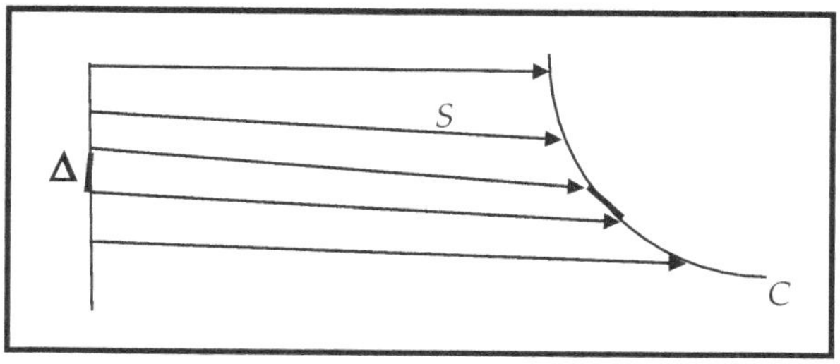

In **SIA** the rigidity of **Δ** is given precise formulation as the

Microaffiness Axiom.[6] *For any map f:* **Δ** → **R,** *there exist unique a, b* ∈ **R** *such that*

$$f(\varepsilon) = a + b\varepsilon$$

for all ε ∈ **Δ.**

This says that any real-valued function on **Δ** is *affine*. If we think of *f* as a graph in the Cartesian plane, the quantity a measures the displacement, and b the rotation undergone by **Δ** under the map *f*.

It follows from the Microaffineness Axiom that any map $f\colon \Delta \to \mathbf{R}^n$ is of the form

$$\varepsilon \mapsto (a_1 + b_1\varepsilon, \ldots, a_n + b_n\varepsilon)$$

for unique $a_1, \ldots, a_n, b_1, \ldots, b_n$. In particular the image of Δ under f is always a straight line.

The infinitesimal object **Δ** can be described in number of remarkable ways:

- As a *generic infinitesimal tangent vector*. For consider any curve C in a space S—that is, the image of a segment of **R** (containing Δ) under a map f into S (Fig. 10.1). Then the image of **Δ** under *f* may considered as a short straight line segment lying along C, that is, as part of the tangent to the curve at its contact point *actually lying in the curve*. Thus curves in S can be considered as infinitesimally straight, or *microstraight*. This is the **Principle of Microstraightness in Space**.
- As an *intensive magnitude* possessing only location and direction, and so which, considered (*per impossibile*) as an extension, cannot be "bent" or "broken".
- As a domain of *nilsquare infinitesimals*. For by considering the curve in **R** × **R** given by $f(x) = x^2$, we see that **Δ** is the intersection of the curve $y = x^2$ with the x-axis[7] (Fig. 10.2):

[6] Also known as the *Kock-Lawvere axiom* after its formulators, Anders Kock and F. W. Lawvere.

[7] This feature of **SIA** brings to mind Protagoras's claim (as reported by Aristotle (in *Metaphysics* III, 2) that "the circle touches the ruler not at a point, but along a line."

Fig. 10.2

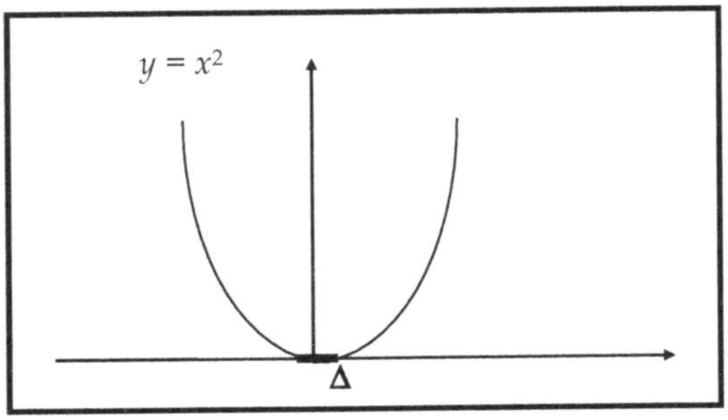

Accordingly

$$\Delta = \{x \in \mathbf{R} : x^2 = 0\}.$$

We see then that **Δ** consists of nilsquare infinitesimals, the species of infinitesimal introduced in the seventeenth century by Nieuwentijdt in opposition to Leibniz's conception.[8] Such infinitesimals will also be called *microquantities.*[9] As we show below, the axioms we shall introduce for smooth infinitesimal analysis will ensure that **Δ** is nondegenerate, i.e. does not reduce to {0}, so that **Δ** may be considered an *infinitesimal neighbourhood* or *microneigbourhood of* 0.

- As an *infinitesimal generator,* or *microgenerator of spaces.* For consider the space $\mathbf{\Delta}^{\mathbf{\Delta}}$ of self-maps of **Δ**. It follows from the Microaffineness Axiom that the subspace $(\mathbf{\Delta}^{\mathbf{\Delta}})_0$ of $\mathbf{\Delta}^{\mathbf{\Delta}}$ consisting of maps vanishing at 0 is isomorphic to **R.**[10] Accordingly **R**, and hence all Euclidean spaces, may be seen as being "generated" by the infinitesimal object **Δ**.

The space $\mathbf{\Delta}^{\mathbf{\Delta}}$ is a monoid[11] under composition which may be regarded as acting on Δ by evaluation: for $f \in \mathbf{\Delta}^{\mathbf{\Delta}}$, $f \cdot \varepsilon = f(\varepsilon)$. Its subspace $(\mathbf{\Delta}^{\mathbf{\Delta}})_0$ is a submonoid naturally identified as the space of *ratios of microquantities*. The isomorphism between $(\mathbf{\Delta}^{\mathbf{\Delta}})_0$ and **R** noted above is easily seen to be an isomorphism of monoids (where **R** is considered a monoid under its usual multiplication.) It follows that **R** itself may be regarded as the space of ratios of microquantities. This was essentially the view of Euler, who, as we have seen,[12] regarded (real) numbers as

[8]See Chap. 2. It should be pointed out, however, that Nieuwentijdt's infinitesimals differ from those of **SIA** in that that the product of any two of the former vanishes, while this is not the case for the latter.

[9]We shall use letters ε, η, ζ to denote arbitrary microquantities.

[10]This can be seen by noting that for any $f \in \mathbf{R}_0$, the Mcroaffineness Axiom ensures that there is a unique $b \in \mathbf{R}$ for which $f(\varepsilon) = b\varepsilon$ for all ε, and conversely each $b \in \mathbf{R}$ yields the map $\varepsilon \mapsto b\varepsilon$ in $\mathbf{R}_0$.

[11]A *monoid* is a multiplicative system (not necessarily commutative) with an identity element.

[12]See Chap. 3.

representing the possible results of calculating the ratio 0/0. For this reason Lawvere has suggested that **R** be called the space of *Euler reals.*[13]

As we have said, **Δ** may be viwed as an intensive magnitude of a certain kind. The microquantities that make up **Δ** may then be termed *intensive quantities*. If we think of **R** as the domain of *extensive* quantities, then an intensive quantity may be identified as an extensive microquantity, and (by the remarks in the paragraph above) an extensive quantity as the ratio of two intensive quantities.[14]

If we think of a smooth world as a model of the natural world, then the Principle of Microstraightness entails not just Leibniz's Principle of Continuity—that natural processes occur continuously, but also the *Principle of Microuniformity,* namely, the assertion that any such process may be considered as taking place at a constant rate over any sufficiently small period of time- a Barrovian timelet. For example, if the process is the motion of a particle, the Principle of Microuniformity entails that over an extremely short period the particle undergoes no accelerations. This idea, although rarely explicitly stated, is freely employed in a heuristic capacity in classical mechanics and the theory of differential equations. The virtual equivalence between the Principles of Microuniformity and Microstraightness becomes manifest when natural processes—the motions of bodies, for example, are represented as curves correlating dependent and independent variables. For then, microuniformity of the process is represented by microstraightness of the associated curve.

The Principle of Microstarightness yields an intuitively satisfying account of *motion*. For it entails that infinitesimal parts of (the curve representing a) motion are not points at which, as Aristotle observed, no motion is detectable—or, indeed, even possible. Rather, infinitesimal parts of the motion are nondegenerate spatial segments just large enough for motion through each to be discernible. On this reckoning a state of motion is to be accorded an intrinsic status, and is not, as Russell claimed, merely to be identified with its result—the successive occupation of a series of distinct positions. Rather, a state of motion is represented by the smoothly varying straight microsegment, the infinitesimal tangent vector, of its associated curve. This straight microsegment may be thought of as an infinitesimal "rigid rod", just long enough to have a slope—and so, like a speedometer needle, to indicate the presence of motion—but too short to bend, and so also too short to indicate a rate of change of motion.

This analysis may also be applied to the mathematical representation of *time*. Classically, time is represented as a succession of discrete instants, isolated "nows" at which time has, as it were, stopped. The Principle of Microstraightness, however, suggests that time be instead regarded as a plurality of smoothly overlapping Barrovian timelets each of which may be held to represent a "now" or "specious present" and over which time is, as it were, still passing. This conception of the

[13]Lawvere (2011).

[14]This would seem to be consonant with Hermann Cohen's conception of the infinitesimal as mentioned at the end of Chap. 4.

nature of time is similar to that proposed by Aristotle to refute Zeno's paradox of the arrow[15]; it is also closely related to Peirce's ideas on time.[16]

10.2 Elementary Differential Geometry in a Smooth World

There is a very simple way of constructing the tangent bundle of a space in a smooth world. Let us start with the real line **R.** Intuitively, the tangent bundle **Tan**S of a space S is the assemblage of infinitesimally short straight paths in S. In a smooth world such a path may be taken to be a map from the generic tangent vector $\boldsymbol{\Delta}$ to S. accordingly the tangent bundle S should be identified with the exponential $S^{\boldsymbol{\Delta}}$.

Let us check the compatibility of this definition of **Tan**S with the usual one in the case of Euclidean spaces $\mathbf{R}^n$. Now $\mathbf{R}^n$ has tangent bundle $\mathbf{R}^n \times \mathbf{R}^n$. But from the Microaffineness Axiom it may be immediately inferred that the map $\mathbf{R}^{\boldsymbol{\Delta}} \to \mathbf{R} \times \mathbf{R}$ which assigns to each $f \in \mathbf{R}^{\boldsymbol{\Delta}}$ the pair ($f(0)$, slope of f) is an isomorphism. It follows that

$$\mathbf{Tan}\,(\mathbf{R}^n) = (\mathbf{R}^n)^{\Delta} \cong \left(\mathbf{R}^{\Delta}\right)^n \cong (\mathbf{R} \times \mathbf{R})^n \cong \mathbf{R}^n \times \mathbf{R}^n.$$

Elements of $S^{\boldsymbol{\Delta}}$ are called *tangent vectors* to S. Thus a *tangent vector* to S at a point $p \in S$ is just a map t: $\boldsymbol{\Delta} \to S$ with $t(0) = p$. That is, a tangent vector at p is a *micropath in S with base point p*. The *base point map* π: $TS \to S$ is defined by $\pi(t) = t(0)$. For $p \in S$, the fibre $\pi^{-1}(p) = \mathbf{Tan}_p S$ is the *tangent space to S at p*.

Observe that, if we identify each tangent vector with its image in S, then *each tangent space to S may be regarded as lying in S*. In this sense each smooth space is "infinitesimally flat".

The assignment $S \mapsto \mathbf{Tan}S = S^{\Delta}$ can be turned into a functor in the natural way—the *tangent bundle functor*. (For f: $S \to T$, **Tan**f: **Tan**$S \to$ **Tan**T is defined by (**Tan**f)$t = f \circ t$ for $t \in$ **Tan**S.)

Synthetic differential geometry turns on the fact that *the tangent bundle functor is rendered representable*: **Tan**S is "represented" as the space of all maps from some fixed object—in this case $\boldsymbol{\Delta}$)—to S. (Classically, this is impossible.) This in turn simplifies a number of fundamental definitions in differential geometry.

For instance, a *vector field* on a smooth space S is an assignment of a tangent vector to S at each point in it, that is, a map ξ: $S \to \mathbf{Tan}S = S^{\Delta}$ such that $\xi(x)(0) = x$ for all $x \in S$. This means that $\pi \circ \xi$ is the identity on S, so that a *vector field is a section of the base point map.*

---

[15]See Chap. 1.

[16]See Chap. 5.

Now we have required that **Space** be a topos. In particular, for any pair S, T of smooth spaces, **Space** must also contain their *product* $S \times T$ and their *exponential* T^S, the space of all (smooth) maps $S \to T$. These are connected in the following way: for any smooth spaces S, T, U, there is a natural correspondence of maps

$$\frac{S \to T^U}{S \times U \to T}$$

(speaking category-theoretically, the product functor is left adjoint to the exponentiation functor). In the usual function-argument notation, this correspondence is given by:

$$(f : S \times U \to T) \mapsto \left(f^{\#} : S \to T^U\right) \text{ with } f^{\#}(s)(u) = f(s, u) \text{for } s \in S, u \in U.$$

This gives rise to a correspondence between *vector fields* on S and what we shall call *microflows* on S:

$$\frac{\xi : S \to S^{\Delta} \quad \text{(vector fields on } S)}{\xi^{\wedge} : S \times \Delta \to S \quad \text{(microflows on } S),}$$

with

$$\xi^{\wedge}(x, \varepsilon) = \xi(x)(\varepsilon).$$

Notice that then $\xi^{\wedge}(x, 0) = x$.

We also get, in turn, a bijective correspondence between microflows on S and *micropaths in* S^S *with the identity map as base point:*

$$\frac{\xi : S \times \Delta \to S \quad \text{(microflows on } S)}{\xi* : \Delta \to S^S \quad \left(\text{micropaths in } S^S\right),}$$

with

$$\xi * (\varepsilon)(x) = \xi^{\wedge}(x, \varepsilon) = \xi(x)(\varepsilon).$$

Thus, in particular,

$$\xi * (0)(x) = \xi(x)(0) = x,$$

so that $\xi*(0)$ is the identity map on S. Each $\xi*(\varepsilon)$ is a microtransformation of S into itself which is "very close" to the identity map.

Accordingly, in **Space,** or **SDG,** *vector fields, microflows, and micropaths are equivalent.*[17] Classically, this is a metaphor at best.

There is another remarkable feature of the microgenerator $\mathbf{\Delta}$ that should be mentioned. First, as Lawvere has emphasized, in smooth categories the tangent bundle functor has an "amazing" right adjoint; that is, for any spaces *S, T,* there is a natural bijection of maps

$$\frac{S^{\Delta} \to T}{S \to T^{1/\Delta}}$$

Maps $S^{\mathbf{\Delta}} \to \mathbf{R}$ are *differential forms* on *S*, so the existence of this right adjoint allows differential forms to be represented as maps $S \to \mathbf{R}^{\frac{1}{\Delta}}$ with values in the bigger algebraic structure $\mathbf{R}^{\frac{1}{\Delta}}$. This feature has led Lawvere to call Δ an "a.t.o.m.": an "amazingly tiny object model". Classically, the only objects having this feature are singletons.

10.3 The Calculus in Smooth Infinitesimal Analysis

In the usual development of the calculus, for any differentiable function f on the real line $\mathbf{R}$, $y = f(x)$, it follows from Taylor's theorem that the increment $\delta y = f(x + \delta x) - f(x)$ in y attendant upon an increment δx in x is determined by an equation of the form

$$\delta y = f'(x)\delta x + A(\delta x)^2, \tag{10.1}$$

where $f'(x)$ is the derivative of $f(x)$ and A is a quantity whose value depends on both x and δx. Now if it were possible to take δx a *nilsquare infinitesimal* or *microquantity,* then (10.1) would assume the simple form

$$f(x + \delta x) - f(x) = \delta y = f'(x)\delta x. \tag{10.2}$$

In **SIA** "sufficient" microquantities are present to ensure that Eq. (10.2) holds *nontrivially* for *arbitrary* functions f: $\mathbf{R} \to \mathbf{R}$. (Of course (10.2) holds trivially in standard mathematical analysis because there 0 is the sole microquantity in this sense.) The meaning of the term "nontrivial" here may be explicated in following way. If we replace δx by the letter ε standing for an arbitrary microquantity, (10.2) assumes the form

[17]Note that, with the appropriate choice of maps, each of these constitute the objects of a further topos, the *topos of first-order differential structures over objects in* **S.**

$$f(x+\varepsilon) - f(x) = \varepsilon f'(x). \tag{10.3}$$

Ideally, we want the validity of this equation to be independent of ε, that is, given *x*, for it to hold for *all* infinitesimal ε. In that case the derivative $f'(x)$ may be *defined* as the unique quantity *D* such that the equation

$$f(x+\varepsilon) - f(x) = \varepsilon D$$

holds for all microquantities ε.

Setting $x = 0$ in this equation, we get in particular

$$f(\varepsilon) = f(0) + \varepsilon D, \tag{10.4}$$

for all ε. Writing, as before, **Δ** for the set of microquantities, that is,

$$\Delta = \{x : x \in \mathbf{R} \wedge x^2 = 0\},$$

we require that, for any f: $\mathbf{\Delta} \to \mathbf{R}$, there is a *unique* $D \in \mathbf{R}$ such that Eq. (10.4) holds for all ε. This says that the graph of *f* is a straight line passing through (0, *f*(0)) with slope *D*. Thus any function on **Δ** is required to be *affine*. In **SIA** this is guaranteed by the Axiom of Microaffiness, as stated above.

If we think of a function $y = f(x)$ as defining a curve, then, for any *a*, the image under *f* of the "microinterval" **Δ** + *a* obtained by translating **Δ** to *a* is straight and coincides with the tangent to the curve at $x = a$ (see Fig. 10.3). In this sense, as in **Space**, each curve is "microstraight".[18] This as the **Principle of Microstraightness** in **SIA.**

From the Microaffiness Axiom we deduce the.

Principle of Microcancellation *If* $\varepsilon a = \varepsilon b$ *for all* ε, *then* $a = b$.

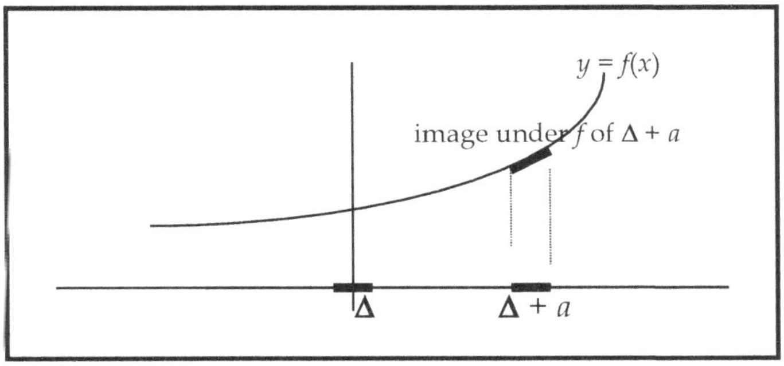

Fig. 10.3

[18] And closed curves can be treated as infinilateral polygons, as they were by Galileo and Leibniz.

For the premise asserts that the graph of the function g: $\mathbf{\Delta} \to \mathbf{R}$ defined by $g(\varepsilon) = a\varepsilon$ has both slope a and slope b: the uniqueness condition in the Microaffineness Axiom then gives $a = b$. The Principle of Microcancellation supplies the exact sense in which there are "enough" infinitesimals in **SIA**.

From the Microaffineness Axiom a prercise version of the *Principle of Continity* can be deduced, namely, that in **SIA** *all functions on* **R** *are continuous*, in the sense of *sending neighbouring points to neighbouring points*. (Here two points x, y on **R** are said to be *neighbours* if $x - y$ is in $\mathbf{\Delta}$, that is, if x and y differ by a microquantity.) To see this, given f: $\mathbf{R} \to \mathbf{R}$ and neighbouring points x, y, note that $y = x + \varepsilon$ with ε in $\mathbf{\Delta}$, so that

$$f(y) - f(x) = f(x + \varepsilon) - f(x) = \varepsilon f'(x).$$

But clearly any multiple of a microquantity is also a microquantity, so $\varepsilon f'(x)$ is a microquantity, and the result follows.

Since Eq. (10.3) holds for any f, it also holds for its derivative f'; it follows that functions in smooth infinitesimal analysis are differentiable arbitrarily many times, thereby justifying the use of the term "smooth".

Let us derive a basic law of the differential calculus, Leibniz'z *product rule:*

$$(fg)' = f'g + fg'.$$

To do this we compute[19]

$$\begin{aligned}
(fg)(x+\varepsilon) &= (fg)(x) + \varepsilon(fg)'(x) = f(x)g(x) + \varepsilon(fg)'(x), \\
(fg)(x+\varepsilon) &= f(x+\varepsilon)g(x+\varepsilon) = [f(x) + \varepsilon f'(x)] \cdot [g(x) + \varepsilon\, g'(x)] \\
&= f(x)g(x) + \varepsilon\big(f'g + fg'\big) + \varepsilon^2 f'g' \\
&= f(x)g(x) + \varepsilon(f'g + fg'),
\end{aligned}$$

since $\varepsilon^2 = 0$. Therefore $\varepsilon(fg)' = \varepsilon(f'g + fg')$, and the result follows by microcancellation. This calculation is depicted in Fig. 10.4.

Fig. 10.4

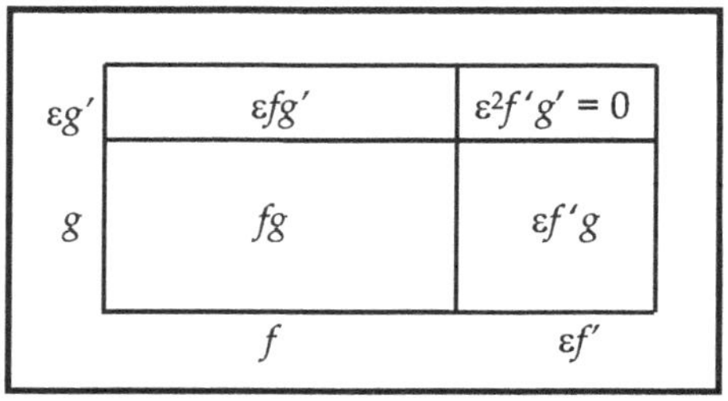

[19]What follows is surely the prettiest demonstration of the product rule ever devised. Leibniz would have found it delightful.

Fig. 10.5

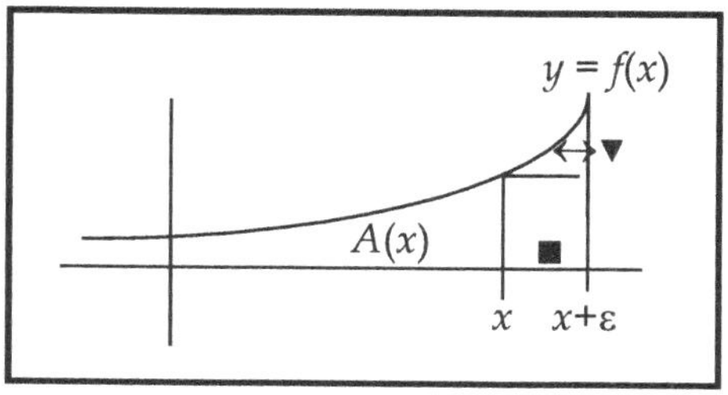

Next, we derive the *Fundamental Theorem of the Calculus* (Fig. 10.5).

Let J be a closed interval $\{x: a \leq x \leq b\}$ in **R** and $f: J \rightarrow$ **R**; let $A(x)$ be the area under the curve $y = f(x)$ as indicated above. Then, using Eq. (10.3),

$$\varepsilon A'(x) = A(x+\varepsilon) - A(x) = \blacksquare + \blacktriangledown = \varepsilon f(x) + \blacktriangledown.$$

Now by Microstraightnesss ▼ is a triangle[20] of area $\frac{1}{2}\varepsilon.\varepsilon\, f'(x) = 0$. Hence $\varepsilon A'(x) = \varepsilon f(x)$, so that, by Microcancellation,

$$A'(x) = f(x),$$

which is the fundamental theorem of the calculus.

Following Fermat,[21] in **SIA** a *stationary point a* in **R** of a function f: **R** → **R** is defined to be a point in whose vicinity microvariations fail to change the value of f, that is, for which $f(a + \varepsilon) = f(a)$ for all ε. This means that $f(a) + \varepsilon f'(a) = f(a)$, so that $\varepsilon f'(a) = 0$ for all ε, from which it follows by microcancellation that $f'(a) = 0$. So a stationary point of a function is precisely a point at which the derivative of the function vanishes.

In classical analysis, if the derivative of a function is identically zero, the function is constant. This fact is the source of the following postulate concerning stationary points adopted in **SIA**:

Constancy Principle. If every point in an interval J is a stationary point of $f: J \rightarrow$ **R** (that is, if f' is identically 0), then f is constant.

It follows from the Constancy Principle that two functions with identical derivatives differ by at most a constant. Thus if we define an *antiderivative* of a function f: $J \rightarrow$ **R** to be a function $g: J \rightarrow$ **R** such that $g' = f$, then any function with an antiderivative has a *unique* antiderivative. This observation, combined with the Principles of Microstraightness and Microcancellation, yields simple derivations of basic equations of mathematics and physics.[22] We illustrate the method by deriving *the equation of a tractrix.*

---

[20]▼ is in fact the *characteristic triangle* of seventeenth century analysis (see Chap. 2). As will be seen, in **SIA** its area reduces to zero.

[21]See Chap. 2.

[22]See Bell (1998).

Fig. 10.6

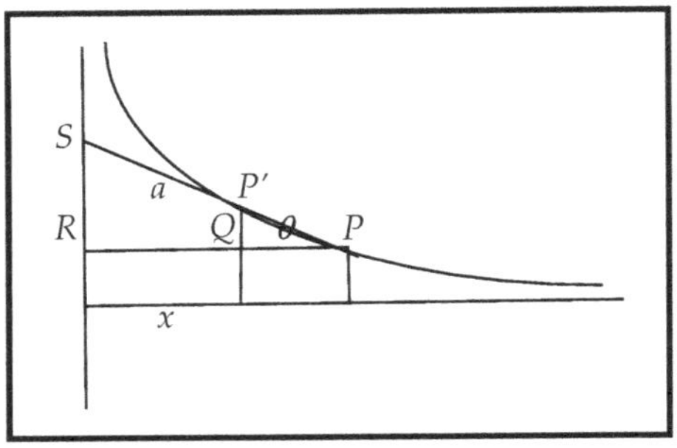

The tractrix has the property that the length of the tangent to the curve from an arbitrary point on it to the y-axis is constant. It is the curve traced out by an object dragged, under the influence of friction, by a string attached to a pulling point that moves at a right angle to the initial line between the object and the puller.

Referring to Fig. 10.6, let P, P' be two neighbouring points on the tractrix curve $y = y(x)$ with coordinates (x, y) and $(x - \varepsilon, y(x - \varepsilon))$ respectively. Let a be the constant length of the tangent from a point of the curve to the y-axis. Then by Microstraightness the tangent PS to the curve passes thfrough P'. Write $\underline{L}$ for the length of a line segment L. Then we have

$$\underline{P'Q} = y(x - \varepsilon) - y(x) = -\varepsilon y'(\mathrm{x}). \qquad (*)$$

But also, writing θ for the angle RPS, we have.

$$\underline{P'Q} = \underline{PQ}\tan\theta = \varepsilon\tan\theta = \varepsilon\underline{RS}/\underline{\mathrm{PR}} = \varepsilon\left(a^2 - x^2\right)^{1/2}/x.$$

Equating this with (∗) gives.

$$\varepsilon y'(\mathrm{x}) = -\varepsilon\left(a^2 - x^2\right)^{1/2}/x.$$

Using Microcancelletion to cancel ε on both sides of this equation, we get.

$$y'(\mathrm{x}) = -\left(a^2 - x^2\right)^{1/2}/x. \qquad (**)$$

Accordingly $y(x)$ is the antiderivative of the function $-(a^2 - x^2)^{1/2}/x$. But the usual computation in the differential calculus (which holds in **SIA**) shows that antiderivative to be the function $a\log[(a + (a^2 - x^2)^{1/2})/x)] - (a^2 - x^2)^{1/2}$. Since, by the Constancy Principle, antiderivatives are unique, it follows that.

$$y = a\log\left[\left(a + \left(a^2 - x^2\right)^{1/2}\right)/x\right)] - \left(a^2 - x^2\right)^{1/2}$$

This is the equation of the tracrix.

This sort of argument is typical in **SIA.** The pattern is this. Suppose we want to determine the explicit form of a function $f(x)$. We subject x to a microvariation by adding an infiniresimal ε, and consider the "increment" in f, $\delta f = f(x + \varepsilon) - f(x) = \varepsilon f'(x)$. Then we use geometric or other methods "in the small" (i.e. employing microstraightness and the nilsquare propery of ε) to present δf in the form $\varepsilon g(x)$, where $g(x)$ is an explicit function with explicit antiderivative h. Since $\varepsilon f'(x) = \varepsilon g(x)$, microcancellation gives $f'(x) = g(x)$. Accordingly f is also an antiderivative of g, and so must be identical with h. This latter function is the explicit form of f.

Put succinctly, the Constancy Principle asserts that "universal infinitesmal (or "local") constancy implies global constancy", or "infinitesimal behaviour determines global behaviour" The Constancy Principle brings into sharp focus the difference in **SIA** between points and infinitesimals. For if in the Constancy Principle one replaces "infinitesimal constancy" by "constancy at a point" the the resulting "Principle" is false because *any function whatsoever* is constant at every point. But since in **SIA** all functions on **R** are microaffine and hence smooth, the Constancy Principle embodies the idea that for such functions local constancy is sufficient for global constancy, that a nonconstant smooth function must be somewhere nonconstant over arbitrarily small intervals.

The Constancy Principle may be seen as furnishing the "missing link" between the infinitesimal and the world "in the large", the lack of which, as we have already observed, Hermann Weyl believed doomed the idea of infinitesimal, and led to its inevitable replacement by thde limit concept:

> [In its struggle with the infinitely small] *the limiting process was victorious. For the* limit *is an indispensable concept, whose importance is not affected by the acceptance or rejection of the infinitely small. But once the limit concept has been grasped, it is seen to render the infinitely small superfluous. Infinitesimal analysis proposes to draw conclusions by integration from the behavior in the infinitely small, which is governed by elementary laws, to the behavior in the large; for instance, from the universal law of attraction for two material "volume elements" to the magnitude of attraction between two arbitrarily shaped bodies with homogeneous or non-homogeneous mass distribution. If the infinitely small is not interpreted 'potentially' here, in the sense of the limiting process, then the one has nothing to do with the other, the process in infinitesimal and finite dimensions become independent of each other, the tie which binds them together is cut.*[23]

In a smooth world the Constancy Principle reconnects the infinitesimal and the extended. Behaviour "in the large" is completely determined by behaviour "in the infinitely small".

---

[23]Weyl (1949) 44–5. Yet we also recall Weyl's observation:

> *The principle of gaining knowledge of the external world from the behaviour of its infinitesimal parts is the mainspring of the theory of knowledge in infinitesimal physics as in Riemann's geometry, and, indeed, the mainspring of all the eminent work of Riemann* (Weyl 1950, p. 92).

10.4 The Internal Logic of a Smooth World Is Intuitionistic

The correctness of the Principle of Continuity in **SIA** induces a subtle, but significant change of logic there: from classical to *intuitionistic*. For, in the first place, we have only to observe that, if the law of excluded middle held without qualification, then each real number x would either be equal to 0 or unequal to 0, in which case the correlation $0 \mapsto 0, x \mapsto 1$ for $x \neq 0$ (the well-known "blip" function) would define a map from the space **R** of real numbers to the set $2 = \{0, 1\}$; but it is evidently discontinuous, contradicting the Principle of Continuity. From this we see that the Principle of Continuity mplies that the law of excluded middle cannot be universally affirmed in **SIA**. In fact, this informal argument shows that the statement.

$$\textit{for any real number } x\textit{, either } x = 0 \textit{ or } x \neq 0. \qquad (*)$$

is *refutable* in **SIA**.

Here is a rigorous refutation[24] Of $(*)$ in **SIA** using the Principle of Microcancellation. To begin with, if $x \neq 0$, then $x^2 \neq 0$, so that, if $x^2 = 0$, then necessarily not $x \neq 0$. This means that.

$$\textit{for all infinitesimal } \varepsilon\textit{, not } \varepsilon \neq 0. \qquad (**)$$

Now suppose that $(*)$ were to hold. Then we would have, for any ε, either $\varepsilon = 0$ or $\varepsilon \neq 0$. But $(**)$ allows us to eliminate the second alternative, and we infer that, for all ε, $\varepsilon = 0$. This may be written.

$$\textit{for all } \varepsilon,\ \varepsilon.1 = \varepsilon.0,$$

from which we derive by Microcancellation the falsehood $1 = 0$. The refutability of $(*)$ follows.

For simplicity we sgall refer to the refutability of the assertion $(*)$ as *the refutability of the law of excluded middle in* **SIA**.

The Principle of Continuity also implies that propositional functions, or predicates, cannot be taken as being merely "bipolar" in Wittgenstein's sense, that is, representable in terms of assuming just two "truth values" within the set $2 = \{\text{true, false}\} = \{1, 0\}$. For let Ω be the domain of truth values in a world in which the Principle holds. Then, as usual, for any object X, parts of X correspond to predicates on X, that is "propositional functions" on X, in other words maps $X \rightarrow \Omega$. Now if X is a (connected) continuum, it presumably does have proper nonempty parts. But there are only two *continuous* maps $X \rightarrow 2$, namely the constant ones corresponding to the whole of X and the empty part of X, because a nonconstant continuous such map on X would yield a "splitting" of X into two nontrivial disconnected pieces. Thus: X has

[24] We givde another refutation of $(*)$ below.

more than two parts; these correspond to maps $X \to \Omega$., so there are more than two of these; but there are just two maps $X \to 2$; whence $\Omega \neq 2$.

It is of interest to recall[25] in this connection Peirce's awareness, even before Brouwer, of the fact that a faithful account of the truly continuous would involve abandoning the unrestricted applicability of the law of excluded middle:

> *Now if we are to accept the common idea of continuity...we must either say that a continuous line contains no points...or that the law of excluded middle does not hold of these points. The principle of excluded middle applies only to an individual...but places being mere possibilities without actual existence are not individuals.*

The prescience shown by Peirce here is all the more remarkable since in **SIA** the law of excluded middle does, in a certain sense, apply to *individuals.* This follows from the fact that, despite its failure for arbitrary predicates, the law of excluded middle can be shown to hold in SIA for arbitrary closed *sentences.*[26] So if P is any predicate and a any particular real number, $P(a) \vee \neg P(a)$ will be true. Also for any particular real numbers a, b the statement $a = b \vee a \neq b$ holds. Note, however, that in **SIA** the truth of this last statement for each pair of particular real numbers does not imply the truth of the universal generalization.

$$\forall x \in \mathbf{R} \forall y \in \mathbf{R} \, (x = y \vee x \neq y).$$

Indeed, we have seen that this is refutable in **SIA:** in short, *equality on* **R** *is undecidable.* This may be taken as indicating that the smooth real line is a genuine continuum in being, unlike a discrete set, more than the mere "sum" of its elements.

The "internal" logic of **SIA** is accordingly not full classical logic. It is, instead, *intuitionistic* logic, that is, the logic derived from the constructive interpretation of mathematical assertions.[27] This "change of logic" is not noticed in the development of basic calculus because there arguments are in the main constructive, proceeding by direct computation.

The refutability of the law of excluded middle in **SIA** leads to the refutability of an important principle of set theory, the *Axiom of Choice.* This is the assertion.

> **(AC)** for any family $\mathcal{A}$ of inhabited[28] sets, there is a *choice function on* $\mathcal{A}$, that is, a function $f: \mathcal{A} \to \bigcup \mathcal{A}$ for which $f(X) \in X$ whenever $X \in \mathcal{A}$.

Now the law of excluded middle can be derived merely from the very special case of the Axiom of Choice which asserts merelythat any *doubleton* $\{U, V\}$ has a choice function. For let p be any proposition, define

---

[25]See Chap. 5.

[26]To be precise, this condition can be shown to hold in a number of models of **SIA**, See McLarty (1988).

[27]See Chap. 9.

[28]A set is said to be *inhabited* if it can be constructively shown to have a member. In intuitionistic logic this is a stronger condition than the assertion that the set be nonempty.

$$U = \{x \in 2 : x = 0 \vee p\}\ V = \{x \in 2 : x = 1 \vee p\}.$$

and let f be a choice function on $\{U, V\}$. Writing $a = f(U)$, $b = f(V)$, we have $a \in U$, $b \in V$, i.e.,

$$[a = 0 \vee p] \wedge [b = 1 \vee p].$$

It follows that

$$[a = 0 \wedge b = 1] \vee p,$$

whence

$$a \neq b \vee p. \qquad (*)$$

Now clearly

$$p \rightarrow U = V = 2 \rightarrow a = b,$$

whence

$$a \neq b \rightarrow \neg p,$$

But this and $(*)$ together imply $\neg p \vee p$.

Thus the law of excluded middle is derivable from this special case of the Axiom of Choice. Since the law of excluded middle is refutable in **SIA**, so equally, then, is **AC**.

The refutability of the Axiom of Choice in **SIA**, and hence its incompatibility with the Principle of Continuity which prevails in smooth worlds, is not surprising in view of the Axiom'swell-known "paradoxical" consequences. One of these is the famous *Banach-Tarski paradox*[29] which asserts that any solid sphere can be decomposed into finitely many pieces which can themselves be reassembled to form two solid spheres each of the same size as the original, or into one solid sphere of any preassigned size. Paradoxical decompositions such as these become possible only when continuous geometric objects are, in Dedekind's words,[30] "dissolved to atoms ... [through a] frightful, dizzying discontinuity" into discrete sets of points which the axiom of choice then allows to be rearranged in an arbitrary (discontinuous) manner. Such procedures are inadmissible in smooth worlds.

In this connection, it should also be mentioned that the classical *intermediate value theorem*, often taken as expressing an "intuitively obvious" property of continuous functions, is *false* in smooth worlds. The intermediate value theorem is

[29]See Wagon (1993).

[30]See Chap. 4.

the assertion that, for any $a, b \in \mathbf{R}$ such that $a < b$, and any continuous f: $[a, b] \to \mathbf{R}$ such that $f(a) < 0 < f(b)$, there is $x \in [a, b]$ for which $f(x) = 0$. In fact this fails in **SIA** even for polynomial functions, as the following informal argument shows. Suppose, for example, that the intermediate value theorem were true in **SIA** for the polynomial function $f(x) = x^3 + tx + u$. Then the value of x for which $f(x) = 0$ would have to depend smoothly on the values of t and u. To be precise, there would have to exist a smooth map g: $\mathbf{R}^2 \to \mathbf{R}$ such that

$$g(t, u)^3 + tg(t, u) + u = 0.$$

A geometric argument can be given to prove that no such smooth map exists.[31]

10.5 Smooth Infinitesimal Analysis as an Axiomatic Theory. Consequences for the Continuum

SIA can be *axiomatized* as a theory formulated within higher-order intuitionistic logic. Here are the basic axioms of the theory.[32]

Axioms for the continuum, or smooth real line R. These include the usual axioms for a commutative ring with unit expressed in terms of two operations + and **.**, (we usually write xy for $x \,.\, y$) and two distinguished elements $0 \neq 1$.[33] In addition we stipulate that **R** is an *intuitionistic field*, i.e., satisfies the following axiom:

$$x \neq 0 \text{ implies} \exists y (xy = 1).$$

Axioms for the strict order relation < on R. These are:

O1. $a < b$ and $b < c$ implies $a < c$.
O2. $\neg(a < a)$.
O3. $a < b$ implies $a + c < b + c$ for any c.
O4. $a < b$ and $0 < c$ implies $ac < bc$.
O5. either $0 < a$ or $a < 1$.
O6. $a \neq b$ implies $a < b$ or $b < a$.[34]
O7. $0 < x$ implies $\exists y\ (x = y^2)$.

Arithmetical Axioms. These govern the set **N** of *Archimedean* (or smooth) *natural numbers*, and read as follows:

---

[31] Moerdijk and Reyes (1991), Remark VII.2.14.

[32] Moerdijk and Reyes (1991)

[33] Here $a \neq b$ stands for $\neg a = b$.

[34] It should be pointed out that axiom 6 is omitted in some presentations of **SIA**, e.g. those in Kock (1981) and McLarty (1992).

1. **N** is a cofinal or Archimedean subset of **R,** i.e. $\mathbf{N} \subseteq \mathbf{R}$ and $\forall x \in \mathbf{R}\ \exists n \in \mathbf{N}\ x < n$.
2. *Peano axioms*:

$$0 \in \mathbf{N}$$
$$\forall x \in \mathbf{R}(x \in \mathbf{N} \rightarrow x + 1 \in \mathbf{N})$$
$$\forall x \in \mathbf{R}(x \in \mathbf{N} \rightarrow x + 1 \neq 0)$$

3. *Restricted Induction scheme.* For every formula $\varphi(x)$ involving just $=, \wedge, \vee, \mathrm{T}, \perp, \exists$[35]

$$\varphi(0) \quad \forall x \in \mathbf{N}[\varphi(x) \rightarrow \varphi(x+1)] \rightarrow \forall x \in N\varphi(x).$$

Using restricted induction it follows that

- **N** has *decidable equality*, i.e. $\forall x \in \ \forall y \in \mathrm{N}(x = y \vee x \neq y)$
- **N** is *linearly ordered*, i.e. $\forall x \in \mathrm{N}\ \forall y \in \mathrm{N}(x < y \vee x = y \vee y < x)$.
- **N** satisfies *decidable induction*: for any formula $\varphi(x)$,

$$\forall x \in \mathbf{N}(\varphi(x) \vee \neg\varphi(x)) \rightarrow \left[\left[\varphi(0) \wedge \forall x \in \mathbf{N}\big(\varphi(x) \rightarrow \varphi(x+1)\big)\right] \rightarrow \forall x\varphi(x)\right].$$

The relation $\leq$ on **R** is defined by $a \leq b \Leftrightarrow \neg b < a$. The open interval (a, b) and closed interval $[a, b]$ are defined as usual, viz. $(a, b) = \{x\colon a < x < b\}$ and $[a, b] = \{x\colon a \leq x \leq b\}$; similarly for half-open, half-closed, and unbounded intervals.

We have written $\boldsymbol{\Delta}$ for the subset $\{x\colon x^2 = 0\}$ of **R** consisting of (nilsquare) *infinitesimals* or *microquantities.* As before, we use the letter ε as a variable ranging over Δ. Δ is subject to the.

Microaffineness Axiom. *For any map* $g\colon \boldsymbol{\Delta} \rightarrow \mathbf{R}$ *there exist unique* $a, b \in \mathbf{R}$ *such that, for all* ε, *we have*

$$g(\varepsilon) = a + b\varepsilon.$$

In **SIA** one also assumes the.

Constancy Principle. *If* $A \subseteq \mathbf{R}$ *is any closed interval on* **R,** *or* **R** *itself, and* $f\colon A \rightarrow \mathbf{R}$ *satisfies* $f(a + \varepsilon) = f(a)$ *for all* $a \in A$ *and* $\varepsilon \in \boldsymbol{\Delta}$, *then* f *is constant.*

It follows easily from the Microaffineness Axiom that $\boldsymbol{\Delta}$ is *nondegenerate,* i.e. $\boldsymbol{\Delta} \neq \{0\}$.[36] For if $\boldsymbol{\Delta} = \{0\}$, then the identity map $i\colon \boldsymbol{\Delta} \rightarrow \boldsymbol{\Delta}$ can be represented as $i(\varepsilon) = b\varepsilon$ for any b, in violation of the uniqueness condition on b.

[35] Here, T, $\perp$ are symbols denoting, respectively, the *true* and the *false*.

[36] It should be noted that, while $\boldsymbol{\Delta}$ does not reduce to $\{0\}$, nevertheless 0 is the only *explicitly nameable* element of Δ. For it is easily seen to be inconsistent to assert that $\boldsymbol{\Delta}$ actually contains an element $\neq 0$.

From the nondegeneracy of **Δ** we can also (again) refute the law of excluded middle in **SIA**, more particularly, we can prove.

$$\neg\forall\varepsilon(\varepsilon = 0 \lor \varepsilon \neq 0). \qquad (*)$$

For we have, for $\varepsilon \in \mathbf{\Delta}$, $\varepsilon^2 = 0$, whence $\neg(\varepsilon \neq 0)$, and $(*)$ would give $\varepsilon = 0$. So **Δ** would be degenerate, contrary to fact. It follows from $(*)$ that, using x and y as variables ranging over **R,**

$$\neg\forall x\forall y(x = y \lor x \neq y).$$

In a word, the identity relation is *undecidable* on **R.**

Call a binary relation S on **R** *stable* if it satisfies

$$\forall x\forall y(\neg\neg xRy \rightarrow xRy).$$

Then the nondegeneracy of **Δ** implies that, in **SIA** the equality relation is *unstable.* For suppose that $=$ were stable. Then, for any ε, it would be the case that $\neg\varepsilon \neq 0 \rightarrow \varepsilon = 0$. But we have already shown above that $\neg(\varepsilon \neq 0)$, so it would follow that $\varepsilon = 0$. This being the case for any ε, **Δ** would be degenerate.

Except for the presence of intuitionistic logic, we note that the *algebraic structure* on **R** in **SIA** differs little from the classical situation. In **SIA**, **R** is equipped with the usual addition and multiplication operations under which it is a field. In particular, **R** satisfies the condition that each $x \neq 0$ has a multiplicative inverse. Notice, however, that since in **SIA** no microquantity (apart from 0 itself) is provably $\neq$ 0, microquantities are not required to have multiplicative inverses (a requirement which would lead to inconsistency). From a strictly algebraic standpoint, **R** in **SIA** differs from its classical counterpart only in being required to satisfy the Principle of Microcancellation.

The situation is otherwise, however, as regards the *order structure* of **R** in **SIA**. Since microquantities do not have multiplicative inverses, and **R** is an intuitionistic field, it must be the case that $\forall\varepsilon\neg(\varepsilon \neq 0)$, whence

$$\forall\varepsilon\neg(\varepsilon < 0 \lor \neg\, \varepsilon > 0),$$

or equivalently

$$\forall\varepsilon(\varepsilon \leq 0 \land \varepsilon \geq 0).$$

It follows easily from this and the nondegeneracy of Δ that

$$\neg\forall x\forall y(x < y \lor y < x \lor x = y).$$

In other words the order relation < on **R** in **SIA** *fails to satisfy the trichotomy law*; it is a *partial,* rather than a *total* ordering.

The axioms of **SIA** entail that **R** differs in certain key respects from its counterpart in *constructive analysis* **CA**.[37] For example, in **CA** the equality relation is stable, while we have ashown above that in **SIA** iit is unstable. Also in **CA** the ordering relation < satisfies

$$\neg(x < y \vee y < x) \rightarrow x = y; \qquad (*)$$

and this is incompatible with the axioms of **SIA**. For (∗) implies.

$$\forall x \neg(x < 0 \vee 0 < x) \rightarrow x = 0. \qquad (**)$$

But in **SIA** it is easy to derive

$$\forall x \in \Delta \ \neg(x < 0 \vee 0 < x),$$

and this, together with (∗∗), would give $\mathbf{\Delta} = \{0\}$, contradicting the nondegeneracy of $\mathbf{\Delta}$.

In **CA** the object $\mathbf{\Delta}$ is degenerate while the nondegeneracy of Δ in **SIA** is one of its characteristic features.

Axiom O6 of **SIA**, together with the transitivity and irreflexivity of <, implies that < is stable. This may be seen as follows. Suppose $\neg\neg a < b$. Then certainly $a \neq b$, since $a = b \rightarrow \neg a < b$ by irreflexivity. Therefore $a < b$ or $b < a$. The second disjunct together with $\neg\neg a < b$ and transitivity gives $\neg\neg a < a$, which contradicts $\neg a < a$. Accordingly we are left with $a < b$. Hence < is stable. But the stability of < cannot be deduced in **CA**.[38]

10.6 Cohesiveness of the Continuum and Its Subsets in SIA

In oclassical analysis the continuum and its closed intervals are connected in the sense that they cannot be split into two nonempty subsets neither of which contains a limit point of the other. In **SIA** the Constancy Principle ensures that these have the vastly stronger property of *cohesiveness*,[39] that is, they cannot be split into two

---

[37]See Chap. 9.

[38]In **CA**, the stability of < can be shown to entail, for certain predicates A, the corresponding instance of *Markov's Principle*, namely

$$\forall x(A(x) \vee \neg A(x)) \wedge \neg\forall x \neg A(x) \rightarrow \exists x A(x).$$

Markov's principle is not generally accepted in constructive analysis. See Dummett (1977) and Bridges and Richman (1987).

[39]See Appendix A.

disjoint nonempty subsets *in any way whatsoever.* This is clearly equivalent to saying that any map $A \to \{0, 1\}$ is constant.

To see this, let A be **R** or any closed interval, and suppose $A = U \cup V$ with $U \cap V = \varnothing$. Define $f: A \to \{0, 1\}$ by $f(x) = 1$ if $x \in U$, $f(x) = 0$ if $x \in V$. We claim that f is constant. For we have

$$(f(x) = 0 \text{ or } f(x) = 1) \ \& \ (f(x+\varepsilon) = 0 \text{ or } f(x+\varepsilon) = 1).$$

This gives four possibilities:

(i) $f(x) = 0 \ \& f(x + \varepsilon) = 0$
(ii) $f(x) = 0 \ \& f(x + \varepsilon) = 1$
(iii) $f(x) = 1 \ \& f(x + \varepsilon) = 0$
(iv) $f(x) = 1 \ \& f(x + \varepsilon) = 1$

Possibilities (ii) and (iii) may be ruled out because f is continuous. This leaves (i) and (iv), in either of which $f(x) = f(x + \varepsilon)$. So f is locally, and hence globally, constant, that is, constantly 1 or 0. In the first case $V = \varnothing$, and in the second case $U = \varnothing$.

From the cohesiveness of closed intervals it can be inferred[40] that in **SIA** *all intervals in* **R** *are cohesive.*

In **SIA** cohesive subsets of **R** correspond, *grosso modo,* to connected subsets of $\mathbb{R}$ in classical analysis, that is, to intervals. This is borne out by the fact that any puncturing of **R** is *decomposable*, for it follows immediately from Axiom O6 that

$$\mathbf{R} - \{a\} = \{x : x > a\} \cup \{x : x < a\}.$$

The set **Q** of (smooth) *rational numbers* is defined as usual to be the set of all fractions of the form m/n with $m, n \in \mathbf{N}$, $n \neq 0$. The fact that **N** is cofinal in **R** ensures that **Q** is dense in **R**.

The set **R – Q** of *irrational* numbers is decomposable as

$$\mathbf{R} - \mathbf{Q} = [\{x : x > 0\} - \mathbf{Q}] \cup [\{x : x < 0\} - \mathbf{Q}\}.$$

This is in sharp contrast with the situation in *intuitionistic analysis* that is, **CA** augmented by Kripke's scheme, Brouwer's Continuity Principle, and bar induction. For we have observed[41] that in intuitionistic analysis not only is any puncturing of **R** cohesive, but that this is even the case for the irrational numbers. This would seem to indicate that in some sense the continuum in **SIA** is considerably less "syrupy"[42] than its counterpart in **SIA.**

---

[40]Bell (2001).

[41]Chap. 9.

[42]It should be emphasized that this phenomenon is a consequence of axiom O6: it cannot necessarily be affirmed in versions of **SIA** not including this axiom.

Fig. 10.7

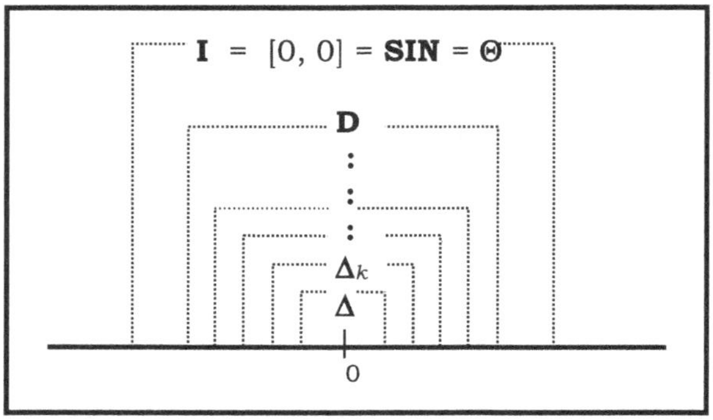

It can also be shown that the various *infinitesimal neighbourhoods* of 0 are cohesive[43] (see Fig. 10.7). The cohesiveness of the first of these infinitesimal neighbourhoods, $\mathbf{\Delta}$ itself, can be established as follows. Suppose f: $\mathbf{\Delta} \to \{0, 1\}$. Then by Microaffineness there are unique a, $b \in \mathbf{R}$ such that $f(\varepsilon) = a + b\varepsilon$ for all ε. Now $a = f(0) = 0$ or 1; if $a = 0$, then $b\varepsilon = f(\varepsilon) = 0$ or 1, and clearly $b\varepsilon \neq 1$. So in this case $f(\varepsilon) = 0$ for all ε. If on the other hand $a = 1$, then $1 + b\varepsilon = f(\varepsilon) = 0$ or 1; but $1 + b\varepsilon = 0$ would impl $b\varepsilon = -1$ which is again impossible. So in this case $f(\varepsilon) = 1$ for all ε. Therefore f is constant and $\mathbf{\Delta}$ cohesive.

In **SIA** *nilpotent infinitesimals* are defined to be the members of the sets.

$$\mathbf{\Delta}_k = \left\{x \in \mathbf{R} : x^{k+1} = 0\right\},$$

for $k = 1, 2, \ldots$, each of which may be considered an infinitesimal neighbourhood of 0. These are subject to the.

Micropolynomiality Principle. *For any* $k \geq 1$ *and any* g: $\mathbf{\Delta}_k \to \mathbf{R}$, *there exist unique* $a, b_1, \ldots, b_k \in \mathbf{R}$ *such that for all* $\delta \in \mathbf{\Delta}_k$ *we have.*

$$g(\delta) = a + b_1\delta + b_2\delta^2 + \ldots + \mathrm{b_k}\,\delta^{\mathrm{k}}.$$

Micropolynomiality implies that no $\mathbf{\Delta}_k$ coincides with $\{0\}$.

An argument similar to that establishing the cohesiveness of $\mathbf{\Delta}$ does the same for each $\mathbf{\Delta}_k$. Thus let f: $\mathbf{\Delta}_k \to \{0, 1\}$; Micropolynomiality implies the existence of $a, b_1, \ldots, b_k \in \mathbf{R}$ such that $f(\delta) = a + \zeta(\delta)$, where $\zeta(\delta) = b_1\delta + b_2\delta^2 + \ldots + b_k\delta^k$. Notice that $\zeta(\delta) \in \mathbf{\Delta}_k$, that is, $\zeta(\delta)$ is nilpotent. Now $a = f(0) = 0$ or 1; if $a = 0$ then $\zeta(\delta) = f(\delta) = 0$ or 1, but since $\zeta(\delta)$ is nilpotent it cannot $=1$. Accordingly in this case $f(\delta) = 0$ for all $\delta \in \mathbf{\Delta}_k$. If on the other hand $a = 1$, then $1 + \zeta(\delta) = f(\delta) = 0$ or 1, but

[43]Notice that in **CA** all of these infinitesimal objects are degenerate. This makes it difficult to formulate a satisfactory theory of infinitesimals in any extension of **CA**, in particular, in intuitionistic analysis.

$1 + \zeta(\delta) = 0$ would imply $\zeta(\delta) = -1$ which is again impossible. Accordingly f is constant and $\mathbf{\Delta}$ cohesive.

The union $\mathbf{D}$ of all the $\mathbf{\Delta}_k$ is the *set of nilpotent infinitesimals,* another infinitesimal neighbourhood of 0. The cohesiveness of $\mathbf{D}$ follows readily from that of each $\mathbf{\Delta}_k$.

The next infinitesimal neighbourhood of 0 is [0, 0], which, as a closed interval, is cohesive. It is easily shown that [0, 0] includes $\mathbf{D}$, so that it does not coincide with $\{0\}$.

It can be shown that [0, 0] coincides with the set $\mathbf{\Theta}$ of *noninvertible* elements of $\mathbf{R}$, as well as with the sets

$$\mathbf{SIN} = \left\{x \in \mathbf{R} : \forall n \in \mathbf{N}\left(-\frac{1}{n+1} < x < \frac{1}{n+1}\right)\right\}.$$

of *strict infinitesimals*, as well as the set

$$\mathbf{I} = \{x \in \mathbf{R} : \neg x \neq 0\}$$

of elements of $\mathbf{R}$ *indistinguishable from* 0. $\mathbf{I}$ is an *ideal*, in fact a *maximal* ideal in the ring $\mathbf{R}$.

Finally, we observe that the sequence of infinitesimal neighbourhoods of 0 generates a strictly ascending sequence of *decomposable* subsets containing $\mathbf{R} - \{0\}$, namely:

$$\begin{aligned}\mathbf{R} - \{0\} \subset (\mathbf{R} - \{0\}) \cup \{0\} \subset (\mathbf{R} - \{0\}) \cup \Delta_1 \subset (\mathbf{R} - \{0\}) \cup \Delta_2 \\ \subset \ldots (\mathbf{R} - \{0\}) \cup \mathbf{D} \subset (\mathbf{R} - \{0\}) \cup [0, 0].\end{aligned}$$

10.7 Comparing the Smooth and Dedekind Real Lines in SIA

A *Dedekind real* is a pair $(U, V) \in \mathscr{P}\mathbf{Q} \times \mathscr{P}\mathbf{Q}$[44] satisfying the conditions:

$$\exists x \exists y (x \in U \wedge y \in V)$$

$$U \cap V = \varnothing$$

$$\forall x (x \in U \leftrightarrow \exists y \in U . x < y)$$

[44] Here $\mathscr{P}A$ is the power set of a set A.

$$\forall x(x \in V \leftrightarrow \exists y \in V.y < x)$$
$$\forall x \forall y(x < y \rightarrow x \in U \vee y \in V).$$

The set $\mathbf{R}_d$ of Dedekind reals can be turned into an ordered ring.[45] This ring is always *constructively complete*, that is, satisfies the condition: Let A be an inhabited subset of $\mathbf{R_d}$ that is bounded above. Then sup A exists if and only if for all $x, y \in \mathbf{R_d}$ with $x < y$, either y is an upper bound for A or there exists $a \in A$ with $x < a$. (A real number b is called a *supremum*, or *least upper bound*, of A if it is an upper bound for A and if for each $\varepsilon > 0$ there exists $x \in A$ with $x > b - \varepsilon$.)

Although $\mathbf{R_d}$ is constructively complete, it is not conditionally complete in the classical sense because of the failure of the logical law $\neg p \vee \neg\neg p$.[46] But $\mathbf{R_d}$ shares some features of the constructive reals not possessed by $\mathbf{R}$, e.g.

$$\neg\neg x = y \rightarrow x = y$$
$$x \leq y \wedge y \leq x \rightarrow x = y$$
$$x^n = 0 \rightarrow x = 0.$$

There is a natural order preserving homomorphism φ: $\mathbf{R} \rightarrow \mathbf{R_d}$ given by.

$$\varphi(r) = (\{q \in \mathbf{Q} : q < r\}, \{q \in \mathbf{Q} : q > r\})'$$

This is injective on $\mathbf{Q}$, and embeds $\mathbf{Q}$ as the rational numbers in $\mathbf{R_d}$. Moreover, the kernel of φ coincides with the ideal $\mathbf{I}$ of strict infinitesimals in $\mathbf{R}$, so φ induces an embedding of the quotient ring $\mathbf{R/I}$ into $\mathbf{R_d}$. $\mathbf{R/I}$ is $\mathbf{R}$ *shorn of its nilpotent infinitesimals*: it is both an intuitionistic field and an integral domain, that is, satisfies

$$\forall x(x \neq 0 \rightarrow x \textit{ is invertible}) \quad \forall x \forall y[xy = 0 \rightarrow x = 0 \vee y = 0].$$

It can be shown that φ is *surjective*—so that $\mathbf{R/I} \cong \mathbf{R_d}$—precisely when $\mathbf{R}$ is constructively complete in the sense above. In that event $\mathbf{R_d}$ is both an intuitionistic field and an integral domain, properties that the ring of Dedekind reals in a topos does not always possess.

In any model of **SIA** the usual *open interval topology* can be defined on $\mathbf{R_d}$. It can be shown[47] that with this topology $\mathbf{R_d}$ is always *connected* in the sense that it cannot

45 See Johnstone (1977). In the topos **Shv**(X) of sheaves over a topological space X, $\mathbf{R}_d$ is the sheaf of continuous real-valued functions on open subsets of X.

46 The failure of this law in **SIA** follows immediately from the cohesiveness of **R** by considering the predicate $x \neq 0$. As originally shown by Johnstone, conditional completeness of $\mathbf{R_d}$ is actually equivalent to this logical law $\neg p \vee \neg\neg p$: in Shv(X), the law holds iff X is extremally disconnected, that is, the closure of every open set is open.

47 L. Stout, *Topological properties of the real numbers object in a topos*. Cahiers Topologie Géom. Différentielle **17**, no. 3, (1976), pp. 295-326.

be partitioned into two disjoint inhabited open subsets. In **SIA** $\mathbf{R_d}$ actually inherits a stronger cohesiveness property from **R.** To see this, call a subset X of a set A *detachable*[48] if there is a subset Y of A such that $X \cap Y = \varnothing$, $X \cup Y = A$. Now we can show that, if X is a detachable subset of $\mathbf{R_d}$, then $\varphi\,[\mathbf{R}] \subseteq X$ or $X \cap \varphi[\mathbf{R}] = \varnothing$. For suppose $X \subseteq \mathbf{R_d}$ detachable and define f: $\mathbf{R_d} \to 2$ by $f(x) = 1$ if $x \in X$, $f(x) = 0$ if $x \notin X$. Then $f \circ \varphi$: $\mathbf{R} \to 2$ must be constant since **R** is cohesive. If $f \circ \varphi$ is constantly 1, then $\varphi[\mathbf{R}] \subseteq X$; if constantly 0, then $A \cap \varphi[\mathbf{R}] = \varnothing$. It follows easily that if φ is surjective then $\mathbf{R_d}$ is itself cohesive.

10.8 Nonstandard Analysis in SIA

In certain models of **SIA** the system of *natural numbers* possesses some intriguing features which make it possible to introduce another type of infinitesimal—the so-called *invertible* infinitesimals—resembling those of nonstandard analysis, whose presence engenders yet another infinitesimal neighbourhood of 0 properly containing all those introduced above.

We recall that the set **N** of smooth natural numbers is required to satisfy not the full principle of mathematical induction for arbitrary properties but only the weaker restricted induction scheme. This raises the possibility that **N** may not coincide with the set $\mathbb{N}$ of *standard natural numbers,* which is defined to be the *smallest* subset of **R** containing 0 and closed under the operation of adding 1. Now, models of **SIA** have been constructed[49] in which $\mathbb{N}$ is a *proper subset* of **N**; accordingly the members of $\mathbf{N} - \mathbb{N}$ may be considered *nonstandard integers.* Multiplicative inverses of nonstandard integers are infinitesimals, but, being themselves invertible, they are of a different type from the (necessarily noninvertible) nilpotent infinitesimals which are basic to **SIA.**

Proceeding formally, we define the set $\mathbb{N}$ of *standard* natural numbers to be the intersection of all *inductive* subsets of **N**, i.e.,

$$\mathbb{N} = \{n \in \mathbf{N}{:}\forall X \subseteq \mathbf{N}[0 \in X \wedge \forall m(m \in X \to m + 1 \in X) \to n \in X]\}$$

$\mathbb{N}$ evidently satisfies *full induction*:

$$\forall X \subseteq \mathbf{N}[0 \in X \wedge \forall m(m \in X \to m + 1 \in X) \to X = \mathbf{N}].$$

The space of *infinitesimals* is the set

[48]It is easily shown that a subset X of a set A is detachable if and only if the property of being a member of X is *decidable*, that is, if $\forall x{\in}A(x \in X$ or $x \notin X)$.

[49]See Moerdijk and Reyes (1991).

$$\mathbf{IN} = \{x \in \mathbf{R} : \forall n \in \mathbb{N}(-1/(n+1) < x < 1/(n+1)\}.$$

This is the largest infinitesimal neighbourhood of zero in **SIA**: it contains the space $\boldsymbol{\Theta}$ of noninvertible infinitesimals as well as the space of *invertible* or *Robinsonian infinitesimals*

$$\mathrm{I} = \{x \in \mathbf{IN} : x \textit{ is invertible}\}.$$

As inverses of "infinitely large" reals (i.e. reals r satisfying $\forall n \in \mathbb{N}\,.\,n < r \vee \forall n \in \mathbb{N}\,.\,r < -n$) invertible infinitesimals are the counterparts in **SIA** of the infinitesimals of nonstandard analysis.[50] Invertible infinitesimals are strictly larger than their noninvertible cousins in that

$$\forall x \forall y[x \in \boldsymbol{\Theta} \wedge y \in \mathrm{I}\; y > 0 \rightarrow x < y].$$

To assert the existence of invertible infinitesimals is to assert that I is inhabited[51]: this is equivalent to asserting that the set $\mathbf{N} - \mathbb{N}$ of *nonstandard integers* is inhabited, or equivalently, that the following holds:

$$\exists n \in \mathbf{N} \forall m \in \mathbb{N} m < n.$$

When this condition is satisfied, as it is in certain models of **SIA,** we shall say that nonstandard integers, or invertible infinitesimals, are *present.* Notice that while it is perfectly consistent to assert the presence of invertible infinitesimals, i.e., that I be inhabited, it is inconsistent to assert the "presence" of nonzero noninvertible infinitesimals, i.e. that $\boldsymbol{\Theta} - \{0\}$ be inhabited.[52]

One may also postulate the condition

$$\forall n \in \mathbf{N}[\forall x \in \mathbf{N} - \mathbb{N}(x > n) \rightarrow n \in \mathbb{N}],$$

i.e. "a natural number which is smaller than all nonstandard natural numbers must be standard". This is in fact equivalent to the condition that $\mathbb{N}$ be a stable subset of **N,** i.e. $\mathbf{N} - (\mathbf{N} - \mathbb{N}) = \mathbb{N}$. Assuming that nonstandard integers are present, this latter may be understood as asserting that as many as possible of these are present.

---

[50]**IN** may accordingly be seen as accommodating both the invertible infinitesimals of Leibniz and the noninvertible nilsquare infinitesimals of Nieuwentijdt.

[51]Recall that a set is A is *inhabited* if it is nonempty in the strong sense of actually possessing an element, as opposed to the constructively weaker sense of the assertion that it is empty being refutable.

[52]On the other hand it follows from the nondegeneracy of Δ that it is also inconsistent to assert that $\boldsymbol{\Theta}$ reduces to $\{0\}$.

In the presence of invertible infinitesimals $\mathbf{R_d}$ is a nonstandard model of the reals lacking nilpotent elements. The passage via φ from **R** to $\mathbf{R_d}$ eliminates the nilpotent elements but preserves invertible infinitesimals. When φ is onto, $\mathbf{R_d}$ is a cohesive *nonstandard model of the reals.*

Within **R** we have the subring of *accessible reals*

$$\mathbf{R_{acc}} = \{x \in \mathbf{R} : \exists n \in \mathbb{N}(-n < x < n)\},$$

in which I is an ideal. Since each open interval in **R** is cohesive, $\mathbf{R_{acc}}$ satisfies the condition of being an inhabited set which includes, for each pair x, y of its members, a cohesive subset I for which $\{x, y\} \subseteq I$. It follows from this that $\mathbf{R_{acc}}$ *is cohesive.*

Within $\mathbf{R_d}$ the subring of *finite reals* may be identified:

$$\mathbf{R_{fin}} = \{x \in \mathbf{R_d} : \exists n \in \mathbb{N}(-n < x < n)\}.$$

Clearly φ carries $\mathbf{R_{acc}}$ into $\mathbf{R_{fin}}$. Since $\mathbf{R_{acc}}$ is cohesive, $\mathbf{R_{fin}}$ inherits a cohesiveness property analogous to that possessed by $\mathbf{R_d}$, namely, if X is a detachable subset of $\mathbf{R_{fin}}$, then $\varphi[\mathbf{R_{acc}}] \subseteq X$ or $X \cap \varphi[\mathbf{R_{acc}}] = \varnothing$.

We observe that $\mathbf{R_{fin}}$ can only be a detachable subset of $\mathbf{R_d}$ when $\mathbf{N} = \mathbb{N}$, or equivalently, when $\mathbf{R_{acc}}$ and **R** coincide, or to put it another way, when no invertible infinitesimals are present. For if $\mathbf{R_{fin}}$ is detachable in $\mathbf{R_d}$, then, as above, either $\varphi[\mathbf{R}] \subseteq \mathbf{R_{fin}}$, or $\mathbf{R_{fin}} \cap \varphi[\mathbf{R}] = \varnothing$. The latter being obviously false, it follows that $\varphi[\mathbf{R}] \subseteq \mathbf{R_{fin}}$. But then $\varphi[\mathbf{N}] \subseteq \mathbf{R_{fin}} \cap \varphi[\mathbf{N}] = \varphi[\mathbb{N}]$, whence $\mathbf{N} = \mathbb{N}$.

10.9 Contrasting Nonstandard Analysis with Smooth Infinitesimal Analysis

Smooth infinitesimal analysis shares with nonstandard analysis the feature that continuity is represented by the idea of "preservation of infinitesimal closeness". Nevertheless, there are a number of differences between the two approaches:

- In models of **SIA**, only smooth maps between objects are present. In models of **NSA,** all set-theoretically definable maps (including, in particular, discontinuous ones) appear.
- The logic of **SIA** is intuitionistic, making possible the nondegeneracy of the infinitesimal neigbourhoods **Δ**, **D** and **SIN**. The logic of **NSA** is classical,[53] causing all these neighbourhoods to collaose to $\{0\}$.
- In **SIA**, all curves are microstraight, and closed curves infinilateral polygons. Nothing resembling this is present in **NSA**.

[53]It should be pointed out, however, that constructive versions of **NSA** have been developed. See Palmgren (1998).

- The nilpotency of the infinitesimals of **SIA** reduces the differential calculus to simple algebra. In **NSA** the use of infinitesimals is a disguised form of the classical limit method.
- The hyperreal line in **NSA** is obtained by augmenting the classical real line with infinitesimals (and infinite numbers), while the smooth real line **R** comes already equipped with infinitesimals.
- In any model of **NSA**, the hyperreal line $\mathbb{R}^{\star}$ has exactly the same set-theoretically expressible properties as does the classical real line: in particular $\mathbb{R}^{\star}$ is an archimedean field in the sense of that model. This means that the infinitesimals (and infinite numbers) of **NSA** are not intrinsically so in the sense of the model in which they "live", but only relative to the "standard" model with which the construction began. That is, speaking figuratively, an inhabitant of a model of **NSA** would be unable to detect the presence of infinitesimals or infinite numbers in $\mathbb{R}^{\star}$. This contrasts with **SIA** in two respects. First, in models of **SIA** containing invertible infinitesimals, the real line is nonarchimedean with respect to the set of standard natural numbers, which is itself an object of the model. In other words, the presence of (invertible) infinitesimals and infinite numbers would be perfectly detectable by an inhabitant of the model. And secondly, the characteristic property of nilpotency possessed by the microquantities of a model of **SIA** is an intrinsic property, perfectly identifiable within the model. In **NSA** the hyperreals have precisely the same algebraic properties as do the classical real numbers, but the smooth reals in **SIA** do not.

The differences between **NSA** and **SIA** arise because the former is essentially a theory of infinitesimal numbers designed to provide a succinct formulation of the limit concept, while the latter is, by contrast, a theory of infinitesimal geometric objects, designed to provide an intrinsic formulation of the concept of differentiability.

10.10 Smooth Infinitesimal Analysis and Physics

In the past physicists showed no hesitation in employing infinitesimal methods,[54] the use of which in turn relied on the implicit assumption that the (physical) world is smooth, or at least that the maps encountered there are differentiable as many times

---

[54]In this connection we recall, for the last time, the words of Hermann Weyl:

> *The principle of gaining knowledge of the external world from the behaviour of its infinitesimal parts is the mainspring of the theory of knowledge in infinitesimal physics as in Riemann's geometry and, indeed, the mainspring of all the eminent work of Riemann* (1922, p. 92).

Hilbert declared set theory to be "Cantor's Paradise"; In the same spirit, **SIA** could be dubbed "Riemann's Paradise". The one, the Paradise of the Discrete; the other, the Paradise of the Continuous.

as needed. For this reason smooth infinitesimal analysis provides an ideal framework for the rigorous derivation of results in classical physics.[55] We present two here.

First, we derive the *equation of continuity for fluids*, whose original derivation by Euler was outlined in Chap. 3. The derivation in **SIA** will follow Euler's very closely, but the use of nilsquare infinitesimals and the Microcancellation Axiom will render the argument entirely rigorous.

Before we begin we require a few observations on *partial derivatives* in **SIA**. Given a function f: $\mathbf{R}^n \to \mathbf{R}$ of n variables $x_1, ..., x_n$, the partial derivative $\frac{\partial f}{\partial x_i}$ is defined as usual to be the derivative of the function $f\,(a_1, ..., x_i, ..., a_n)$ obtained by fixing the values of all the variables apart from x_i. In that case, for an arbitrary microquantity ε, we have

$$f(x_1,\ldots,x_i+\varepsilon,\ldots,x_n)+\varepsilon\frac{\partial f}{\partial x_i}(x_1,\ldots,x_n). \tag{10.5}$$

Using the fact that $\varepsilon^2 = 0$, it is then easily shown that

$$f(x_1+a_1\varepsilon,\ldots,x_n+a_n\varepsilon)=f(x_1,\ldots,x_n)+\varepsilon\sum_{i=1}^{n}a_i\frac{\partial f}{\partial x_i}(x_1,\ldots,x_n). \tag{10.6}$$

These equations are pivotal in deriving the equation of continuity. Here we are given a inviscid fluid of varying density flowing smoothly in space. At any point $O = (x, y, z)$ in the fluid and at any time t, the fluid's density ρ and the components u, v, w of the fluid's velocity are given as functions of x, y, z, t. Following Euler, we consider the elementary volume element **E**—a microparallelepiped—with origin O and edges OA, OB, BC of microlengths ε, η, ζ and so of mass εηζρ:

Fig. 10.8

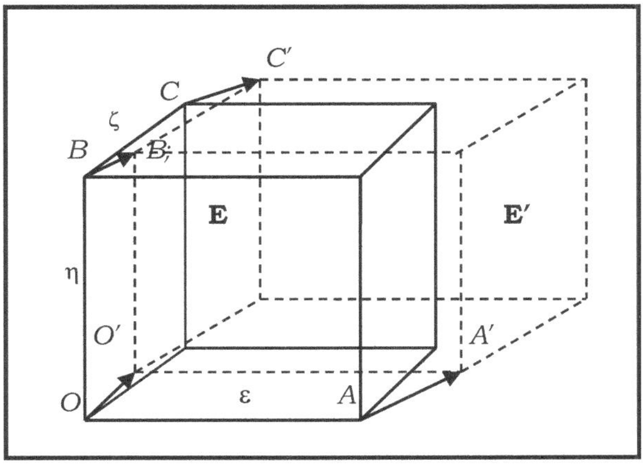

[55]A number of these are derived in Bell (1998).

Fluid flow during the microtime τ transforms the volume element **E** into the microparallelepiped **E**$'$ (Fig. 10.8) with vertices O', A', B', C'. We first calculate the length of the side $O'A'$. Now, using (10.5), the rate at which A is moving away from O in the x-direction is

$$u(x+\varepsilon, y, z, t) - u(x, y, z, t) = \varepsilon \frac{\partial u}{\partial x}.$$

The change in length of OA during the microtime τ is thus $\varepsilon\tau \frac{\partial u}{\partial x}$, so that the length of $O'A'$ is $\varepsilon + \varepsilon\tau \frac{\partial u}{\partial x} = \varepsilon\left(1 + \tau \frac{\partial u}{\partial x}\right)$. Similarly, the lengths of $O'B'$ and $O'C'$ are, respectively,

$$\eta\left(1 + \tau \frac{\partial v}{\partial y}\right), \zeta\left(1 + \tau \frac{\partial w}{\partial z}\right).$$

The volume of **E**$'$ is the product of these three quantities, which, using the fact that $\tau^2 = 0$, comes out as

$$\varepsilon\eta\zeta\left[1 + \tau\left(\frac{\partial u}{\partial x} + \frac{\partial v}{\partial y} + \frac{\partial w}{\partial z}\right)\right]. \tag{10.7}$$

Since the coordinates of O' are $(x + u\tau,\ y + v\tau,\ z + w\tau)$, the fluid density ρ' there at time$t + \tau$ is, using (10.6),

$$\rho + \tau\left(\frac{\partial \rho}{\partial t} + u\frac{\partial \rho}{\partial x} + v\frac{\partial \rho}{\partial y} + w\frac{\partial \rho}{\partial z}\right). \tag{10.8}$$

The mass of **E**$'$ is then the product of (10.7) and (10.8), which, again using the fact that that $\tau^2 = 0$, comes out as

$$\varepsilon\eta\zeta\rho + \varepsilon\eta\zeta\tau\left(\frac{\partial \rho}{\partial t} + \rho\frac{\partial u}{\partial x} + \rho\frac{\partial v}{\partial y} + \rho\frac{\partial w}{\partial z} + u\frac{\partial \rho}{\partial x} + v\frac{\partial \rho}{\partial y} + w\frac{\partial \rho}{\partial z}\right). \tag{10.9}$$

Now by the principle of conservation of mass, the masses of the fluid in **E** and **E**$'$ are the same, so equating the mass $\varepsilon\eta\zeta\rho$ of **E** to the mass of **E**$'$ given by (10.9) yields

$$\varepsilon\eta\zeta\tau\left(\frac{\partial \rho}{\partial t} + \rho\frac{\partial u}{\partial x} + \rho\frac{\partial v}{\partial y} + \rho\frac{\partial w}{\partial z} + u\frac{\partial \rho}{\partial x} + v\frac{\partial \rho}{\partial y} + w\frac{\partial \rho}{\partial z}\right) = 0.$$

Microcancellation gives

$$\frac{\partial \rho}{\partial t} + \rho\frac{\partial u}{\partial x} + \rho\frac{\partial v}{\partial y} + \rho\frac{\partial w}{\partial z} + u\frac{\partial \rho}{\partial x} + v\frac{\partial \rho}{\partial y} + w\frac{\partial \rho}{\partial z} = 0,$$

i.e.,

$$\frac{\partial \rho}{\partial t} + \frac{\partial}{\partial x}(\rho u) + \frac{\partial}{\partial y}(\rho v) + \frac{\partial}{\partial z}(\rho w) = 0,$$

Euler's equation of continuity.

Next, we derive the *Kepler-Newton areal law of motion under a central force*. We suppose that a particle executes plane motion under the influence of a force directed towards some fixed point O. If P is a point on the particle's trajectory with coordinates x, y, we write r for the length of the line PO and θ for the angle that it makes with the x-axis OX. Let A be the area of the sector ORP, where R is the point of intersection of the trajectory with OX. We regard x, y, r, θ as functions of a time variable t: thus

$$x = x(t), y = y(t), r = r(t), \theta = \theta(t), A = A(t).$$

Now let Q be a point on the trajectory at which the time variable has value $t + \varepsilon$, with ε in $\mathbf{\Delta}$ (Fig. 10.9). Then by Microstraightness the sector OPQ is a triangle of base $r(t + \varepsilon) = r + \varepsilon r'$ and height

$$r \sin[\theta(t+\varepsilon) - \theta(t)] = r \sin \varepsilon\theta' = r\varepsilon\theta'.$$[56]

The area of OPQ is accordingly

$$2 \text{ base} \times \text{height} = 2(r + \varepsilon r')r\varepsilon\theta' = 2\left(r^2\varepsilon\theta' + \varepsilon^2 r r'\theta'\right) = 2r^2\varepsilon\theta'.$$

Therefore

$$\varepsilon A'(t) = A(t+\varepsilon) - A(t) = \textit{area } OPQ = 2\varepsilon r^2\theta',$$

so that, cancelling ε,

Fig. 10.9

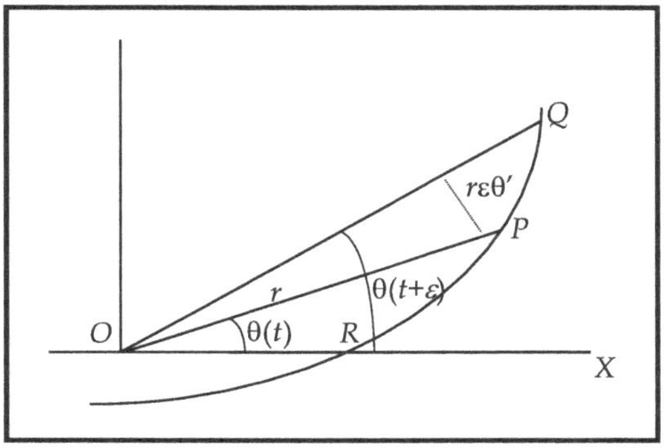

$$A'(t) = 2r^2\theta'. \quad (*)$$

Now let $H = H(t)$ be the acceleration towards O induced by the force. Resolving the acceleration along and normal to OX, we have

$$x'' = H\cos\theta \qquad y'' = H\sin\theta.$$

Also $x = r\cos\theta$, $y = r\sin\theta$. Hence

$$yx'' = Hy\cos\theta = Hr\sin\theta\cos q \qquad xy'' = Hx\sin\theta = Hr\sin\theta\cos\theta,$$

from which we infer that

$$(xy' - yx')' = xy'' - yx'' = 0.$$

Hence

$$xy' - yx' = k, \quad (**)$$

where k is a constant.

Finally, from $x = r\cos\theta$, $y = r\sin\theta$, it follows in the usual way that

$$xy' - yx' = r^2\theta',$$

and hence, by $(**)$ and $(*)$, that

$$A'(t) = 2k.$$

Assuming $A(0) = 0$, we conclude that

$$A(t) = 2kt.$$

Thus the radius vector joining the body to the point of origin sweeps out equal areas in equal times (Kepler's law).

Here is an appropriate place to remark on an intriguing use of infinitesimals in Einstein's celebrated 1905 paper *On the Electrodynamics of Moving Bodies,*[56] in which the special theory of relativity is first formulated. In deriving the Lorentz

[56]Reprinted in English translation in Einstein et al. (1952). It should be noted, however, that in subsequent presentations of special relativity Einstein avoided the use of infinitesimals

transformations from the principle of the constancy of the velocity of light Einstein obtains the following equation for the time coordinate $\tau(x', y, z, t)$ of a moving frame[57]:

$$\frac{1}{2}\left[\tau(0,0,0,t)+\tau\left(0,0,0,t+\frac{x'}{c-v}+\frac{x'}{c+v}\right)\right]=\tau\left(x',0,0,t+\frac{x'}{c-v}\right). \quad (10.10)$$

He continues:

Hence, if x' be chosen infinitesimally small,

$$\frac{1}{2}\left(\frac{1}{c-v}+\frac{1}{c+v}\right)\frac{\partial\tau}{\partial t}=\frac{\partial\tau}{\partial x'}+\frac{1}{c-v}\frac{\partial\tau}{\partial t}, \quad (10.11)$$

or

$$\frac{\partial\tau}{\partial x'}+\frac{v}{c^2-v^2}\frac{\partial\tau}{\partial t}=0.$$

Now the derivation of Equation (10.11) from equation (10.10) can be simply and rigorously carried out in **SIA** by *choosing* x' *to be a microquantity* ε. For then (10.10) becomes

$$\frac{1}{2}\left[\tau(0,0,0,t)+\tau\left(0,0,0,t+\varepsilon\left(\frac{1}{c-v}+\frac{1}{c+v}\right)\right)\right]=\tau\left(\varepsilon,0,0,t+\frac{\varepsilon}{c-v}\right).$$

From this we get, using equation (1) above,

$$\tau(0,0,0.t)+\frac{1}{2}\varepsilon\left(\frac{1}{c-v}+\frac{1}{c+v}\right)\frac{\partial\tau}{\partial t}=\tau(0,0,0,t)+\varepsilon\left(\frac{\partial\tau}{\partial x'}+\frac{1}{c-v}\frac{\partial\tau}{\partial t}\right).$$

So

$$\frac{1}{2}\varepsilon\left(\frac{1}{c-v}+\frac{1}{c+v}\right)\frac{\partial\tau}{\partial t}=\varepsilon\left(\frac{\partial\tau}{\partial x'}+\frac{1}{c-v}\frac{\partial\tau}{\partial t}\right),$$

and (ii) follows by microcancellation.

Spacetime metrics have some arresting properties in **SIA**. In a spacetime the metric can be written in the form.

$$ds^2=\Sigma g_{\mu\nu}dx_\mu dx_\nu \qquad \mu,\nu=1,2,3,4. \quad (*)$$

[57]Here x' is simply a symbol for the x-coordinate of the moving frame, not to be confused with the derivative of x

In the classical setting (∗) is in fact an abbreviation for an equation involving derivatives and the "differentials" ds and dx_μ are not really quantities at all. What form does this equation take in **SIA**? Notice that the "differentials" cannot be taken as microquantities since all the squared terms would vanish. But the equation does have a very natural form in terms of microquantities. Here is an informal way of obtaining it.

We think of the dx_μ as being multiples $k_\mu e$ of some small quantity e. Then (∗) becomes

$$ds^2 = e^2 \Sigma g_{\mu\nu} k_\mu k_\nu,$$

so that

$$\mathrm{d}s = e\sqrt{\Sigma g_{\mu\nu} k_\mu k_\nu}.$$

Now replace e by a microquantity ε. Then we obtain the metric relation: in **SIA**

$$\mathrm{d}s = \varepsilon\sqrt{\Sigma g_{\mu\nu} k_\mu k_\nu}.$$

This tells us that the "infinitesimal distance" ds between a point P with coordinates (x_1, x_2, x_3, x_4) and an infinitesimally near point Q with coordinate ($x_1 + k_1\varepsilon$, $x_2 + k_2\varepsilon$, $x_3 + k_3\varepsilon$, $x_4 + k_4\varepsilon$) is $\varepsilon\sqrt{\Sigma g_{\mu\nu} k_\mu k_\nu}$. Here a curious situation arises. For when the "infinitesimal interval" ds between P and Q is timelike (or lightlike), the quantity $\Sigma g_{\mu\nu} k_\mu k_\nu$is nonnegative, so that its square root is a real number. In this case ds may be written as εd, where d is a real number. On the other hand, if ds is spacelike, then $\Sigma g_{\mu\nu} k_\mu k_\nu$is negative, so that its square root is imaginary. In this case, then, ds assumes the form iεd, where d is a real number (and, of course i $= \sqrt{-1}$). On comparing these we see that, if we take ε as the "infinitesimal unit" for measuring infinitesimal *timelike* distances, then iε serves as the "imaginary infinitesimal unit" for measuring infinitesimal *spacelike* distances.

For purposes of illustration (Fig. 10.10), let us restrict the spacetime to two dimensions (x, t), and assume that the metric takes the simple form ds^2 = dt^2 – dx^2. The infinitesimal light cone at a point P divides the infinitesimal neighbourhood at P into a timelike region T and a spacelike region S bounded by the null lines l and l′ respectively (see Fig. 10.9). If we take P as origin of coordinates, a typical point Q in this neighbourhood will have coordinates (aε, bε) with a and b real numbers: if $|b| > |a|$, Q lies in T; if $a = b$, P lies on l or l′; if $|a| < |b|$, P lies in S. If we write $d = \sqrt{|\,a^2 - b^2\,|}$, then in the first case, the infinitesimal distance between P and Q is εd, in the second, it is 0, and in the third it is iεd.

Minkowski introduced "ict" to replace the "t" coordinate so as to make the metric of relativistic spacetime positive definite. This was purely a matter of formal

Fig. 10.10

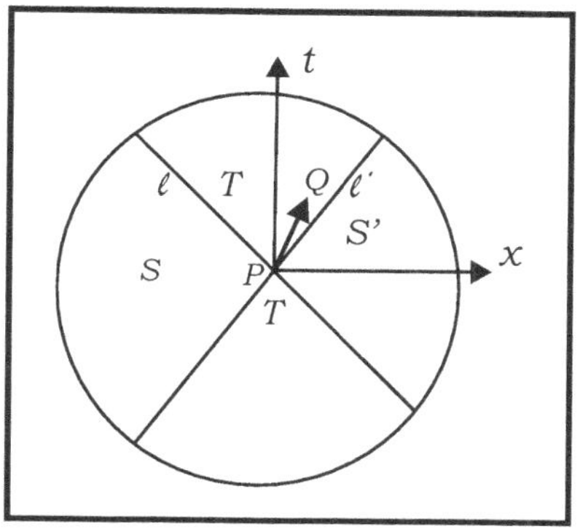

convenience and was later rejected by (general) relativists.[58] In conventional physics one never works with nilpotent quantities so it is always possible to replace formal imaginaries by their (negative) squares. But spacetime theory in **SIA** *forces* one to use imaginary units, since, infinitesimally, one can't "square oneself out of trouble". This being the case, it would seem that, infinitesimally, Misner, Thorne and Wheeler's[59] dictum *Farewell to ict* needs to be replaced by

Vale ict, ave iε!

To quote Misner, Thorne and Wheeler again,

> *Another danger in curved spacetime is the temptation to regard ... the tangent space as lying in spacetime itself. This practice can be useful for heuristic purposes but is incompatible with complete mathematical precision.*[60]

The consistency of **SIA** shows that, on the contrary, yielding to this temptation *is* compatible with complete mathematical precision: there tangent spaces may indeed be regarded as lying in spacetime itself.

We conclude this section with a speculation. Observe that the microobject **Δ** is "tiny" in the order-theoretic sense. For, using ε, η, as variables ranging over **Δ**, it is easily seen that that.

$$\forall\varepsilon\forall\eta\neg(\varepsilon < \eta \vee \eta < \varepsilon). \qquad (*)$$

whence

$$\forall\varepsilon\forall\eta(\varepsilon \leq \eta \wedge \eta \leq \varepsilon).$$

[58]See, for example Box 2.1 of Misner et al. (1972).

[59]See footnote immediately above.

[60]*Op. cit.*, p. 205.

In particular, the members of **Δ** are all simultaneously ≤ 0 and ≤ 0 but cannot (because of the nondegeneracy of **Δ**) be shown to coincide with zero.

In his book *Just Six Numbers*[61] the astrophysicist Martin Rees comments on the microstructure of space and time, and the possibility of developing a theory of quantum gravity. In particular he says:

> *Some theorists are more willing to speculate than others. But even the boldest acknowledge the "Planck scales" as an ultimate barrier. We cannot measure distances smaller than the Planck length [about 10^{19} times smaller than a proton]. We cannot distinguish two events (or even decide which came first) when the time interval between them is less than the Planck time (about 10^{-43} seconds).*

On this account, Planck scales seem very similar in certain respects to **Δ**. In particular, the sentence (∗) above seems to be an exact embodiment of the idea that we cannot decide of two "events" in **Δ** which came first; in fact it makes the stronger assertion that actually neither comes "first".

Could **Δ** serve as a suitable model for "Planck scales"? While **Δ** is unquestionably small enough to play the role, it inhabits a domain in which everything is smooth and continuous, while Planck scales live in the quantum world which, if not outright discrete, is far from being universally continuous. So if Planck scales could indeed be modelled by microneighbourhoods in **SIA**, then one might begin to suspect that the quantum microworld, the Planck regime—smaller, in Rees's words, "than atoms by just as much as atoms are smaller than stars"—is not, like the world of atoms, discrete, but instead continuous like the world of stars. This would be a major victory for the Continuous in its long struggle with the Discrete.

10.11 Relating Sets and Smooth Spaces

Considerable light is shed on the nineteenth century arithmetization, or set-theorization, of analysis by examining the relationship that exists between **Space** and the category **Set** of sets.[62]

While the law of excluded middle holds in **Set**, we have seen that it fails in **Space.** In particular the identity relation on the smooth real line **R** in **Space** is not decidable, that is, with variables *x, y* over **R,**

$$\neg\forall x\forall y[x = y \vee x \neq y].$$

This may be understood as saying that elements of **R** cannot always be fully distinguished; **R** is *amorphous* in some degree. While **R** *contains* "well-distinguished" points such as 0, 1, π, etc., it cannot, unlike a discrete set, actually be *made up* of these.

---

[61] Ress (2001).

[62] My account here is based on McLarty's illuminating paper (1988).

Now, as we have seen, this is precisely the view that most mathematicians and philosophers took of the geometric line, and of continua generally, before the nineteenth century set-theorization of analysis. This suggests that we view the objects of **Space** as representing continua as they were conceived before that process took place.[63] It is also reasonable to take the smooth maps in **Space** as representing the functions between continua actually recognized by pre-nineteenth century mathematicians, since these were, before the emergence of the notion of a function as an arbitrary correspondence, mostly smooth in any case. The category **Space** accordingly provides a working "model" of the pre-set-theoretic mathematics of continua.

As we know, the view that a mathematical continuum cannot be composed of points began to change in the nineteenth century, giving way to the arithmetization of analysis and the emergence of a set-theoretic, or discrete, account of the continuum. In effect, this meant the displacement of **Space** by **Set** as the locus of activity in mathematical analysis. Let us investigate the relation between the two.

First, certain objects in **Space** may be identified as "set-like". These are the *discrete* spaces S which consist of well-distinguished elements in that the law of excluded middle in the form

$$\forall x \forall y [x = y \vee x \neq y]$$

holds with variables x, y over S. (It is easily shown, for example that the space $\mathbf{N}$ of natural numbers is discrete.) Since every object in **Set** is discrete in this sense, discrete spaces are the counterparts, in **Space**, of sets.[64]

Next, recall that each topological space or manifold has an *underlying set* of elements or points; we want to extend this idea to a space as an object of **Space**. What should we mean by a *point* of a space? The natural response is to define a point of a space S to be a map (in **Space**) from the terminal object (one-point space) 1 to S. We think of points as being the smallest possible nonempty spaces, so it is natural to stipulate that **Space** satisffy the

Points Axiom. Each space is either empty or has points.

It follows from this axion that, for any points p, q of a space, either $p = q$ or $p \neq q$. In other words, as Aristotle asserted, points either coincide or are totally separate. Clearly 0 is then the only point of $\mathbf{\Delta}$.

To ensure that each space has an underlying set of points we introduce the

[63]In this spirit, we may take the microobjects in **Space** to represent the diverse theories of infinitesimals that were still in place before set theory swept them away.

[64]As mentioned in McLarty (1988), it can be shown that, in the presence of the axioms for **SIA** augmented by the two additional axioms introduced below, discrete spaces together with the maps between them form a category which satisfies the system of axioms characterizing the category **Set.** In this sense **Set** may be seen as the result of "imposing" the law of excluded middle on the objects of **Space**, or more precisely, of discarding those objects of **Space** which fail to satisfy that law. McLarty mentions another method of obtaining **Set** from **Space,** that of passing to double-negation sheaves.

Discrete Subspace Axiom. Each space S has a unique discrete subspace ΓS such that every point of S is in ΓS.

The space ΓS is called the *set of points* of the space S. It is the "arithmetized" or "atomized" version of S. ΓS may be thought of as the set obtained from S by removing the "glue" binding the points of S together. Notice that $\Gamma \mathbf{\Delta} = \{0\}$.

These new axioms (together with those of **SIA**) suffice to ensure that each space S has an underlying set of points and each map between spaces induces an underlying function between the corresponding sets of points. But "the underlying sets and functions are far too weakly axiomatized to supply foundations for geometry".[65] In the case of the smooth real line **R**, for example, the stated axioms ensure only that its underlying set of points $\Gamma\mathbf{R} = \mathbb{R}$ is a field, nothing more than a discrete algebraic structure. For $\mathbb{R}$ to provide an adequate surrogate for **R,** it is necessary to capture the concept of convergence, of capital importance to the arithmetical account of the continuum. Since the amorphousness of **R** undermines the uniqueness of the limit required by the theory of convergence, whatever axioms are introduced to ensure that the usual convergence criteria introduced by Cauchy are satisfied, they must of necessity be formulated for $\mathbb{R}$ rather than **R.** This fact helps to explain why "Cauchy's theory of convergence led towards set-theoretic as opposed to geometric foundations."[66]

We have seen that, following Weierstrass's lead, Cantor and Dedekind laboured to formulate an independent characterization of $\mathbb{R}$ as a discrete set of real numbers. Working as they did "only with discrete collections, using the law of excluded middle",[67] their efforts can be seen as taking place in **Set** rather than **Space**. Each provided a definition of $\mathbb{R}$ as an ordered field, both postulating in addition that "this represented the set of points on the geometric line with [its] arithmetic and order relation."[68] In effect, they "defined $\mathbb{R}$ within **Set** and added an axiom $\Gamma\mathbf{R} = \mathbb{R}$ plus others for arithmetic and order."[69]

We turn next to the space $\mathbf{R}^{\mathbf{R}}$ of maps from **R** to **R** in **Space.** We need to distinguish carefully between $\Gamma(\mathbf{R}^{\mathbf{R}})$, the set of functions corresponding to smooth maps from **R** to **R**, and $\Gamma\mathbf{R}^{\Gamma\mathbf{R}} = \mathbb{R}^{\mathbb{R}}$, the set of arbitrary functions from $\mathbb{R}$ to $\mathbb{R}$. Clearly, however, $\Gamma(\mathbf{R}^{\mathbf{R}}) \subseteq \mathbb{R}^{\mathbb{R}}$; in fact, it can be shown that every function in $\Gamma(\mathbf{R}^{\mathbf{R}})$ has derivatives of all orders.

The set $\Gamma(\mathbf{R}^{\mathbf{R}})$ includes all the real-valued functions known to eighteenth century mathematics, in particular, the polynomial, trigonometric, and exponential functions and all functions obtained by composing these. $\Gamma(\mathbf{R}^{\mathbf{R}})$ may be said to represent functions of "the well-behaved form with which mathematicians [of the day] were familiar".[70] But the work of Fourier on trigonometric series in the early nineteenth

[65]McLarty (1988), p. 83.

[66]*Op. cit.,* p. 84

[67]*Ibid.*

[68]*Op. cit.,* p. 85.

[69]*Ibid.*

[70]Boyer (1968), p. 600.

century stimulated mathematicians to begin to admit as "functions" not of that well-behaved form, that is, functions which are clearly not in $\Gamma(\mathbf{R}^{\mathbf{R}})$.[71] but which we would today recognize as being in $\mathbb{R}^{\mathbb{R}}$. This development provided a further motive for the development of an independent theory of $\mathbb{R}$, so also assisting in leading analysis away from **Space** to **Set.**

The upshot was that "the [geometric] line came to be *defined* as $\mathbb{R}$ with additional structure [and] smooth maps were *defined* to be functions with derivatives of all orders." Set theory came to dominate analysis, and, eventually, geometry as well. In the process, as we have seen, infinitesimals fell by the wayside,[72] not to be returned to active duty for another three-quarters of a century. Even more lamentably,

> *the disappearance of infinitesimals is only a symptom of a deeper loss. The independent reality of spaces and maps almost disappeared from mathematical consciousness as everything was reduced to sets.*[73]

It is fortunate that today, through category theory and intuiutionistic logic, the means for reviving the geometric vision are at hand.

[71]For example, Dirichlet's function r: $\mathbf{R} \to \mathbf{R}$ defined by $r(x) = 1$ for x rational and $r(x) = 0$ for x irrational.

[72]In the passage from **Space** to **Set,** nonzero infinitesimals sink without trace, since the application of Γ reduces microobjects such as $\mathbf{\Delta}$ to singletons such as $\{0\}$.

[73]McLarty (1988), p. 87.

Appendices

Appendix A: The Cohesiveness of Continua

It is, as we have observed, characteristic of a continuum that it is "gapless" or "all of one piece", in the sense of not being *actually separated* into two or more disjoint proper parts. On the other hand, as we have also noted, it has been taken for granted from antiquity that continua are *limitlessly divisible,* or *separable* into parts in the sense that any part of a continuum can be "divided", or "separated" into two or more disjoint proper parts. Now there is a traditional conceptual difficulty in seeing just how the parts of a continuum obtained by separation – assumed disjoint – "fit together" exactly so as to reconstitute the original continuum. This difficulty is simply illustrated by considering the case in which a straight line X is divided into two segments L, R by cutting it at a point p. What happens to p when the cut is made? On the face of it, there are four possibilities (not all mutually exclusive): (*i*) p is neither in L nor in R; (*ii*) p may be identified as the right-hand endpoint p_L of L: (*iii*) p may be identified as the left-hand endpoint p_R of R; (*iv*) p may be identified as *both* the right-hand endpoint of L *and* the left-hand endpoint of R. Considerations of symmetry suggest that there is nothing to choose between and (*iii*), so that if either of the two holds, then so does the other. Accordingly we are reduced to possibilities (*i*) and (*iv*). In case (*i*), L and R are disjoint, but since neither contains p, they together fail to cover X; while in case (*iv*), L and R together cover X, but since each contains p, they are not disjoint. This strongly suggests that a (linear) continuum *cannot* be separated, or decomposed, into two disjoint proper parts *which together cover it*[1] Herein lies the germ of the idea of cohesiveness.

[1]That is, in the words of Weyl [1925], "A continuum cannot be put together out of parts".

J. L. Bell, *The Continuous, the Discrete and the Infinitesimal in Philosophy and Mathematics*, The Western Ontario Series in Philosophy of Science 82,
https://doi.org/10.1007/978-3-030-18707-1

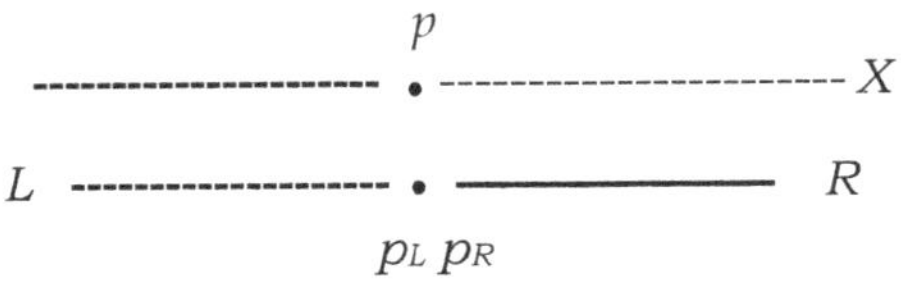

Of course, this analysis is quite at variance with the standard set-theoretic (Cantor-Dedekind) account of the linear continuum as a discrete linearly ordered set $\mathbb{R}$ of real numbers. "Cutting" $\mathbb{R}$ (or any interval thereof) at a point p amounts to partitioning it into the pairs of subsets $(\{x : x \leq p\}, \{x : p < x\})$ or $(\{x : x < p\}, \{x : p \leq x\})$: the first and second of these correspond, respectively, to cases (*ii*) and (*iii*) above. Now in the discrete case, one cannot appeal to symmetry as before: consider, for instance, the partitions of the set of natural numbers into the pairs of subsets $(\{n : n \leq 1\}, \{n : 1 < n\})$ and $(\{n : n < 1\}, \{n : 1 \leq n\})$. The first of these is $(\{0, 1\}, \{2, 3, \ldots\})$ and the second $(\{0\}, \{1, 2, \ldots\})$. Here it is manifest that the symmetry naturally arising in the continuous case does not apply: in the first partition 1 is evidently a member of its first component and in the second partition, of its second. In sum, when a discrete linearly ordered set X is "cut", no ambiguity arises as to which segment of the resulting partition the cut point is to be assigned, so that the segments of the partition can be considered disjoint while their union still constitutes the whole of X.[2]

Acknowledging the fact that the set-theoretic continuum, as a discrete entity, *can* be separated into disjoint parts, set theory proceeds to capture the characteristic "gaplessness" of a continuum by restricting the *nature* of the parts into which it can be so separated. In set-theoretic topology this is done by confining "parts" to *open* (or *closed*) subsets, leading to the standard topological concept of *connectedness*. Thus a space S is defined to be connected if it cannot be partitioned into two disjoint nonempty open (or closed) subsets[3] – or equivalently, given any partition of S into two open (or closed) subsets, one of the members of the partition must be empty. It is a standard topological theorem that the space $\mathbb{R}$ of real numbers and all of its intervals are connected in this sense.

But now let us return to our original analysis. This led to the idea that a continuum cannot be decomposed into disjoint parts. Let us take the bull by the horns and attempt to turn this idea into a definition. We shall call a space S *cohesive* or *indecomposable*,[4] or a (genuine) *continuum* if, for *any* parts, or subsets U and V of

---

[2]Even so, as Michael White remarks (White 1992, p. 20), "if we take a (supposedly continuous) physical object and cut it into two pieces, it would seem strange to say that one piece contains its limit (at the place of bisection) but that the other piece does not contain such a limit – that the end where it has been cut, although obviously limited, does not *contain* its terminus in the way that the other piece does." While certainly not inconsistent, this violation of intuition is part of the price that must be paid for treating the continuous as discrete.

[3]We note that the partitions obtained above by cutting $\mathbb{R}$ at p each consist of an open set and a closed set.

[4]The term *unsplittable* is also used. We shall use the term *decomposable* for "not cohesive". It should be noted that in topos theory the term "indecomposability" sometimes receives the

S, whenever $U \cup V = S$ and $U \cap V = \varnothing$, then one of U, V must be empty, or, equivalently, one of U, V must coincide with S. Put succinctly, a space is cohesive if it cannot be partitioned into two nonvoid parts. Another form of cohesiveness, slightly weaker than the version just stated, is: for any subsets U, V each of which contains at least one point, if $U \cup V = S$, then $U \cap V$ cannot be empty.

Cohesiveness may also be phrased in the following way. Recall that a subset U of S is said to be *detachable* if a "complementary" subset V of S exists satisfying $U \cup V = S$ and $U \cap V = \varnothing$. Then S is cohesive precisely when its only detachable subsets are $\varnothing$ and S itself.

Cohesiveness can be furnished with various "logical" formulations. Namely, S is cohesive in the first, stronger sense, if and only if, for any property P defined on S, the following implication holds:

$$(*)\forall x \in S[P(x) \vee \neg P(x)] \rightarrow [\forall x \in S\, P(x) \vee \forall x \in S \neg P(x)].$$

And S is cohesive in the second, weaker sense if and only if, for any properties P, Q defined on S:

$$\begin{aligned}&[\forall x \in S\,[P(x) \vee Q(x)] \wedge \exists x \in S\,P(x)\\ &\quad \wedge \exists x \in S\,Q(x)] \rightarrow [\neg\forall x \in S \neg[P(x) \wedge Q(x)]^5\end{aligned}$$

We observe that the law of excluded middle of classical logic confines cohesive spaces to the trivial empty space and one-point spaces. For nontrivial cohesive spaces to become admissible, therefore, it is necessary to *abandon the law of excluded middle.*[5] In fact the existence of nontrivial cohesive spaces is, as we have seen in Chaps. 9 and 10, compatible with *intuitionistic* logic. Indeed, much more can be said: it is compatible with intuitionistic logic that *every space which is connected in the usual topological sense is cohesive.*[6] We get a clear picture of how this can come about by reformulating our definitions in terms of *maps*, rather than parts. If we denote by **2** the two-element discrete space, then connectedness of a space S is equivalent to the condition that any *continuous* map $S \rightarrow$ **2** is constant, and cohesiveness of S to the condition that any map $S \rightarrow$ **2** *whatsoever* is constant.

---

considerably stronger meaning of *irreducibility* (see, e.g. Bell (1988) or Lambek and Scott (1986)). A space S is said to be *irreducible* if, for any subsets U and V of S for which $U \cup V = S$, we have $U = S$ or $V = S$, even when U and V *are not assumed to be disjoint.* (Notice that $\mathbb{R}$ can never be irreducible.) The "logical" condition for irreducibility is:

$$\forall x \in S\,[P(x) \vee Q(x)] \rightarrow [\forall x \in S P(x) \vee \forall x \in S Q(x)].$$

for arbitrary properties P and Q.

[5]It is easy to see that the law of excluded middle in the form $\forall x \forall y[x = y \vee x \neq y]$ must fail for any cohesive space S with at least two elements. For if that law held in S, then, for $a \in S$, the sets $\{a\}$, $S \setminus \{a\}$ would constitute a partition of S into two nonempty parts.

[6]This x is x the case in **SDG** (Chap. 10).

Supposing S to be connected and to possess more than one point, then from the law of excluded middle it follows that there exist nonconstant – and hence discontinuous – maps $S \to \mathbf{2}$. But the situation would be decidedly otherwise if *all* maps defined on S were continuous, for then, clearly, the connectedness of S would immediately yield its cohesiveness. So if S could be conceived as inhabiting a universe $\mathbf{U}$ in which all maps defined on S are continuous, then, within $\mathbf{U}$, S would be both nontrivial *and* cohesive. As we have seen, such universes $\mathbf{U}$ can in fact be constructed within category theory as *toposes*. Their underlying logic is intuitionistic, and within them the law of excluded middle fails in just the way necessary to allow for the presence of nontrivial cohesive spaces.

Tracing the Idea of Cohesiveness: Aristotle, Veronese, Brentano

We have observed in Chap. 5 that Brouwer and Weyl explicitly acknowledged cohesiveness as being a characteristic feature of the continuum. But related ideas in regard to continuous entities can be traced back to antiquity. As we have remarked, Anaxagoras, for example, asserted around 450 B.C. that

> *The things in the one world-order are not separated one from the other nor cut off with a hatchet, neither the hot from the cold nor the cold from the hot.*

Here the "one world-order" is the homogeneous continuum supposed by Anaxagoras to constitute the world.

In Chap. 1 it was pointed out it was Aristotle who first undertook the systematic analysis of continuity and discreteness. Let us summarize Aristotle's account of the of the continuum as presented there. Aristotle held that all change is continuous and that continuous variation of quality, of quantity and of position are inherent features of perception and intuition. He considered it evident that a continuum cannot consist of points. Any pair of unextended points, he observes, are such that they either touch or are totally separated: in the first case, they yield just a single unextended point, in the second, there is a definite gap between the points. Aristotle held that any continuum – a continuous path, say, or a temporal duration, or a motion – may be divided *ad infinitum* into other continua but not into what might be called "discreta" – parts that cannot themselves be further subdivided. Accordingly, paths may be divided into shorter paths, but not into unextended points; durations into briefer durations but not into unextended instants; motions into smaller motions but not into unextended "stations". Nevertheless, this does not prevent a continuous line from being divided at a point constituting the common border of the line segments it divides. But such points are, according to Aristotle, just *boundaries*, and not to be regarded as actual *parts* of the continuum from which they spring. If two continua have a common boundary, that common border unites them into a single continuum. Such boundaries exist only *potentially*, since they come into being when they are, so to speak, marked out as connecting parts of a continuum; and the parts in their turn are similarly dependent as parts upon the existence of the continuum.

Aristotle identified continuity (and discreteness) as attributes applying to the category of Quantity. As examples of continuous quantities he offers lines, planes, solids (i.e., solid bodies), extensions, movement, time and space; among discrete quantities he includes number and speech. Recall his definition of the continuous from *Physics, V, 3:*

> *I mean by one thing being continuous with another that those extremities of the two things in virtue of which they are in contact with each other become one and the same thing and (as the very name indicates) are "held together", which can only be if the two limits do not remain two but become one and the same.*

Aristotle indicates that a single continuous whole can be brought into existence by "gluing together" two things which have been brought into contact, which suggests that the continuity of a whole should derive from the way its *parts* "join up".

For Aristotle, constituent parts of continuous quantities "always join together at a common boundary". This suggests that we call a quantity *A continuous in the Aristotelian sense* or an *Aristotelian continuum,* if any pair of (nonvoid) "constituent parts" of A have a "common boundary", that is, whenever $U, V \subseteq A$ are such that $U \neq \emptyset$ and $V \neq \emptyset$ and $U \cup V = A$, then $U \cap V \neq \emptyset$. This corresponds (essentially) to the weaker version of cohesiveness formulated above.

Of particular relevance to our discussion is an observation Aristotle makes in connection with the joining and division of bodies:

> *But points, lines and planes, although they exist at one time and at another do not, cannot be in the process of being either generated or destroyed; for whenever bodies are joined or divided, at one time, when they are joined, one surface is instantaneously produced, and at another, when they are divided, two. Thus, when the bodies are combined the surface does not exist, but has perished; and when they are divided, surfaces exist which did not exist before. (The indivisible point is of course never divided into two.)*[7]

Aristotle's view is accordingly that the actual division of a body produces bounding surfaces, and so, analogously, in the words of the contemporary scholar Michael White, that the "actual bisection of an interval results in two distinct points, a limit or terminus of each sub-interval, where there was formerly one 'position'".[8] Moreover, when the subintervals are rejoined, the two distinct endpoints "become"

[7]Aristotle (1996a, b), III. Compare also the following passage in *Physics*, VIII, in which Aristotle comments, in connection with Zeno's paradoxes of motion, on the division of space and movement:

> *For whoever divides the continuous into two halves thereby confers a double function upon the point of division, for he makes it both a beginning and an end. And that is just what the counting man, or the dividing man whose half-sections he counts, is doing; and by the very act of division both the line and the movement cease to be continuous.*

[8]White (1992), 20. In fact, according to White, Aristotle "does not seem to recognize" the "open (or half open) intervals of magnitude", that is, intervals lacking at least one of their end points, generated by cutting the set-theoretic linear continuum.

one. This brings to mind case (iv) of the analysis with which we started and from which the cohesiveness (as we have defined it) of a linear continuum was inferred.

It would of course be grossly anachronistic to infer from all this that Aristotle conceived of continua as being cohesive in the exact technical sense in which it has been defined here. Nevertheless, the quotations do suggest that Aristotle saw something like cohesiveness as the feature distinguishing continuous from discrete quantity.[9]

In the nineteenth century Veronese, whose views on the continuum were discussed in Chap. 5 also comes close to the idea that the continuum is cohesive. This emerges on reconsidering his two "thought experiments" concerning the linear continuum (see Figs. 2). Let us recall these. In the first it is supposed that

> *...the part a of the rectilinear object is painted red, the remaining part a′ white, and suppose further that there is no other colour between the white and the red. That which separates the white from the red can be coloured neither white nor red, and therefore cannot be a part of the object, since by assumption all its parts are white or red. And this sign of separation of uniting can be considered as belonging either to the white or to the red, if one considers them independently of one another. If we now abstract from the colours, we can assume that the sign of separation between the parts a and a′ belongs to the object itself.*

Accordingly, a point can belong to a continuum thorough "assignment", but cannot be a part of it.

In the second thought experiment, Veronese

> *cut[s] a very fine thread at the place indicated by X with the blade of an extremely sharp knife,* [so that] *the two parts a and a' separate* [Fig. 1] *and we assume that one can put the thread back together without seeing where the cut was* [Fig. 2], *in other words, without a particle of the thread being lost. One produces this, apparently, if one looks at the thread from a certain distance. If one now considers the part a from right to left as the arrow above indicates, then what one sees of the cut is surely not part of the thread, just as what one sees from a body is not part of the body itself. It happens analogously if one looks at the part a' from left to right. If the sign of separation X of the parts a and a', which by assumption belongs to the thread itself, were part of the thread, then looking at a from right to left, one would not see all of this part, since that which separates the part a from a' is only that which one sees in the way indicated above when one supposes the thread put back together.*

---

[9]White's "principle of non-supervenience of continuity" (*ibid.*, p. 29), which he attributes to Aristotle, in fact amounts essentially to cohesiveness. This principle reads:"

(∗) Each partition of a continuous magnitude into proper parts yields parts each of which is pairwise continuous with at least one other part".

Here the notion of a part being "continuous with" another part is taken in the Aristotelian sense of "the limits of both parts at which they touch are the same", in other words, part of their boundaries coincide. In particular if a continuous magnitude is divided into two parts, (∗) asserts that these parts cannot be disjoint. That is, (∗) asserts that a continuous magnitude is cohesive in the weaker sense.

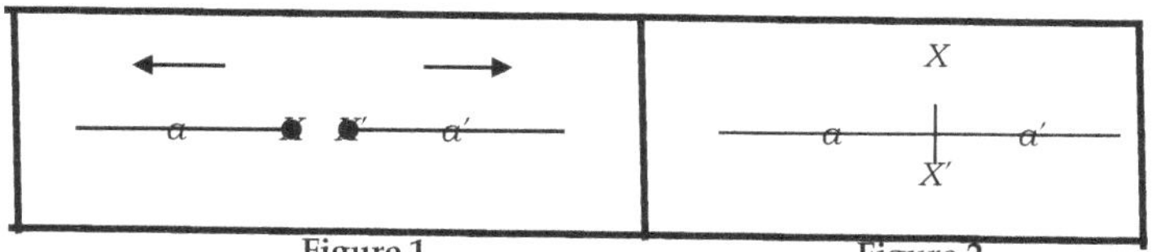

Figure 1 Figure 2

One produces this, apparently, if one looks at the thread from a certain distance. If one now considers the part a from right to left as the arrow above indicates

The premise of Veronese's first thought experiment is that a continuum can be coloured completely by two different colours, which is tantamount to claiming that it is decomposable. He then attempts to get round the difficulty of which segment of the resulting partition of the continuum the boundary ("that which separates") is to be assigned by arguing that it is not actually part of the continuum – it is, in fact, no more than a "sign of separation" which can be considered as belonging to either segment. But he goes on, using an appeal to "abstraction", to assert that the sign of separation, i.e. the boundary between the segments, can be regarded as being part of the continuum itself. Unfortunately, this move reintroduces precisely the difficulty we noted at the beginning, namely, to which segment is the boundary, now conceived as being an actual part of the continuum, to be assigned? As long as the "sign of separation" remains a potentiality, that is, as long as the continuum is not actually separated, this "boundary issue" does not arise. But conceiving of the continuum as being actually separated into two segments inevitably rekindles the boundary issue.[10] Although Veronese does not actually suggest it, this issue could be avoided by simply denying that a continuum is decomposable, that is, affirming that continua are cohesive.

The premise of Veronese's second thought experiment is the assumption that the thread, or linear continuum, can be separated into two parts which, upon being rejoined, reconstitute the continuum in its entirety, i.e., again that the continuum is decomposable. From this assumption Veronese infers that the point, or sign, of separation cannot be part of the continuum. The argument may be clearer in its contrapositive form, namely, if points of separation are parts of (linear) continua, then such continua are *cohesive* in the sense of being inseparable into two parts which, upon being rejoined, reconstitute the continuum in its entirety.

Suppose then that any point of separation of a linear continuum *L* is part of *L*, and suppose that *L* is separated into two parts *a, a'* with point of separation *X*. By assumption *X* is part of *L;* but on the other hand *X* cannot be part either of *a* or *a'*, since if it were part of one it would, by symmetry, have to be part of the other so that the parts *a* and *a'* would overlap contrary to assumption. So rejoining *a* and *a'* would not yield *L*, since the result would lack its part *X*. The cohesiveness of *L* follows.

Veronese's purpose in presenting these thought experiments is to establish that a point cannot be a part of a continuum, and to do this he makes what seems to him the

[10]In fact, Veronese's initial assumption that the given rectilinear continuum is coloured in the way he specifies *already entails* that it is actually separated into segments.

natural assumption that any continuum is decomposable. He would not, perhaps, have taken very seriously the contrapositive version of his argument to the effect that if points can be considered parts of continua, then the latter must be cohesive. But presumably he would have regarded the argument as valid – if unsound.

Finally, we return to Brentano, whose views on the continuum were also discussed in Chap. 5. As we have observed, Brentano, like Aristotle, considered it self-evident that a continuum cannot consist of points. Points are just boundaries, and not to be regarded as actual parts of the continuum from which they spring. If two continua have a common boundary, that common border unites them into a single continuum. Such boundaries exist only potentially, since they come into being when they are, so to speak, marked out as connecting parts of a continuum; and the parts in their turn are similarly dependent as parts upon the existence of the continuum.

As we pointed out, Brentano was unmoved by the efforts of mathematicians to construct the continuum from numbers, or points. In a discussion of Dedekind's construction of the real numbers we read:

> *Dedekind believes that either the number 1/2 forms the beginning of the series 1/2 to 1, so that the series 0 to 1/2 would thereby be spared a final member, i.e. an end point which would belong to it, or conversely. But this is not how things are in the case of a true continuum.*
>
> *Much rather it is the case that, when one divides a line, every part has a starting point, but in half plerosis.*[11]
>
> *That one has... postulated something completely absurd is seen immediately if one splits the supposedly continuous series of fractions between 0 and 1 into two parts at some arbitrary position. One of the two parts will then end with some fraction f, the second however could now start only if there were some fraction in the series which was the immediate neighbour of f, which is however not the case ... We should apparently have something that began but without having any beginning.*[12]

And again:

> *Geometry teaches that a line that is halved is halved in a single point. The line $\overline{a\ b\ c}$ in the point b. And further, that one is able to lay the one half over the other, for example in such a way that cb would come down on ba, the point c coinciding with the point b, the other end coinciding with the point a. According to the doctrine here considered [i.e., the Dedekind construction of the continuum], the divisions of the line would not occur in points, but in some absurd way behind a point and before all others, of which none, however, would stand closest to the cut. One of the two lines into which the line would be split upon division would therefore have an end point, but the other no beginning point. This inference has quite correctly been drawn by Bolzano, who was led thereby to his monstrous doctrine that there would exist bodies with and without surfaces.*[13]

From these quotations it becomes clear that Brentano rejected the idea of "splitting" a continuum into two parts, one of which lacks a boundary. His view is that the boundary is common to both parts, but with a difference in "plerosis" depending on which part the boundary is considered as bounding. Thus, for example, when one

---

[11] Brentano (1988), pp. 40–41.

[12] *Ibid.*, 4.

[13] *Ibid.*, p. 5.

divides a closed interval $[a, b]$ at an intermediate point c, one necessarily obtains the closed intervals $[a, c]$, $[c, b]$, with the common point c. Brentano would maintain that the "plerosis" of the point c is different in its two manifestations: as a right-hand endpoint of the first interval, it is in half-plerosis to the left; in the second, analogously, in half-plerosis to the right. But this does not affect the fact that the point c is common to both intervals. That being the case, Brentano would probably have regarded a continuous line as indecomposable, into disjoint intervals at least.

Concluding Remarks

There is an old philosophical argument, going back to the pre-Socratics, to the effect that, if continuous extended magnitudes are limitlessly divisible, then they must be composed of indivisible atoms. For suppose one starts with such a magnitude, a line, say, then proceeds to divide it in two, then divides each half in two, indefinitely. Imagine this process to be carried out completely. The result is a multitude of parts that cannot be further divided, that is, atoms. But now, the argument continues, since at no stage in the process of division is any part of the original line "lost", that is, at any stage the original line remains the sum of the parts obtained by division, it follows that this must still be the case when the process of division is completed and the parts into which the line has been divided have become atoms. Conclusion: any line is the sum of atoms. Hermann Weyl put the matter thus:

> *The old principle that "one cannot separate that which is not already separated (Gassendi) here again comes into its own. Indeed, Democritus argues with good reason that if I can break a stick, then it was from the outset not a whole. Strictest atomism is the inescapable conclusion of this.*[14]

Now let us grant for the purposes of argument that the process of division can be "completed". Then drawing the conclusion depends crucially on the assumption – call it "*A*" – that *at no stage in the process of division is any part "lost"*. (Without *A* one is left at the completion of the division with a multitude of atoms whose sum does not necessarily coincide with the given line.) Now to posit *A* is tantamount to asserting that *lines (or continua generally) are decomposable.* Consequently (granting the completability of the division process), *if* continua are decomposable, they are sums of atoms, hence discrete. This conclusion is, of course, the central claim of atomism. Those who (like myself) grant the limitless divisibility of continua, are happy to accept the imaginative possibility of "completing" the infinite process of dividing them and grant as well that atoms are the result – but who nevertheless find unpalatable the conclusion that continua are the sums of atoms, and so ultimately discrete – can avoid that conclusion by the simple expedient of denying that continua are decomposable, that is, by asserting that *continua are cohesive.* Cohesiveness forces us to recognize that, in separating a continuum into parts,

---

[14]Weyl (1925), p. 135

something *is* lost thereby – call it the "glue" uniting the parts into the original whole. Before the whole is separated into parts, those parts are implicit, and so the potential existence of the "glue" uniting them lies unrecognized. The loss of that "glue", indeed its very presence within the whole, only becomes apparent once the whole is separated. When the process of dividing a continuum is "completed", one is left with a multitude of separated atoms, but the "glue" uniting them within the whole has vanished without trace. Without that "glue" to hold them together, the atoms fail to sum to the original whole.

It has to be admitted that, even if one accepts the argument just presented, cohesiveness, as defined here, may still seem a bizarre notion. For *of course* a stick can be cut into two pieces, and *of course* a board can be painted half black and half white. But if such objects were truly cohesive, wouldn't it then become impossible to carry out such routine procedures? Perhaps only those few (necessarily philosophers) bent on avoiding atomism would be forced to countenance such a curious idea! But it is not really necessary to invoke the spectre of atomism, or even to be a philosopher, to acknowledge the concept of cohesiveness. For a moment's thought shows that cohesiveness is not such an unreasonable notion after all. In fact, the cohesiveness principle does not make the cutting of a stick in two or the painting of a board half black and half white impossible *per se*; it asserts nothing more than that it is impossible to do such things with *complete exactitude*. In the first case, the two half sticks cannot exactly reconstitute the original stick, and in the second the board cannot be painted with sufficient precision so as to cover it and at the same time avoid an overlapping, however small, of the painted areas. From an empirical standpoint, these facts are quite commonplace. The concept of cohesiveness may thus be seen as the result of elevating certain limitations *in practice* in the handling of continuous objects into a limitation *in principle*. These practical limitations not only make the idea of cohesiveness less offensive to intuition, they actually serve to distinguish continuous from discrete objects in their everyday handling. Thus, while dividing a stick leads to subtleties, none arise in separating a dozen eggs into two half-dozens!

Given universal discreteness, or classical logic, as in classical set theory, cohesiveness collapses into the trivial property of having no nonvoid proper parts. Consequently the counterpart of cohesiveness in set-theoretic topology – connectedness – is not and cannot be an intrinsic property of a space; it is rather a property of the topology imposed on the space: with a different topology, e.g. the discrete topology, a connected space can become highly disconnected. The property of cohesiveness, on the other hand, is intrinsic to a space, marking it as a genuine continuum in itself. A space is guaranteed to be cohesive if all maps on it are continuous. It is remarkable that the use of intuitionistic logic is compatible with the pervasive continuity of maps, hence also with the existence of cohesive spaces and genuine continua, so allowing the latter to assume their rightful place in mathematics and philosophy.

Appendix B: Cohesiveness of Spaces in Topos Models of Intuitionistic Set Theory

We have observed that in classical set theory the only cohesive spaces are trivial. It is a remarkable fact, however, that the existence of a whole range of nontrivial cohesive spaces, including the real line and all of its intervals, is consistent with *intuitionistic* set theory **IST.**[15] This is established by constructing topos models of **IST** in which the existence of such spaces can be demonstrated.

There are a number of toposes which contain nontrivial cohesive objects. In Chap. 10 we described smooth toposes in which the smooth real line and its closed intervals are cohesive. Here we describe four other kinds of topos in which cohesive objects are present.

We recall the classification of toposes into three types given in Chap. 7:

Type I: categories of "sets undergoing variation These include sheaves and presheaves.
Type II: categories of sets with a generalized equality relation.
Type III: categories arising as models of intuitionistic many- sorted theories.

1. **The *topos T of sheaves on the site of topological spaces equipped with the* open cover topology.**[16]

This topos is of type I. Here the "variation" takes place over a suitable small category of topological spaces containing the usual real line $\mathbb{R}$. It can shown that, in *T*, every map from the space $\mathbb{R}_{\mathbf{D}}$ of Dedekind real numbers to itself is continuous (with respect to the usual open-interval topology). Now Stout (1976) has shown that, in **IST**, $\mathbb{R}_{\mathbf{D}}$ is connected in the following sense:

(∗) For all subsets U, V of $\mathbb{R}_{\mathbf{D}}$ open in the usual open-interval topology,
$$[\mathbb{R}_{\mathbf{D}} = U \cup V \& \exists x.x \in U \& \ \exists x.x \in V] \Rightarrow U \cap V \neq \varnothing.$$

It follows that, in *T*, $\mathbb{R}_{\mathbf{D}}$ is cohesive in the weaker sense introduced earlier. For suppose U, V are (arbitrary) subsets of $\mathbb{R}_{\mathbf{D}}$ for which $\mathbb{R}_{\mathbf{D}} = U \cup V$, $U \cap V \neq \varnothing$.and $\exists x.\, x \in U \,\&\, \exists x.\, x \in V$. Define the map $f\colon \mathbb{R}_{\mathbf{D}} \to \mathbf{2}$ by $f(x) = 0$ if $x \in U$, $f(x) = 1$ if $x \in V$. Then, in T, f is continuous, so U and V are open. We infer from (∗) that $U \cap V \neq \varnothing$. The weak cohesiveness of $\mathbb{R}_{\mathbf{D}}$ follows.

2. ***The topos Shv*($\mathbb{R}$) *of sheaves over the real line*** $\mathbb{R}$[17]

This topos is also of type I. Here the "variation" takes place over the category of open subsets of $\mathbb{R}$. It can be shown that in ***Shv***($\mathbb{R}$) every map from a closed interval

[15] For the axions od **IST** see Chap. 7.

[16] This topos receives a detailed discussion in Mac Lane and Moerdijk (1992).

[17] See Scott (1970), Hyland (1979).

of $\mathbb{R}_\mathbf{D}$ to $\mathbb{R}_\mathbf{D}$ is *uniformly* continuous, and it follows easily from this that $\mathbb{R}_\mathbf{D}$ and all of its intervals are cohesive.

3. ***The free topos*** F

This topos is of type III. **F** is the category-theoretic embodiment of the *free* or *minimal* higher-order intuitionistic theory.[18] **F** is minimal in the sense that it is the common part of all such theories. It has been shown by Joyal that, in **F**, all maps $\mathbb{R}_\mathbf{D} \to \mathbb{R}_\mathbf{D}$ are continuous. It follows, as above, that $\mathbb{R}_\mathbf{D}$ is cohesive in the weaker sense.

In **F** power sets exhibit an exotic property we shall call *supercohesiveness.* A space S is *supercohesive* if whenever $\{X_n\colon n \in \mathbb{N}\}$ is a (not necessarily disjoint) covering of S indexed by the set $\mathbb{N}$ of natural numbers, then $S = X_n$ for some n. Obviously an extremely cohesive space is cohesive. Notice that the real line, while it can be cohesive, is never extremely cohesive because it can be expressed as the union of the family of closed intervals $[-n, n]$ for $n \in \mathbb{N}$.

Superohesiveness of S is equivalent to the *uniformity rule* for S, namely,

$$(*)\forall x \in S\exists n \in \mathbb{N}R(x,n) \Rightarrow \exists n \in \mathbb{N}\forall x \in S\, R(x,n).$$

supercohesiveness can also be expressed within infinitary logic: S is supercohesive if, for any countable collection $\{P_n\colon n \in \mathbb{N}\}$ of properties of S, the following implication is valid:

$$\forall x \in S\vee_{n\in\mathbb{N}}\, R_n(x) \Rightarrow \vee_{n\in\mathbb{N}}\forall x \in S\, P_n(x).$$

Clearly, if S is supercohesive, any map $S \to \mathbb{N}$ is constant. Supercohesiveness may be construed in a number of ways:

- As an extreme version of the *pigeonhole principle.* The pigeonhole principle (in a general form) states that, given two sets S and I, if the cardinality of M exceeds that of I, then any I-indexed covering of S has a non- singleton member. The more the cardinality of S exceeds that of I, the larger must one of these non-singleton members be: in particular, it might be S itself. Let us call S *incomparably bigger* than I if any I-indexed covering of S *always* contains S as a member. Clearly, if S is incomparably bigger than I, then any map $S \to I$ is constant. Any set with more than one element is incomparably bigger than any singleton. Extreme cohesiveness is the same as being incomparably bigger than $\mathbb{N}$.
- As a kind of *measurability* in the set-theoretic sense.[19] This follows if one observes that S is incomparably bigger than I exactly when the trivial filter $\{S\}$ on S is I-complete.[20] So S is incomparably bigger than $\mathbb{N}$, i.e. S is supercohesive, if and only if the trivial filter $\{S\}$ on S is countably complete. Compare this with

[18]See Lambek and Scott (1986).

[19]I owe this insight to Jean Petitot.

[20]A filter F on a set M is *I-complete* if whenever the union of any I-indexed family S of subsets of M is a member of F, then at least one member of S is also a member of F.

the definition of a measurable cardinal in the set- theoretic sense: a cardinal K is measurable if it supports a countably complete ultrafilter not generated by a singleton.

- As an extreme version of countable compactness with countable covers replacing open covers.
- As an extreme kind of *amorphousness*. Let us call a space *D discrete* if it satisfies $\forall x \in D\, \forall y \in D\, (x = y \vee x \neq y)$ In any topos the set $\mathbb{N}$ is discrete. It is natural to call a space which is not discrete *amorphous*: elements of an amorphous space may be thought of as being only partially distinguishable from one another. It is easy to see that a cohesive – and *a fortiori* an supercohesive – space is amorphous. Note also that in any non-Boolean topos (i.e., one in which the law of excluded middle fails) the power set of a set with at least one element is always amorphous.

Recall now the uniformity rule (∗) for S. Its premise may be construed as an attempt to coordinate each element of S via the relation R with a natural number, that is, to give a function $f: S \to \mathbb{N}$ for which $\forall x \in S\, \mathrm{R}\, (x, fx)$. If S is sufficiently amorphous, there is no way of distinguishing elements x and y of S so as to make the elements fx, fy of the discrete set $\mathbb{N}$ distinct. This means that f has to be constant, and we infer the consequent of (∗).

4. *The effective topos* EFF[21]

This topos is of type III. The "generalized equality relation" here on a set X is a $\mathcal{P}\mathbb{N}$-valued relation on X satisfying formal versions of symmetry and transitivity formulated in terms of the notion of *recursive realizability*. A map in **EFF** between two sets X, Y equipped with generalized equality relations in this sense is a $P\mathbb{N}$-valued relation R on $X \times Y$ satisfying the corresponding formal version of the condition "R is a single-valued relation with domain X and codomain Y". In **EFF** maps between objects constructed from the natural numbers correspond to recursive functions between them. In particular maps from $\mathbb{N}$ to $\mathbb{N}$ may be considered as being (total) recursive functions on $\mathbb{N}$. Hence, in **EFF**, *Church's thesis* holds in the strong sense that *every* function $\mathbb{N} \to \mathbb{N}$ is recursive. It follows from this that in **EFF** the domain of Cauchy real numbers $\mathbb{R}_\mathbf{C}$ corresponds to the *recursive reals,* that is, the real numbers arising as limits of recursive Cauchy sequences of rationals. (This means that "real analysis" in **EFF** coincides with recursive analysis.) Using the fact from classical recursion theory that recursive maps on the recursive reals are continuous, it follows that, in **EFF**, every map $\mathbb{R}_\mathbf{C} \to \mathbb{R}_\mathbf{C}$ is continuous. So, as above, we infer that $\mathbb{R}_\mathbf{C}$ is cohesive in the weaker sense.[22]

Finally, it has been shown that, in **EFF**, $P\mathbb{N}$ is extremely cohesive.

[21]See Hyland (1982)

[22]Hence $\mathbb{R}_\mathbf{D}$ is also cohesive since in **EFF.** $\mathbb{R}_\mathbf{D}$ can be shown to be isomorphic to $\mathbb{R}_\mathbf{C}$.

Appendix C: Continuity and the Logic of Perception

In his essay *On What is Continuous* of 1914, Brentano makes the following observation:

> *If we imagine a chess-board with alternate blue and red squares, then this is something in which the individual red and blue areas allow themselves to be distinguished from each other in juxtaposition, and something similar holds also if we imagine each of the squares divided into four smaller squares also alternating between these two colours. If, however, we were to continue with such divisions until we had exceeded the boundary of noticeability for the individual small squares which result, then it would no longer be possible to apprehend the individual red and blue areas in their respective positions. But would we then see nothing at all? Not in the least; rather we would see the whole chessboard as violet, i.e. apprehend it as something that participates simultaneously in red and blue.*[23]

Let us think of *attributes* or *qualities* such as "blackness", "hardness", etc. as being *manifested over* or *supported by* parts of a (perceptual) *space*. For instance if the space is my total sensory field, part of it manifests blackness and part manifests hardness and, e.g., a blackboard manifests both attributes. Each attribute α is correlated with a *proposition* (more precisely, a propositional function) of the form "— manifests the attribute α."

Let us assume given a supply of *atomic* or *primitive* attributes, i.e., attributes not decomposable into simpler ones: these will be denoted by *A, B, C*. For each primitive attribute *A* and each space **S** we may consider the total part of **S** which manifests *A*; this will be called the *A-part* of **S** and denoted by $[\![A]\!]_S$. Thus, for instance, if **S** is my visual field and *A* is the attribute "redness", then $[\![A]\!]_S$ is the total part of my sensory field where I see redness: the red part of my visual field.

Attributes may be *combined* by means of the logical operators $\wedge$ (and), $\vee$ (and/or), $\neg$ (not) to form *compound* or *molecular* attributes. The term "attribute" will accordingly be extended to include compound attributes. It follows that (symbols for) attributes may be regarded as the statements of a propositional language **L**—the *language of attributes.*

In order to be able to correlate parts of any given space **S** with compound attributes, i.e., to be able to define the *A*-part of **S** for arbitrary compound *A*, we need to assume the presence of operations $\sqcap$, $\bigsqcup$, $\sim$ corresponding respectively to $\wedge$, $\vee$, $\neg$, on the parts of **S**. For then we will be able to define the **S**-part $[\![A]\!]_S$ for arbitrary attributes α according to the following scheme:

$$\begin{aligned}
[\![\alpha \wedge \beta]\!]_S &= [\![\alpha]\!]_S \sqcap [\![\beta]\!]_S \\
[\![\alpha \vee \beta]\!]_S &= [\![\alpha]\!]_S \bigsqcup [\![\beta]\!]_S \qquad (*) \\
[\![\neg\alpha]\!]_S &= {}^{\sim}[\![\alpha]\!]_S
\end{aligned}$$

Once this is done, we can then define the basic relation $\sqsubseteq_S$ of *inclusion* between attributes over **S**:

[23] Brentano (1988), p. 8.

$$\alpha \sqsubseteq s\beta \Leftrightarrow [\![\alpha]\!]s \subseteq [\![\beta]\!]s,$$

where, as usual, "⊆" denotes the relation of set-theoretic inclusion.

Now the conventional meaning of "∧" dictates that, for any attributes α and β, we should have $\alpha \wedge \beta \sqsubseteq_S \alpha$ and $\alpha \wedge \beta \sqsubseteq_S \beta$ and, for any γ, if $\gamma \sqsubseteq_S \alpha$ and $\gamma \sqsubseteq_S \beta$ then $\gamma \sqsubseteq_S \alpha \wedge \beta$. In other words, $[\![\alpha \wedge \beta]\!]_S$ should be taken to be the largest part (w.r.t. ⊆) of **S** included in both $[\![\alpha]\!]_S$ and $[\![\beta]\!]_S$. By the first equation in (∗) above, the same must be true of $[\![\alpha]\!]_S \sqcap \beta]\!]_S$. Consequently, for any parts *U, V* of **S**, $U \sqcap V$ should be the largest part of **S** included in both *U* and *V*.

Similarly, now using the conventional meaning of "∨", we find that, for any parts *U, V* of **S**, $U \bigsqcup V$ should be the smallest part of **S** which includes both *U* and *V*.

We shall suppose that there is a *vacuous* attribute ⊥ for which $[\![\bot]\!]_S = \varnothing$, the empty part of **S.** In that case, for any attribute α, we have

$$[\![\alpha]\!]_S \sqcap {}^{\sim}[\![\alpha]\!]_S = [\![\alpha]\!]_S \sqcap [\![\neg\alpha]\!]_S = [\![\alpha \wedge \neg\alpha]\!]_S = [\![\bot]\!]_S = \varnothing.$$

Consequently, for any part *U* of **S** we should require that $U \sqcap {\sim}U = \varnothing$, i.e. that *U* and *~U* be *mutually exclusive.*

It follows from these considerations that we should take the parts of a perceptual space **S** to constitute a *lattice of subsets* of (the underlying set of) **S,** on which is defined an operation ~ ('complementation') corresponding to negation or exclusion satisfying the condition of mutual exclusiveness mentioned above. Formally, a *lattice of subsets* of a set *S* is a family $\mathbb{L}$ of subsets of *S* containing ∅ and *S* such that for any $U, V \in \mathbb{L}$ there are elements $U \sqcap V$, $U \bigsqcup V$ of $\mathbb{L}$ such that $U \sqcap V$ is the largest (w.r.t. ⊆) element of $\mathbb{L}$ included in both *U* and *V* and $U \bigsqcup V$ is the smallest (w.r.t. ⊆) element of $\mathbb{L}$ which includes both *U* and *V*. $U \sqcap V$, $U \bigsqcup V$ are called the *meet* and *join,* respectively, of *U* and *V*. A lattice $\mathbb{L}$ of subsets of *S* equipped with an operation $\sim: \mathbb{L} \to \mathbb{L}$ satisfying $U \sqcap {\sim}U = \varnothing$ for all $U \in \mathbb{L}$ is called a ~ – *lattice* of subsets of *S*.

We can now formally define a *perceptual space,* or simply a *space*, to be a pair $\mathbf{S} = (S, \mathbb{L})$ consisting of a set *S* and a ~ – lattice $\mathbb{L}$ of subsets of *S*. Elements of $\mathbb{L}$ are called *parts* of **S**, and $\mathbb{L}$ is called the *lattice of parts* of *S*.

The perceptual spaces that most closely resemble actual perceptual fields are called *proximity spaces.* These in turn are derived from *proximity structures.* A *proximity structure* is a set *S* equipped with a *proximity relation*, that is, a symmetric reflexive binary relation ≈. Here we think of *S* as a field of perception, its points as *locations* in it, and the relation ≈ as representing *indiscernibility of locations,* so that $x \approx y$ means that *x* and *y* are "too close" to one another to be perceptually distinguished. (Caution: ≈ is *not* generally transitive!) For each $x \in S$ we define the *sensum* at *x*, Q_x, by

$$Q_x = \{y \in S : x \approx y\}.$$

We may think of the sensum Q_x as representing the *minimum perceptibilium* at the location x. Unions of families of sensa are called *parts* of S. Parts of S correspond to *perceptibly identifiable subregions* of S. It can be shown that the family Part(S) of parts of S forms a ~ – lattice of subsets of S (actually, a complete ortholattice) in which the join operation is set-theoretic union, the meet of two parts of S is the union of all sensa included in their set-theoretical intersection, and, for $U \in$ Part(S),

$$^{\sim}U = \{y \in S : \exists x \notin U . x \approx y\}.$$

The pair $\mathbf{S} = (S, \text{Part}(S))$ is called a *proximity space.*

The most natural proximity structures (and proximity spaces) are derived from *metrics.* Any metric d on a set S and any nonnegative real number ε determines a proximity relation ≈ given by $x \approx y \Leftrightarrow \mathrm{d}(x, y) \leq \varepsilon$. When $\varepsilon = 0$ the associated proximity relation is the *identity relation* =: the corresponding proximity space is then called *discrete.* It can be shown that, if a proximity space **S** has a *transitive* proximity relation, then it is *almost* discrete in the sense that its lattice of parts is isomorphic to the lattice of parts of a discrete space.

Given a perceptual space $\mathbf{S} = (S, \mathbb{L})$ we define an *interpretation* of the language **L** of attributes to be an assignment, to each primitive attribute A, of a part $[\![A]\!]_{\mathbf{S}}$ of **S.** Then we can extend the assignment of parts of **S** to all attributes as in (∗) above. Given an attribute α and a part U of **S,** we think of the relation $U \subseteq [\![\alpha]\!]_{\mathbf{S}}$ as meaning that U is *covered* by the attribute α. Now there is another relation between parts and attributes the *manifestation relation* $\Vdash_{\mathbf{S}}$—which reflects more closely the way compound attributes are built up from primitive ones. $U \Vdash_{\mathbf{S}} \alpha$, which is read "$U$ *manifests* α" or "A is *manifested over U*" is defined as follows:

$$U \Vdash_{s} A \Leftrightarrow U \subseteq [\![A]\!]_{s} \text{ for primitive A,}$$

$$U \Vdash_{s} \alpha \wedge \beta \Leftrightarrow U \Vdash_{s} \alpha \text{ and } U \Vdash_{s} \beta;$$

$$U \Vdash_{s} \alpha \vee \beta \Leftrightarrow V \Vdash_{s} \alpha \;\&\; W \Vdash_{s} \beta$$
$$\text{for some parts } V, W \text{ of } \mathbf{S} \text{ such that } U = V \cup W,$$

$$U \Vdash_{s} \neg\alpha \Leftrightarrow (\text{for all parts } V \text{ of } \mathbf{S})\; U \Vdash_{s} \alpha \Rightarrow V \subseteq \sim U.$$

Thus U manifests a disjunction $\alpha \vee \beta$ provided there is a "covering" of U by two "subparts" manifesting α and β respectively, and U manifests a negation ¬α provided any part of **S** manifesting α is included in the "complement" of ~U of U.

In general, the manifestation and covering relations fail to coincide in proximity spaces. The reason for this is that, while the latter has a certain *persistence* property, the former, in general, fails to possess this property. By persistence of the covering relation is meant the evident fact that if a part U of a space is covered by an attribute, then this attribute continues to cover any subpart of U. However, as we shall see, this is not the case for the manifestation relation: there are attributes manifested over a part of a space which fail to be manifested over a subpart.

Let us call an attribute α **S**-*persistent* (or persistent over **S**) if for all parts U, V of S we have

$$V \subseteq U \;\&\; U \Vdash_{S} \alpha \Rightarrow V \Vdash_{S} \alpha.$$

(Note that a *primitive* attribute is always persistent. More generally, it is not hard to show that the same is true for any compound attribute not containing occurrences of the disjunction symbol $\vee$.) Let us call a space **S** *persistent* if every attribute is S-persistent (for any interpretation of **L** in **S**). We now give an example of a *nonpersistent* proximity space, a one-dimensional version of Brentano's chessboard:

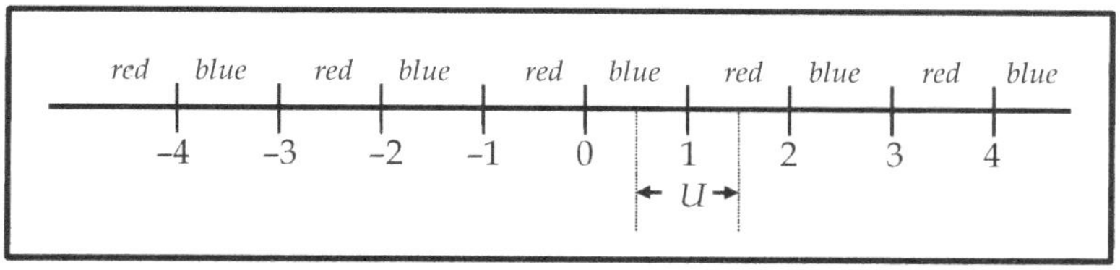

Consider the real line with the proximity relation $\approx$ defined by $x \approx y \Leftrightarrow |x - y| \leq 1/2$, and let **P** be the associated proximity space. The sensum at a point x is then the closed interval of length 1 centred on x. Suppose now we are given two primitive attributes B ('blue') and R ('red'). Let the B-part of **R** be the union of all closed intervals of the form $[2n, 2n + 1]$ and let the R-part of **P** be the union of all closed intervals of the form $[2n - 1, 2n]$. To put it vividly, we "colour" successive unit segments alternately blue and red. Clearly, then, **P** manifests the disjunction $R \vee B$. But if U is the sensum $Q_1 = [1/2, 3/2]$, then $R \vee B$ is *not* manifested over U, since U is evidently not covered by two subparts over which R and B are manifested, respectively—indeed U has no proper subparts.

Thus arises the curious phenomenon that, although we can tell, by surveying a (sufficiently large part of) the whole space **P**, that the part U is covered by redness and blueness, nevertheless U—unlike **P**—does not split into a red part and a blue part. In some sense redness and blueness are conjoined or *superposed* in U: it seems natural then to say that U manifests a *superposition* of these attributes rather than a disjunction. If we take the unit of length on the real line sufficiently small (or equivalently, redefine $x \approx y$ to mean $|x - y| \leq \varepsilon$ for sufficiently small ε) so that each interval of unit length represents the minimum length discernible to human visual perception, we have (essentially) Brentano's chessboard in one dimension. In that case, the "superposition" of the two attributes blue and red turns out to be *violet*, which is what we actually see.

Actually, the covering of our proximity space by parts like U looks like this:

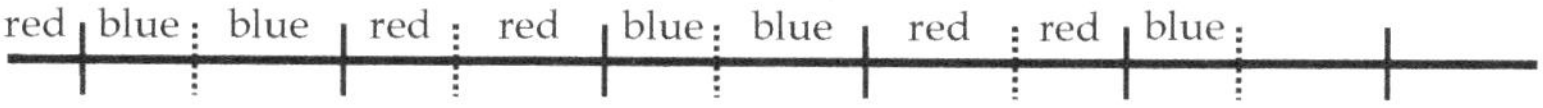

while Brentano's chessboard looks like this:

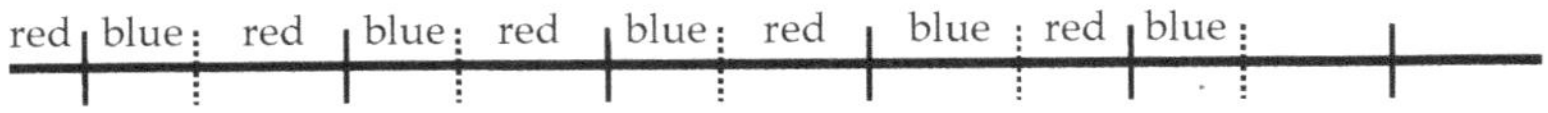

But the two arrangements are evidently isomorphic.

The concept of superposition of attributes admits a very simple rigorous formulation. In the example we have just considered, the part U manifests a superposition of the attributes R and B just when there is a part V of the space which includes U and manifests $R \vee B$ (in this case, V may be taken to be the whole real line). This prompts the following definition. Given a proximity space **S,** an interpretation of **L** in **S** and attributes α, β, we say that a part U of **S** *manifests a superposition of* α *and* β if there is a part V of **S** such that $U \subseteq V$ and $V \Vdash_{\mathbf{S}} \alpha \vee \beta$. Now for any attribute γ, it is readily shown that

$$U \Vdash_{\mathrm{S}} \neg\neg\gamma. \Leftrightarrow V \Vdash_{\mathrm{S}} \gamma \text{ for some part } V \text{ such that } U \subseteq V.$$

So the condition that U manifest a superposition of α and β is just

$$U \Vdash_{\mathrm{S}} \neg\neg(\alpha \vee \beta).$$

It follows that *a superposition is a double negation of a disjunction.* In the human visual field, then, the attribute "violet" is the double negation of the attribute "blue or red". Similarly, the attribute "grey" is the double negation of the attribute "black or white", etc..[24]

These ideas may be linked up with *continuity*. Let us call a proximity structure $(S, \approx)$ *continuous* if for any $x, y \in S$ there exist $z_1, \ldots, z_n$ such that $x \approx z_1$, $z_1 \approx z_2, \ldots, z_{n-1} \approx z_n$, $z_n \approx y$. Continuity in this sense means that any two points can be joined by a finite sequence of points, each of which is indistinguishable from its immediate predecessor.[25] If d is a metric on S such that the metric space (S, d) is connected, then every proximity structure determined by d is continuous. When S is a perceptual field such as that of vision, the fact that it does not fall into separate parts means that it is connected as a metric space with the inherent metric. Accordingly every proximity structure on S determined by that metric is continuous. Note that this continuity emerges even when S is itself an assemblage of discrete "points". This would seem to be the way in which continuity of perception is engendered by an essentially discrete system of receptors.

---

[24]It can also be shown that superposition of states in quantum physics is a special case of the notion of superposition introduced here, See Bell (1986).

[25]This is essentially Poincaré's definition of a perceptual continuum: see Chap. 5 above. In the case of the nonpersistent proximity space we have presented, continuity means that a red segment and a blue segment can always be joined by a violet line provided that the coloured segments are taken to be sufficiently small.

Appendix D: Kant's First Antinomy

Kant's First Antinomy in the *Critique of Pure Reason* concerns the opposition between the boundedness and unboundedness of space and time. He draws conclusions essentially similar to those he arrived at in the Second Antinomy – in this case, the Thesis is that the world has a beginning in time, and is also limited as regards space; and the Antithesis that the world has no beginning, and no limits in space; it is infinite as regards both time and space.

Jonathan Bennett, in his thought-provoking paper, *The Age and Size of the World*,[26] sums up the conclusion of the First Antinomy as follows:

> *Although Kant denies that the world can be infinitely old or large, he thinks that it cannot be finitely large or old either.*

In explicating this assertion, Bennett concludes that what Kant means by "the world is not finite in size" is "no finite amount of world includes all the world there is", or "every finite quantity of world excludes some world". Bennett submits that this last statement "seems to Kant to be a weaker statement than the statement that there is an infinite amount of world." More generally, Bennett suggests that

> *Kant is one of those who think that*
> *Every finite set of F's excludes at least one F*, (1)
> *though it contradicts the statement that there are only finitely many F's, is nevertheless weaker than*
> *There is an infinite number of F's* (2).

Bennett implies that Kant is simply mistaken here, that in fact (1) and (2) are equivalent. But is this right? Let us bring to bear some contemporary mathematical ideas on the matter.

Call a set A *finite* if for some natural number n, all the members of A can be enumerated as a list $a_0, \ldots, a_n$; *potentially infinite* if it is not finite, that is, if, for any natural number n, and any list of n members of A, there is always a member of A outside the list; *actually infinite* if there is a list $a_0, \ldots, a_n, \ldots$(one for each natural number n) of *distinct* members of A; and *Kantian* if it is potentially, but not actually infinite, that is, if it is neither finite nor actually infinite.[27] Now *it is possible for a set to be Kantian*, just as Kant (according to Bennett) thought the actual world was.

For suppose that we are given a potentially infinite set A, and we attempt to show that it is actually infinite by arguing as follows. We start by picking a member a_0 of A; since A is potentially infinite, there must be a member of A different from a_0; pick such a member and call it a_1. Now again by the fact that A is potentially infinite, there is a member of A different from a_0, a_1—pick such and call it a_2. In this way we generate a list a_0, a_1, a_2, ... of distinct members of A; so, we are tempted to

---

[26]Bennett (1971).

[27]In the usual set-theoretic terminology, my term "potentially infinite" corresponds to "infinite"; "actually infinite" to "transfinite" or "Dedekind infinite"; and "Kantian" to "infinite Dedekind finite".

conclude, A is actually infinite. But clearly the cogency of this argument hinges on our presumed ability to "pick", for each n, an element of A distinct from $a_0, \ldots, a_n$—an ability enshrined in the set-theoretic principle known as the *Axiom of Choice*.[28] Now the Axiom of Choice is, as Gödel showed in 1938,[29] a perfectly consistent mathematical assumption. But, as Paul Cohen showed in 1964,[30] its denial is equally consistent. In fact, it can be denied in such a way as to prevent the argument just presented from going through, that is, to allow the presence of potentially infinite sets which are not at the same time actually infinite. That is, the existence of Kantian sets is consistent with the axioms of set theory (and classical logic) *as long as the Axiom of Choice is not assumed.*

The Axiom of Choice may be regarded as a principle ensuring that the universe of sets is "static" in the sense that families of sets "sit still" long enough to enable elements to be extracted from them. Accordingly the existence of "Kantian" sets is compatible with classical set theory, as long as it has not been rendered "overstatic" through the imposition of the Axiom of Choice.

Another, more direct way of obtaining a Kantian set is to allow our "sets" to undergo *explicit variation*, to wit, variation over discrete time (that is, over the natural numbers). For consider the following universe of discourse[31] $\mathcal{U}$. Its objects are all sequences of maps between sets:

$$A_0 \xrightarrow{f_0} A_1 \xrightarrow{f_1} A_2 \xrightarrow{f_2} \cdots A_n \xrightarrow{f_n} A_{n+1} \rightarrow \ldots$$

Such an object may be thought of as a set $\boldsymbol{A}$ "varying over (discrete) time": A_n is its "value" at time n. Now consider the temporally varying set

$$K = (\{0\} \rightarrow \{0,1\} \rightarrow \{0,1,2\} \cdots \rightarrow \{0,1,\ldots,n\} \rightarrow \cdots)$$

in which all the arrows are identity maps. In $\mathcal{U}$, $\boldsymbol{K}$ "grows" indefinitely and hence potentially infinite. On the other hand, at each specific time it is finite and so is not actually infinite. In short, in the universe of "sets through time", $\boldsymbol{K}$ is a Kantian set.

It would seem then that Kant's conclusion in the First Antinomy that space and time are neither finite nor infinite—that is, potentially infinite—*is* coherent (or at least consistent) after all. This applies equally to the Second Antinomy, at least in respect of Kant's contention that ongoing sequences of divisions of appearances cannot be assumed to terminate and are accordingly neither finite nor infinite. But

[28] A straightforward version of the axiom of choice is the following: for any collection $\mathcal{U}$ of nonempty sets no two of which have common elements there is a set having exactly one element in common with each member of $\mathcal{U}$. See Chap. 10.

[29] Gödel (1940).

[30] Cohen (1966).

[31] *Cognoscenti* will recognize $\mathcal{A}$ as the topos of sets varying over the natural numbers. But within this universe not only the Axiom of Choice but also the law of excluded middle has to be abandoned, a course that Kant would likely have found most unpalatable.

these assertions only make sense in a conceptual framework conceived as undergoing some form of variation.[32]

Appendix E: Logic and Variation

The intuitionistic logic associated with Brouwer's conception of the continuum can be seen as arising not only from the opposition between the Continuous and the Discrete which has been the focus of this book, but also from the equally important, and closely related opposition between the *Constant* and the *Variable.*[33]

Tradition took for granted that a single overarching system of reasoning, governed by classical logic, was applicable *pari passu* to all conceivable oppositions. But does a single logic really suffice?

The world as we perceive it is in a perpetual state of flux. But the objects of mathematics are usually held to be eternal and unchanging. How then is the phenomenon of variation to be given mathematical expression? Consider, for example, a fundamental and familiar form of variation: *change of position,* or *motion,* a form of variation so basic that the mechanical materialist philosophers of the eighteenth and nineteenth centuries held that it subsumes all forms of physical variation. Now motion is itself reducible to a still more fundamental form of variation—*temporal* variation.[34] But this reduction can only be effected once the idea of *functional dependence* of spatial locations on temporal instants has been grasped. Lacking an adequate formulation of this idea, the mathematicians of Greek antiquity were unable to produce a satisfactory analysis of motion, or more general forms of variation, although they grappled mightily with the problem. The problem of analyzing motion was compounded by Zeno's paradoxes, which, as we know, were designed to show that motion was impossible, and that in fact the world is a Parmenidean unchanging unity.

It was not until the seventeenth century that motion came to be conceived as a functional relation between space and time, as the manifestation of a dependence of variable spatial position on variable time. This enabled the many forms of spatial variation to be reduced to the one simple fundamental notion of temporal change, and the concept of motion to be identified as the *spatial representation* of temporal change. The "static" version of this idea is that space curves are the "spatial representations" of straight lines.

---

[32]See Chap. 9.

[33]Equally important as driving forces in the history of thought are the oppositions between the One and the Many, the Finite and the Infinite, and the Whole and the Part. For a detailed discussion of oppositions, see Bell (2016)

[34]It may be noted here that according to Whitehead even this is not the ultimate reduction: cf. his notion of "passage of nature". See Whitehead (1920).

Now this account of motion (and its central idea, functional dependence) in no way compels one to conceive of either space or time as being further analyzable into static indivisible atoms, or *points*. All that is required is the presence of two domains of variation—in this case, space and time—correlated by a functional relation. True, in order to be able to establish the correlation one needs to be able to localize within the domains of variation, (e.g. a body is in place x_i at "time" t_i, $i = 0, 1, 2, \ldots$) and it could be held that these domains of variation are just the "ensemble" of all conceivable such "localizations". But even this does not necessitate that the localizations themselves be atomic points—cf. Whitehead's method of "extensive abstraction"[35] and, latterly, the rise of "pointless" topology.[36]

The incorporation of variation into seventeenth century mathematics led to the triumphs of the calculus and mathematical physics, and to the mathematization of nature, with all of which we are familiar. But, as we have seen, difficulties arose in the attempt to define the instantaneous rate of change of a varying quantity—the fundamental concept of the differential calculus. Like the ancient Pythagorean program of reduction of the continuous to the discrete, the attempt by seventeenth century mathematicians to reduce the varying to the static— through the use of infinitesimals—led to outright *contradictions*.

Marx and Engels, and their Marxist successors, thought that the analysis of variation would require the creation of a dialectical logic or a "logic of contradiction". But traditional logic survived in mathematics, largely as a result of the replacement of variation by stasis at the hands of the great nineteenth century arithmetizers Weierstrass, Dedekind and Cantor. As we have seen, Cantor replaced the concept of a varying quantity by that of a completed, static *domain of variation* which may be regarded as an ensemble of atomic individuals—thus, like the Pythagoreans, replacing the continuous by the discrete. He also banished infinitesimals and the idea of geometric objects as being generated by points or lines in motion.

But as we know, certain mathematicians and philosophers raised objections to the idea of "discretizing" or "arithmetizing" the linear continuum. Brentano, for example, rejected the idea that a true continuum can be completely analyzed into a collection of discrete points, no matter how many of them there might be.

It was only with Brouwer, for whom the phenomenon of temporal variation was fundamental, that logic became an issue within mathematics. Rejecting the Cantorian account of the continuum as purely discrete, Brouwer identifies points on the line as entities "in the process of becoming" in a temporal, even subjective sense, that is, as embodying variation generating a potential infinity. He rejects the law of excluded middle for such objects, a move which led, as we have seen, to a new form of logic, *intuitionistic logic*. It is a remarkable fact that this logic is compatible with a very general concept of variation, which embraces all forms of (objective) continuous variation, and which in particular allows the use of

[35]Whitehead (1920).

[36]See Chap. 7.

(continuous) infinitesimals.[37] While its roots lie in the subjective, intuitionistic logic is thus revealed to have an *objective* character.

The application of intuitionistic logic to resolve the contradiction engendered by variation shows that it was not in the end necessary—as claimed by dialectical philosophy—to reject the *law of noncontradiction*$\neg(p \wedge \neg p)$, but rather its dual the *law of excluded middle* $p \vee \neg p$.

It is a characteristic of the intuitionistic conception of mathematical objects undergoing variation that, once a property of a such an object has been established by means of a construction, the property remains established "for all time"; it is, in a word, *unalterable*. This is reflected in the *persistence property* of the semantics of intuitionistic logic: a statement, once "forced" be true, remains true. This suggests that intuitionistic logic can, roughly, be regarded as the logic of the *past tense*: a statement of the form "such and such was the case" once true, remains true forever.[38] This is a particular case of an association among types of variation, philosophical leaning and logic, as presented in the following concordance:

Variation	Philosophy	Logic
Static: no variation: eternal present: objective state of affairs independent of our knowledge	*Platonic realism*	*Classical Logic*
Continuously cumulative: no revision of information at later stages: once known, always known	*Broad constructivism*	*Intuitionistic Logic*
	Kantian idealism	
Noncumulative: possible revision, falsification, or loss of information at later stages.	*Indeterminism*	*Quantum Logic*
	Humean scepticism	

Variation can also be correlated with certain concepts and branches of mathematics:

Variation	Mathematical correlate
Temporal	*Natural numbers (discrete)*
	Real line (continuous)
Positional (motion)	*Real line*
	Differential Calculus
	Mathematical Analysis
Morphological	*Topology*
	Category Theory

The idea of variation plays as fundamental a role in thought as does the concept of continuity to which it is so closely tied—and as does the concept of the discrete to which continuity is in such fundamental opposition.

[37]See Chap. 10.

[38]Provided, of course, that the universe contains no closed time-like lines.

Appendix F: The Construction of Smooth Worlds: Assembling the Continuous from the Discrete

The construction of a smooth world begins with the category **Man** of manifolds.[39] As already noted, **Man** does not contain microobjects such as **Δ.** Nevertheless we can identify **Δ** indirectly through its *coordinate ring*—that is, the ring $\mathbf{R}^{\Delta}$ of smooth maps on **Δ** to **R.** We know that, as a space, $\mathbf{R}^{\Delta}$ is isomorphic to $\mathbf{R} \times \mathbf{R}$. Using this isomorphism to transfer the obvious ring structure of $\mathbf{R}^{\Delta}$ to $\mathbf{R} \times \mathbf{R}$, one finds that $\mathbf{R}^{\Delta}$ is isomorphic to the ring $\mathbf{R}*$ which has underlying set $\mathbf{R} \times \mathbf{R}$ and addition $\oplus$ and multiplication $\otimes$ defined by

$$(a, b) \oplus (c, d) = (a + c, b + d) \quad (a, b) \otimes (c, d) = (ac, ad + bc).$$

This suggests that in order to enlarge **Man** to a category containing microbjects—the first stage in constructing a smooth world—we replace each manifold M by its coordinate ring CM—the ring of smooth functions on M to $\mathbb{R}$—and then adjoin to the result every ring which, like the counterpart $\mathbb{R}*$ in **Set** of $\mathbf{R}*$, is required to be present as the coordinate ring of a microobject.

More precisely, we proceed as follows. Each smooth map $f: M \to N$ of manifolds yields a ring homomorphism $Cf: CN \to CM$ sending each g in CN to the composite $g \circ f$: accordingly C is a (contravariant) functor from **Man** to the category **Ring** of (commutaitive) rings. Now a certain subcategory **A** of **Ring** is selected, the objects of which include all coordinate rings of manifolds, together with all rings which ought to be coordinate rings of microobjects, but whose maps include only those ring homomorphisms which correspond to smooth maps. The contravariant functor C is then an embedding of **Man** into the opposite category $\mathbf{A}^{\mathbf{op}}$ of **A.** Thus $\mathbf{A}^{\mathbf{op}}$ is the desired enlargement of **Man** to a category containing microbjects, and just smooth maps. However, $\mathbf{A}^{\mathbf{op}}$ is not a topos, so it needs to be enlarged to one. The natural first candidate presenting itself here is the topos $\mathbf{Set}^{\mathbf{A}}$of sets varying over **A,** with the Yoneda embedding $Y : \mathrm{A} \to \mathbf{Set}^{\mathbf{A}}$. The composite $i = Y \circ C$ then embeds **Man** in $\mathbf{Set}^{\mathbf{A}}$. In the latter category the role of the smooth line **R** is played by the object $i\,\mathbb{R}$, and that of **Δ** by the object $Y\,\mathbb{R}*$—the images in $\mathbf{Set}^{\mathbf{A}}$of $C\,\mathbb{R}$ and $\mathbb{R}*$, respectively.

Now $\mathbf{Set}^{\mathbf{A}}$is close to being a smooth world. For the truth of most of the axioms of **SIA** therein can be shown to follow from certain established properties of $\mathbb{R}$ (considered as a manifold), or its coordinate ring $C\mathbb{R}$. For instance, that **R** is a commutative ring with identity in the sense of $\mathbf{Set}^{\mathbf{A}}$follows directly from that corresponding fact about $\mathbb{R}$. The correctness of the Microaffineness Axiom for **Δ** can be shown to follow from a result of classical analysis known as Hadamard's theorem, which asserts that, for any smooth map $F: \mathbb{R}^n \times \mathbb{R} \to \mathbb{R}$ there is a smooth map $G: \mathbb{R}^n \times \mathbb{R} \to \mathbb{R}$ such that

[39]More exactly, with the category of manifolds which are Hausdorff and have a countable base. My account here follows Moerdijk and Reyes (1991).

$$F(x,t) = F(x,0) + t\frac{\partial F}{\partial t}(x,0) + t^2G(x,t).$$ [41]

Unfortunately, however, certain key principles of **SIA** do not hold in $\mathbf{Set}^{\mathbf{A}}$. For instance, $\forall x \in \mathbf{R}(x < 1 \vee x > 0)$, that is, the assertion "the intervals $(\leftarrow, 1)$ and $(0, \rightarrow)$ cover **R**" is false in $\mathbf{Set}^{\mathbf{A}}$, although the corresponding principle for $\mathbb{R}$ is evidently true. This may be summed up by saying that the embedding *i* fails to preserve open covers.

To put this right, a suitable Grothendieck topology is imposed on $\mathbf{A}^{\mathbf{op}}$ and $\mathbf{Set}^{\mathbf{A}}$ is replaced by the topos of sheaves with respect to it. This has the effect of trimming $\mathbf{Set}^{\mathbf{A}}$ to those sets varying over **A** which "believe" that open covers in **Man** still cover in $\mathbf{Set}^{\mathbf{A}}$. It can then be shown that the resulting topos **S** of sheaves is a model of all the principles laid down in **SIA—S** is accordingly a smooth world. In **S**, *L***R** is the smooth line, and $L\mathbf{\Delta}$ the domain of microquantities, where L: $\mathbf{Set}^{\mathbf{A}} \rightarrow$ S is the associated sheaf functor.

Suitable refinements of the choice of Grothendieck topology on $\mathbf{A}^{\mathbf{op}}$ lead to toposes of sheaves which can be shown to satisfy the other principles of smooth infinitesimal analysis we have mentioned. For each such smooth world **S** there is a chain of functors

$$\mathbf{Man} \xrightarrow{C} \mathbf{A}^{\mathbf{op}} \xrightarrow{Y} \mathbf{Set}^{\mathbf{A}} \xrightarrow{L} \mathbf{S}$$

whose composite *s* can be shown to have the following properties:

- $s\mathbb{R}$ = the smooth line **R**
- $s(\mathbb{R} - \{0\})$ = the set of invertible elements of **R**
- $s(f') = s(f)'$ for any smooth map f: $\mathbb{R} \rightarrow \mathbb{R}$
- $s(\mathbf{Tan}M) = (sM)^{\mathbf{\Delta}}$ for any manifold *M*.

Such smooth worlds are said to be *well-adapted.*

Let us call the image *sM* of a manifold *M* in a well-adapted smooth world **S** its *representative* in **S.** Now it might be thought that manifolds and their representatives are radically different. For example, classically, satisfies $\forall x(x = 0 \vee x \neq 0)$ holds in the real line $\mathbb{R}$ but, as we know, this is not the case for its representative, the smooth line **R.** However, this difference is less profound than it seems. In fact, on analyzing the meaning of any statement of the internal language of **S** containing a variable ranging over **R**, one finds that it holds in **S** if and only if the corresponding statement is, in addition to being true for all points of $\mathbb{R}$, is also *locally true for all smooth maps to* $\mathbb{R}$. A smooth map f: $\mathbb{R} \rightarrow \mathbb{R}$ may have $f(a) = 0$ for some point *a* and yet not be constantly zero in any neighbourhood of *a*. This means that neither $f = 0$ nor $f \neq 0$ is locally true at *a*, so that "$f = 0 \vee f \neq 0$" fails to be locally true. Similarly, the trichotomy law $\forall x \, \forall \, y(x < y \vee y \vee y < x)$, although true in $\mathbb{R}$, fails for **R**, since for smooth maps f, g: $\mathbb{R} \rightarrow \mathbb{R}$ there may exist points *a* on no single neighbourhood of which does $f < g$ or $f = g$ or $g < f$. On the other hand, since *f* is continuous, each point a either has a neighbourhood on which $0 < f$ or one on which $f < 1$, so that $\forall x\ (0 < x \vee x < 1)$ holds for **R.**

Most well-adapted smooth worlds have the further property that elements of the smooth line **R**, that is, maps $1 \rightarrow$ **R,** correspond to points of $\mathbb{R}$. This means that in passing from $\mathbb{R}$ to **R** no new "genuine" elements are added, but only "potential" ones. As already remarked, it can also be shown that these satisfy the closed law of excluded middle in the sense that $p \vee \neg p$ is true whenever p is a closed sentence.

*

The construction and verification of properties of smooth worlds is, it must be admitted, a laborious business, far more complex than the process of constructing models for nonstandard analysis. But perhaps the situation here can be likened to the use of a complicated film projector to produce a simple image (in the case at hand, an image of ideal smoothness), or to the activity of a brain whose intricate neurochemical structure contrives somehow to present simple images to consciousness. The point is that, although the fashioning of smooth worlds is by no means a simple process, it is designed to embody simple principles. The path to simplicity must on occasion pass through the complex.

Bibliography

Abraham, R., J. Marsden, and T. Ratiu. 1988. *Manifolds, Tensor Analysis, and Applications*. 2nd ed. New York: Springer.

Al-Azm, S. 1972. *The Origins of Kant's Arguments in the Antinomies*. Oxford: Oxford University Press.

Alexander, H.G., ed. 1956. *The Leibniz-Clarke Correspondence*. Manchester: University of Manchester Press.

Aristotle. 1980. *Physics*, 2 vols. Trans. Cornford and Wickstead. Loeb Classical Library, Harvard University Press and Heinemann.

———. 1996a. *Metaphysics, Oeconomica, Magna Moralia*. Vols. 2. Trans. Cooke and Tredinnick. Loeb Classical Library, Harvard University Press.

———. 1996b. *The Categories, On Interpretation, Prior Analytics*. Trans. Cooke and Tredinnick. Loeb Classical Library, Harvard University Press.

———. 2000a. *On the Heavens*. Trans. Guthrie. Loeb Classical Library, Harvard University Press.

———. 2000b. *On Sophistical Refutations, On Coming-to-Be and Passing Away, On the Cosmos*. Trans. Forster and Furley. Loeb Classical Library, Harvard University Press.

Arnauld, A., and P. Nicole. 1996. *Logic or the Art of Thinking*. Trans. Buroker. Cambridge: Cambridge University Press.

Bair, J., P. Blaszczyk, P. Heinig, V. Kanovei, M. Katz, and T. McGaffey.. (to appear). *Cauchy's Work on Integral Geometry, Centers of Curvature, and Other Applications of Infinitesimals.*

Ballard, J.G. 1993. *You and me and the continuum*. In *The Atrocity Exhibition*. London: Flamingo.

Barnes, J. 1986. *The Presocratic Philosophers*. London: Routledge.

Baron, M.E. 1987. *The Origins of the Infinitesimal Calculus*. New York: Dover.

Bayle, P. 1965. *Historical and Critical Dictionary; Selections*. Trans. Popkin. Bobbs-Merrill.

Beeson, M.J. 1985. *Foundations of Constructive Mathematics*. Berlin: Springer.

Bell, E.T. 1945. *The Development of Mathematics*. 2nd ed. New York/London: McGraw-Hill.

———. 1965. *Men of Mathematics*. Vol. 2. Harmondsworth/Middlesex: Penguin Books.

Bell, J.L. 1986. *A new approach to quantum logic. British Journal for the Philosophy of Science* 37: 83–99.

———. 1988. *Toposes and Local Set Theories: An Introduction*, Oxford Logic Guides 14. Oxford: Clarendon Press. Reprinted by Dove, 2008.

———. 1998. *A Primer of Infinitesimal Analysis*. Cambridge: Cambridge University Press. Second edition 2oo8.

———. 2000. *Continuity and the logic of perception. Tiranscendent Philosophy* 1 (2): 1–7.

J. L. Bell, *The Continuous, the Discrete and the Infinitesimal in Philosophy and Mathematics*, The Western Ontario Series in Philosophy of Science 82,
https://doi.org/10.1007/978-3-030-18707-1

———. 2001. In *The Continuum in Smooth Infinitesimal Analysis*, In Schuster, Berger, and Osswald (2001), 19–24.
———. 2003. *Hermann Weyl's later philosophical views: His divergence from Husserl*. In *Husserl and the Sciences*, ed. R. Feist. Ottawa: University of Ottawa Press.
———. 2009. *Cohesiveness*. *Intellectica* 41: 145–168.
———. 2014. *Intuitionistic Set Theory*. London: College Publications.
———. 2016. *Oppositions and Paradoxes: Philosophical Perplexities in Science and Mathematics*. Peterborough: Broadview Press.
Bell, J.L., and M. Machover. 1977. *A Course in Mathematical Logic*. Amsterdam/New York/ Oxford: North-Holland.
Benacerraf, P., and H. Putnam. 1977. *Philosophy of Mathematics: Selected Readings*. 2nd ed. Cambridge: Cambridge University Press.
Bennett, J. 1971. *The age and size of the world*. *Synthese* 23: 127–146.
Berkeley, G. 1960. *Principles of Human Knowledge*. New York: Doubleday.
Bishop, E. 1967. *Foundations of Constructive Analysis*. New York: McGraw-Hill.
Bishop, E., and D. Bridges. 1985. *Constructive Analysis*. Berlin: Springer.
Bolzano, B. 1950. *Paradoxes of the Infinite*. Trans. Prihovsky. Routledge & Kegan Paul.
Bos, H. 1974. *Differentials, higher order differentials and the derivative in the Leibnizian calculus*. *Archive for History of Exact Sciences* 14: 1–90.
Bourbaki, N. 1990. *Éléments de mathématique. Topologie générale*. Paris: Masson.
Boyer, C. 1959. *The History of the Calculus and Its Conceptual Development*. New York: Dover.
———. 1968. *A History of Mathematics*. Hoboken: Wiley.
Boyer, C., and U. Merzbach. 1989. *A History of Mathematics*. 2nd ed. Hoboken: Wiley.
Brentano, F. 1974. *Psychology from an Empirical Standpoint*. New York: Humanities Press.
———. 1988. *Philosophical Investigations on Space, Time and the Continuum*. Trans. Smith, B. Croom Helm.
Bridges, D. 1994. *A constructive look at the real line*. In *Real Numbers, Generalizations of the Reals, and Theories of Continua*, ed. P. Ehrlich, 29–92. Dordrecht: Springer.
———. 1999. *Constructive mathematics: A foundation for computable analysis*. *Theoretical Computer Science* 219: 95–109.
Bridges, D., and F. Richman. 1987. *Varieties of Constructive Mathematics*. Cambridge: Cambridge University Press.
Brouwer, L.E.J. 1975. In *Collected Works, 1*, ed. A. Heyting. New York: North-Holland.
Burns, C.D. 1916. *William of Ockham on continuity*. *Mind* 25: 506–512.
Busemann, H. 1955. *The Geometry of Geodesics*. New York: Academic.
Cajori, F. 1919. *A Short History of the concepts of Limits and Fluxions*. Chicago: Open Court.
Cantor, G. 1961. *Contributions to the Founding of the Theory of Transfinite Numbers*. New York: Dover.
Carnot, L. 1832. *Reflexions sur la Métaphysique du Calcul Infinitesimal*. Trans. Browell. Parker, Oxford.
Chevalier, G., et al. 1929. *Continu et Discontinu*. *Cahiers de la Nouvelle Journée* 15.
Child, J. 1916. *The Geometrical Lectures of Isaac Barrow*. Chicago: Open Court.
Cohen, P.J. 1966. *Set Theory and the Continuum Hypothesis*. New York: W. A. Benjamin Inc.
Cohn, P.M. 1957. *Lie Groups*. Cambridge: Cambridge University Press.
Cusanus, N. 1954. *Of Learned Ignorance*. Trans. Heron. London: Routledge and Kegan Paul.
D' Alembert, J., and D. Diderot. 1966. *Encyclopédie, ou, Dictionnaire raisonné des sciences, des arts et des métiers/(mis en ordre et publié par Diderot, quant à la partie mathématique, par d'Alembert.)* Cannstatt = Frommann, Stuttgart.
Dauben, J. 1979. *Georg Cantor: His Mathematics and Philosophy of the Infinite*. Cambridge: Harvard University Press.
Dedekind, R. 1963. *Essays on the Theory of Numbers*. New York: Dover.
Descartes, R. 1927. *Discourse on method, meditations, and principles of philosophy*. In *Everyman's Library*. London: Dent.

Dugas, R. 1988. *A History of Mechanics*. New York: Dover.

Dummett, M. 1977. *Elements of Intuitionism*. Oxford: Clarendon Press.

Ehrlich, P., ed. 1994. *Real Numbers, Generalizations of the Reals, and Theories of Continua*. Dordrecht: Kluwer.

———. 1994a. *All numbers great and small*. In , ed. Ehrlich, 239–258. Dordrecht: Kluwer Academic Publishers.

———. 2006. *The rise of non-Archimedean mathematics and the roots of a misconception. I. The emergence of non-Archimedean systems of magnitudes. Archive for History of Exact Sciences* 60 (*1*): 1–121.

Einstein, A., et al. 1952. *The Principle of Relativity : a Collection of Original Memoirs on the Special and General Theory of Relativity*. Trans. Perrett and Jeffrey. Dover. Reprint of 1923 Methuen edition.

Euler, L. 1843. *Letters of Euler on Different Subjects in Natural Philosophy : Addressed to a German Princess; with Notes, and a Life of Euler / by David Brewster; Containing a Glossary of Scientific Term[s] with Additional Notes, by John Griscom*. New York: Harper and Brothers. Reprint by Forgotten Books, 2012.

———. 1990. *Introduction to Analysis of the Infinite*. Trans. Blanton. New York: Springer.

Evans, M. 1955. *Aristotle, Newton, and the theory of continuous magnitude. Journal of the History of Ideas* 16: 548–557.

Ewald, W. 1999. *From Kant to Hilbert, A Source Book in the Foundations of Mathematics*. Vol. I and II. Oxford: Oxford University Press.

Fisher, G. 1978. *Cauchy and the infinitely small. Historia Mathematica* 5: 313–331.

———. 1981. *The infinite and infinitesimal quantities of du Bois-Reymond and their reception. Archive for History of Exact Sciences* 24 (2): 101–163.

———. 1994. *Veronese's non-Archimedean linear continuum. In Ehrlich* 1994: 107–146.

Folina, J. 1992. *Poincaré and the Philosophy of Mathematics*. New York: St. Martin's Press.

Fourman, M., and J. Hyland. 1979. *Sheaf models for analysis*. In , ed. Fourman et al., 280–302. Berlin: Springer.

Fourman, M., C. Mulvey, and D. Scott., eds. 1979. *Applications of Sheaves. Proc. L.M.S. Durham Symposium* 1977. Springer Lecture Notes in Mathematics 753.

Fraenkel, A. 1976. *Abstract Set Theory*. Fourth edition of first (1953) edition, revised by A. Levy. North-Holland.

Fraenkel, A., Y. Bar-Hillel, and A. Levy. 1973. *Foundations of Set Theory*. 2nd ed. Amsterdam: North-Holland.

Frege, G. 1980. *Foundations of Arithmetic*. Trans. Austin. Northwestern University Press.

———. 1984. In *Collected Papers on Mathematics, Logic and Philosophy*, ed. B. McGuinness. Oxford: Blackwell.

Fuchs, L. 1963. *Partially Ordered Algebraic Systems*. Oxford: Pergamon Press.

Furley, D. 1957. *Two Studies in the Greek Atomists*. Princeton: Princeton University Press.

———. 1982. *The Greek commentators' treatment of Aristotle's theory of the continuous. In Kretzmann* 1982: 17–36.

———. 1987. *The Greek Cosmologists*. Cambridge: Cambridge University Press.

Furley, D., and R. Allen, eds. 1970. *Studies in Presocratic Philosophy, Vol. I*. London: Routledge and Kegan Paul.

Galilei, G. 1954. *Dialogues Concerning Two New Sciences*. Trans. Crew and De Salvio. Dover.

Gangopadhyaya, M. 1980. *Indian Atomism*. Calcutta: Bagchi.

Gödel, K. 1940. *The Consistency of the Axiom of Choice and of the Generalized Continuum-Hypothesis with the Axioms of Set Theory*, Annals of Mathematics Studies 3.. Revised edition, 1951. Princeton: Princeton University Press.

Grant, E., ed. 1974. *A Source Book in Medieval Science*. Cambridge, MA: Harvard University Press.

Gray, J. 1973. *Ideas of Space: Euclidean, Non-Euclidean, and Relativistic*. Oxford: Clarendon Press.

Gregory, J. 1931. *A Short History of Atomism*. London: A. & C. Black.
Grothendieck, A. 1986. *Récoltes et Semailles*. Trans. Lisker. http: // www. fermentmagazine.org/rands/recoltes1.html
Grünbaum, A. 1967. *Zeno's Paradoxes and Modern Science*. London: Allen and Unwin.
Hallett, M. 1984. *Cantorian Set Theory and Limitation of Size*. Oxford: Clarendon Press.
Hannequin, A. 1899. *Essai Critique sur l'Hypothese des Atomes*. Paris: Alcan.
Hardy, G.H. 1910. *Orders of Infinity*. Cambridge: Cambridge University Press.
———. 1960. *A Course of Pure Mathematics*. Cambridge: Cambridge University Press.
Heath, T. 1949. *Mathematics in Aristotle*. Oxford: Oxford University Press.
———. 1981. *A History of Greek Mathematics*. Vol. 2. New York: Dover.
Hegel, G. 1961. *Science of Logic*. Trans. Johnston & Struthers. Allen & Unwin.
Hellman, G., and S. Shapiro. 2018. *Varieties of Continua: From Regions to Points and Back*. Oxford: Oxford University Press.
Heyting, A. 1956. *Intuitionism: An Introduction*. Amsterdam: North-Holland.
Hobson, E.W. 1957. *The Theory of Functions of a Real Variable*. New York: Dover.
Hocking, J.G., and G.S. Young. 1961. *Topology*. Reading: Addison-Wesley.
Holden, T. 2004. *The Architecture of Matter*. Oxford: Oxford University Press.
Houzel, C., et al. 1976. *Philosophie et Calcul de l'Infini*. Paris: Maspero.
Hume, D. 1962. *A Treatise of Human Nature, Book I*. London: Collins.
Hyland, J. 1979. *Continuity in spatial toposes*. In Fourman et al. (1979), pp. 442–465.
———. 1982. *The effective topos*. In *The L.E.J. Brouwer Memorial Symposium*, ed. A. Troelstra and D. van Dalen. North-Holland. Publ. Co.
Jesseph, D. 1993. *Berkeley's Philosophy of Mathematics*. Chicago: Chicago University Press.
Johnstone, P.T. 1977. *Topos Theory*. London: Academic.
———. 1982. *Stone Spaces*, Cambridge Studies in Advanced Mathematics 3. Cambridge: Cambridge University Press.
———. 1983. *The point of pointless topology. Bulletin of American Mathematical Society (N.S.)* 8 (1): 41–53.
———. 2002. *Sketches of an Elephant: A Topos Theory Compendium, Vols. I and II*, Oxford Logic Guides Vols. 43 and 44. Oxford: Clarendon Press.
Kahn, C. 2001. *Pythagoras and the Pythagoreans: A Brief History*. Indianapolis: Hackett.
Kant, I. 1964. *Critique of Pure Reason*. London: Macmillan.
———. 1970. *Metaphysical Foundations of Natural Science*. Trans. Ellington. Bobbs-Merrill.
———. 1977. *Prolegomena to Any Future Metaphysics*. Trans. Carus, revsd. Ellington. Hackett.
———. 1992. In *Theoretical Philosophy 1755–1770*, ed. Watford and Meerbote. Cambridge: Cambridge University Press.
Keisler, H. 1994. *The hyperreal line. In Ehrlich* 1994: 207–238.
Kelley, J.L. 1960. *General Topology*. Princeton: Van Nostrand.
Kirk, G.S., J.E. Raven, and M. Schofield. 1983. *The Presocratic Philosophers*. 2nd ed. Cambridge: Cambridge University Press.
Kline, M. 1972. *Mathematical Thought from Ancient to Modern Times*. Vol. 3 volumes. Oxford: Oxford University Press.
Kneebone, G.T. 1963. *Mathematical Logic and the Foundations of Mathematics: An Introductory Survey*. London: Van Nostrand.
Kock, A. 1981. *Synthetic Differential Geometry*. Cambridge: Cambridge University Press.
Körner, S. 1955. *Kant*. Harmondsworth: Penguin Books.
———. 1960. *The Philosophy of Mathematics*. London: Hutchinson.
Kretzmann, N., ed. 1982. *Infinity and Continuity in Ancient and Medieval Thought*. Ithaca: Cornell University Press.
Kuratowski, K., and A. Mostowski. 1968. *Set Theory*. Amsterdam: North-Holland.
Lambek, J., and P.J. Scott. 1986. *Introduction to Higher-Order Categorical Logic*. Cambridge: Cambridge University Press.

Lawvere, F.W. 1971. *Quantifiers and sheaves*. In *Actes du Congrès Intern. Des Math. Nice 1970, tome I*, 329–334. Paris: Gauthier-Villars.
———. 2011. *Euler's continuum functorially vindicated*. In *Vintage Enthusiasms: Essays in Honour of John L. Bell*, ed. D. Devidi, P. Clark, and M. Hallett. Dordrecht: Springer.
Leibniz, G. 1951. In *Selections*, ed. P. Wiener. New York: Scribners.
———. 1961. *Philosophical Writings*. Trans. Morris. Everyman Library, Dent Dutton.
———. 2001. *The Labyrinth of the Continuum : Writings on the Continuum Problem, 1672–1686*. Trans. Arthur. Yale University Press.
Lucretius. 1955. *The Nature of the Universe*. Trans. Latham. Penguin Books.
Mac Lane, S. 1986. *Mathematics: Form and Function*. New York: Springer.
Mac Lane, S., I. Moerdijk, and I. 1992. *Sheaves in Geometry and Logic: A First Introduction to Topos Theory*. New York: Springer.
Maimonides, M. 1956. *The Guide for the Perplexed*. Trans. Friedländer. Dover.
Mancosu, P. 1996. *Philosophy of Mathematics and Mathematical Practice in the Seventeenth Century*. New York/Oxford: Oxford University Press.
———. 1998. *From Brouwer to Hilbert: The Debate on the Foundations of Mathematics in the1920s*. Oxford: Clarendon Press.
Martin-Löf, P. 1990. *Mathematics of Infinity*. In *Colog-88 Computer Logic*, ed. P. Martin-Löf and G. Mints. Berlin: Springer.
Marx, K. 1983. *Mathematical Manuscripts*. Clapham/London: New Park Publishers.
McLarty, C. 1988. *Defining sets as sets of points of spaces*. *Journal of Philosophical Logic* 17: 75–90.
———. 1992. *Elementary Categories, Elementary Toposes*. Oxford: Oxford University Press.
Miller, F. 1982. *Aristotle against the atomists. In Kretzmann* 1982: 87–111.
Misner, C., K. Thorne, and J. Wheeler. 1972. *Gravitation*. San Francisco: Freeman.
Moerdijk, I. 1995. *A model for intuitionistic non-standard arithmetic*. *Annals of Pure and Applied Logic* 73: 297–325.
Moerdijk, I., and G.E. Reyes. 1991. *Models for Smooth Infinitesimal Analysis*. New York: Springer.
Moore, A.W. 1990. *The Infinite*. London: Routledge.
Moynahan, G. 2003. *Hermann Cohen's Das Prinzip der Infinitesimal-Methode und Seine Geschichte, Ernst Cassirer, and the Politics of Science in Wilhelmine Germany*. *Perspectives in Science* 11 (1): 35–75.
Murdoch, J. 1957. *Geometry and the Continuum in the 14th Century: A Philosophical Analysis of Th. Bradwardinne's Tractatus de Continuo*. Ph.D. thesis, University of Wisconsin.
———. 1982. *William of Ockham and the logic of infinity and continuity. In Kretzmann* 1982: 165–206.
Needham, J. 1954. *Science and Civilisation in China*. Vol. 4. Cambridge: Cambridge University Press.
Newton, I. 1952. *Opticks (1730)*. New York: Dover.
———. 1962. *Principia, Vols, I, II*. Trans. Motte, revsd. Cajori. University of California Press.
Nicholas of Autrecourt. 1971. *The Universal Treatise*. Trans. Kennedy et al. Milwaukee: Marquette University Press
Oresme, N. 1968. *Le Livre du Ciel et du Monde*. Trans. Menut. Madison: University of Wisconsin Press.
Palmgren, E. 1998. *Developments in constructive nonstandard analysis*. *The Bulletin of Symbolic Logic* 4: 233–272.
———. 2001. *Unifying constructive and nonstandard analysis. In Schuster, Berger and Osswald* 2001: 167–183.
Pascal, B. 1966. *Pensées*. Trans. Krailsheimer. Penguin Books.
Peirce, C.S. 1976. In *The New Elements of Mathematics*, ed. Carolyn Eisele, vol. III. Atlantic Highlands: Mouton Publishers and Humanities Press.
———. 1992. In *Reasoning and the Logic of Things*, ed. Kenneth Laine Ketner. Cambridge: Harvard University Press.

Peters, F. 1967. *Greek Philosophical Terms*. New York: New York University Press.
Plato. 1961. In *The Collected Dialogues*, ed. E. Hamilton and H. Cairns. Princeton: Princeton University Press.
Poincaré, H. 1946. *Foundations of Science*. Trans. Halstead. Science Press.
———. 1963. *Mathematics and Science: Last Essays*. New York: Dover.
Pycior, H. 1987. Mathematics and philosophy: Wallis, Hobbes, Barrow, and Berkeley. *Journal of the History of Ideas* 48 (2): 265–286.
Pyle, A. 1997. *Atomism and Its Critics*. Bristol: Thoemmes Press.
Rees, M. 2001. *Just Six Numbers: The Deep Forces that Shape the Universe*. New York: Basic Books.
Rescher, N. 1967. *The Philosophy of Leibniz*. Englewood Cliffs: Prentice-Hall.
Robinson, A. 1996. *Non-Standard Analysis*. Princeton: Princeton University Press.
Russell, B. 1945. *A History of Western Philosophy*. New York: Simon and Schuster.
———. 1958. *A Critical Exposition of the Philosophy of Leibniz*. 2nd ed. London: Allen & Unwin.
———. 1964. *The Principles of Mathematics*. London: Allen and Unwin.
———. 1995. *Introduction to Mathematical Philosophy*. London: Routledge.
Sambursky, S. 1963. *The Physical World of the Greeks*. London: Routledge and Kegan Paul.
———. 1971. *Physics of the Stoics*. London: Hutchinson.
Schuster, P., U. Berger, and H. Osswald. 2001. *Reuniting the Antipodes—Constructive and Nonstandard Views of the Continuum*. Dordrecht: Kluwer.
Smith, D.E. 1959. *A Source Book in Mathematics*. New York: Dover.
Sorabji, R. 1982. *Atoms and time atoms. In Kretzmann* 1982: 37–86.
———. 1983. *Time, Creation and the Continuum*. Ithaca: Cornell University Press.
Spivak, M. 1975. *A Comprehensive Introduction to Differential Geometry*. Houston: Publish or Perish Press.
Stokes, M. 1971. *One and Many in Presocratic Philosophy*. Cambridge: Harvard University Press.
Stones, G.B. 1928. *The atomic view of matter in the XVth, XVIth and XVIIth centuries. Isis* 10: 444–465.
Struik, D. 1948. *A Concise History of Mathematics*. New York: Dover.
Tokaty, G. 1994. *A History and Philosophy of Fluid Mechanics*. New York: Dover.
Tolstoy, L. 1961. *War and Peace*. Trans. Edmonds, R.. Penguin Books.
Truesdell, C. 1972. *Leonard Euler, supreme geometer*. In *Irrationalism in the Eighteenth Century*, ed. H. Pagliaro. Cleveland: Press of Case Western Reserve University.
Van Atten, M., D. van Dalen, and R. Tieszen. 2002. *The phenomenology and mathematics of the intuitive continuum. Philosophia Mathematica* 10: 203–236.
Van Cleve, J. 1981. *Reflections on Kant's 2nd antinomy. Synthese* 47: 1147–1150.
Van Dalen, D. 1995. *Hermann Weyl's intuitionistic mathematics. The Bulletin of Symbolic Logic* 1 (2): 145–169.
———. 1997. *How connected is the intuitionistic continuum? Journal of Symbolic Logic* 62: 1147–1150.
———. 1998. *From a Brouwerian point of view. Philosophia Mathematica* 6: 209–226.
van Heijenoort, J. 1981. *From Frege to Gödel: A Sourcebook in Mathematical Logic 1879 – 1931*. Cambridge, MA: Harvard University Press.
Van Melsen, A. 1952. *From Atomos to Atom*. Trans. Koren. Pittsburgh: Duquesne University Press.
Vesley, R. 1981. *An intuitionistic infinitesimal calculus*. In *Lecture Notes in Mathematics 873*, 208–212. Berlin: Springer.
Wagon, S. 1993. *The Banach-Tarski Paradox*. Cambridge: Cambridge University Press.
Walker, E. 1932. *A Study of the Traité des Indivisibles of Gilles Personne de Roberval*. New york: Bureau of Publications, Teachers College, Columbia Unversity.
Weyl, H. 1921. *On the new foundational crisis in mathematics*. (English translation of 'Über der neue Grundslagenkrise der Mathematik,' Mathematische Zeitschrift 10, 1921, 37–79.) In Mancosu (1998), pp. 86–122.

———. 1929. *Consistency in Mathematics*. Rice Institute Pamphlet 16, 245–265. Reprinted in Weyl (1968) II, pp. 150–170.
———. 1932. *The Open World: Three Lectures on the Metaphysical Implications of Science*. New Haven: Yale University Press.
———. 1940. *The ghost of modality*. In *Philosophical Essays in Memory of Edmund Husserl*. Cambridge: Harvard University Press. Reprinted in Weyl (1968) III, pp. 684–709.
———. 1949. *Philosophy of Mathematics and Natural Science*. Princeton: Princeton University Press. (An expanded Engish version of *Philosophie der Mathematik und Naturwissenschaft*, Leibniz Verlag, 1927.).
———. 1950. *Space-Time-Matter*. Trans. Brose, H. Dover. (English translation of *Raum, Zeit, Materie*, Springer, 1918.)
———. 1954. *Address on the unity of knowledge*. Columbia University Bicentennial Celebration. Reprinted in Weyl (1968) IV, pp. 623–630.
———. 1968. In *Gesammelte Abhandlungen*, ed. K. Chandrasehharan, vol. I-IV. Berlin: Springer.
———. 1969. *Insight and reflection*. (Lecture delivered at the University of Lausanne, Switzerland, May 1954. Translated from German original in *Studia Philosophica*, 15, 1955.) In T.L. Saaty and F.J. Weyl, eds., *The Spirit and Uses of the Mathematical Sciences*, pp. 281–301. McGraw-Hill, 1969.
———. 1985. *Axiomatic versus constructive procedures in mathematics.*, ed. T. Tonietti. *Mathematical Intelligencer* 7, no. 4, pp. 10–17, 38.
———. 1987. *The Continuum: A Critical Examination of the Foundation of Analysis*, Trans. S. Pollard and T. Bole. Thomas Jefferson University Press. (English translation of *Das Kontinuum*, Leipzig: Veit, 1918.)
———. 1998a. *On the New Foundational Crisis in Mathematics*. (English translation of 'Über der neue Grundslagenkrise der Mathematik,' *Mathematische Zeitschrift* 10, 1921, pp. 37–79.) In Mancosu (1998), pp. 86–122.
———. 1998b. *On the Current Epistemological Situation in Mathematics*. English translation of 'Die Heutige Erkenntnislage in der Mathematik, *Symposion* 1, 1925–27, pp. 1–32.) In Mancosu (1998), pp. 123–142.
White, M.J. 1992. *The Continuous and the Discrete: Ancient Physical Theories from a Contemporary Perspective*. Oxford: Clarendon Press.
Whitehead, A.N. 1920. *The Concept of Nature*. Cambridge: Cambridge University Press.
Whyte, L. 1961. *Essay on Atomism from Peano to 1960*. Middletown: Wesleyan University Press.
Wike, V. 1982. *Kant's Antinomies of Reason: Their Origin and Resolution*. Washington: University Press of America.
Wilson, C. 2015. *Leibniz's Metaphysics: A Historical and Comparative Study*. Princeton: Princeton University Press.

Index

J. L. Bell, *The Continuous, the Discrete and the Infinitesimal in Philosophy and Mathematics*, The Western Ontario Series in Philosophy of Science 82,
https://doi.org/10.1007/978-3-030-18707-1

D

The manufacturer's authorised representative in the EU is Springer Nature Customer Service Centre GmbH, Europaplatz 3, 69115 Heidelberg, Germany. If you have any concerns regarding our products, please contact ProductSafety@springernature.com

Printed and bound by CPI Group (UK) Ltd, Croydon, CR0 4YY
06/10/2026
02228459-0004